CENTIGRADE AND FAHRENHEIT EQUIVALENTS

$$F = \frac{9}{5} C + 32 \qquad C = \frac{5}{9} (F - 32)$$

CENTIGRADE	FAHRENHEIT
100°	212° ← water boils at sea level
95°	203°
90°	194°
85°	185°
80°	176°
75°	167°
70°	158°
65°	149°
60°	140°
55°	131°
50°	122°
45°	113°
40°	104°
35°	95°
comfort { 30°	comfort { 86°
range { 25°	range { 77°
20°	68°
15°	59°
10°	50°
5°	41°
0°	32° ←

ARITHMETIC
FOR COLLEGE STUDENTS

ARITHMETIC FOR COLLEGE STUDENTS

Third Edition

D. FRANKLIN WRIGHT

Cerritos College

D. C. HEATH AND COMPANY
Lexington, Massachusetts Toronto

This book is dedicated to the memory of
Sara M. Wright

PREFACE

This new edition contains the basic features and topics of the first and second editions, with the following important changes:

■ Most sections contain a Self Quiz consisting of three to five problems similar to those in the exercises. The answers are provided beside the problems. If the students cover these answers and do the problems in class, the instructor can judge the effectiveness of his lecture and spot potential difficulties and misunderstandings before the students leave the classroom.

■ Each chapter (except Chapter 7) contains a chapter summary of definitions and key ideas presented in the chapter..These summaries, along with the Chapter Reviews, should be very helpful to the students in preparing for examinations.

■ Chapter 6 on percents contains more detailed explanations and many more applications, including separate sections on simple interest and compound interest.

■ Chapter 7 on the metric system has been rearranged and simplified by the addition of a simple system of charts for changing metric units.

■ The work with base two and base five numerals has been placed in Appendix II. These topics are still valuable for students interested in computers.

Chapter 1 (Whole Numbers) begins with a discussion of exponents so that the student can develop an understanding of the place value concept used in the decimal system. Addition, subtraction, multiplication, and division with whole numbers are presented with plenty of examples and practice exercises and with an attempt to explain just why these operations work as they do.

Chapter 2 (Prime Factors) makes use of the knowledge of exponents from Chapter 1 and forms a basis for the work with rational numbers in Chapter 3 (Rational Numbers). Skills with rational numbers are developed thoroughly in Chapter 3, and mixed numbers are treated as rational numbers. The term *improper fraction* is presented, but there is no emphasis on this idea.

Chapter 4 (Decimal Numbers) treats decimal numbers as rational numbers with powers of ten as denominators. Reading and writing decimal numbers correctly is emphasized. The last section contains a discussion of real numbers and relates rational numbers to infinite repeating decimals.

Chapter 5 (Ratio and Proportion) forms the foundation for the development of percent in Chapter 6 (Percent). The explanation and exercises in Chapter 6 have been expanded, and the topic of compound interest (annual, semiannual, and quarterly) has been included.

Chapter 7 (Metric System) should be particularly valuable for many adults who are not familiar with the basic metric units. The empahsis is on changing units within the metric system, and a technique of using charts is developed. Included are formulas for area and volume and tables of English–metric equivalents, although there are no exercises relating English and metric units.

Chapter 8 (Square Root) contains two methods of finding square roots and a discussion of the Pythagorean Theorem.

Chapter 9 (Negative Numbers) and Chapter 10 (Equations and Inequalities) constitute the algebra sections and are especially valuable for those students who will continue in mathematics. Word problems are presented at a basic level, and the solution of inequalities is a new topic that ends the text.

Each chapter contains a set of Review Questions and a Chapter Summary (except Chapter 7). The Answer Key in the back of the book has the answers to all questions in the exerccises, except for multiples of four, and the answers to all the questions in the reviews. The remaining answers, along with sample tests, are in the accompanying *Instructor's Manual.*

Enough material is included for a three- or four-semester unit course. The appendixes contain information and exercises on ancient numeration systems and the base two and base five systems. A table of powers and roots is on the inside back cover.

Thanks to Pat Wright for such patience, understanding, and a terrific job of reading my handwriting.

D. Franklin Wright

CONTENTS

Lesson 1 & Lesson 2

Lesson 3
Stress Definations

① Test chap 1 & 2
Lesson 4
Lesson 5
Lesson 6

Lesson 7 Review
Board work
② Test chap ③

4

DECIMAL NUMBERS   124

LESSON 8

5

RATIO AND PROPORTION   154

LESSON 9

③ TEST CHAP 4&5

6

PERCENT   168

LESSON 10

LESSON 11 START PG 179 PROB 21

7

MEASUREMENT: THE METRIC SYSTEM   198

SKIP

ARITHMETIC
FOR COLLEGE STUDENTS

WHOLE NUMBERS

1

So you want to improve in arithmetic! Have you ever questioned why $40 + 50 = 90$? Why doesn't $40 + 50 = 900$; or $40 + 50 = 4050$; or $40 + 50 = 120$? (In some systems $40 + 50$ does equal 120!) Improvement in arithmetic usually follows when the student begins to ask a lot of questions.

For the Romans, $XL + L = XC$ was the same as $40 + 50 = 90$ is for us. Why don't we still use the Roman system? Why have we adopted another system given to us by the Hindu-Arabic tribes about 800 A.D.? The answer is that our system allows us to add, subtract, multiply, and divide much more easily and faster than any of the ancient number systems, including the Roman system. (A discussion of several ancient number systems is found in Appendix I.)

What do you know about our **decimal system?** Do you know that it is a **place value system** that uses the number ten as a base? This chapter will give you a foundation for understanding the decimal system so that you may **understand the reasons** for some of the rules and procedures used in arithmetic.

1.1   EXPONENTS

Suppose we want to multiply 5 times 3. One notation that is sometimes used is a raised dot, as in $5 \cdot 3$. Thus, $\mathbf{5 \cdot 3 = 15}$ is read *five times three equals fifteen*. In the equation $5 \cdot 3 = 15$, 15 is called the **product,** and 5 and 3 are called **factors** of the product.

Similarly, since $6 \cdot 2 = 12$, 6 and 2 are factors of the product 12. Are 6 and 2 the only factors of 12? The answer is *no*, because $4 \cdot 3 = 12$ and $1 \cdot 12 = 12$. We see then that 6, 2, 4, 3, 1, and 12 are all factors of 12.

A number may have a factor that is repeated, as the following examples illustrate.

EXAMPLES

1. $7 \cdot 7 = 49$
2. $3 \cdot 3 = 9$
3. $2 \cdot 2 \cdot 2 \cdot 2 = 16$
4. $5 \cdot 5 \cdot 3 = 75$
5. $10 \cdot 10 \cdot 10 = 1000$

Mathematicians have invented a shorthand notation to indicate repeated factors. With this notation, Examples 1–5 can be rewritten as follows:

1'. $7^2 = 49$
2'. $3^2 = 9$
3'. $2^4 = 16$
4'. $5^2 \cdot 3 = 75$
5'. $10^3 = 1000$

The small numbers written to the right and above the other numbers are called **exponents.** The number to the left and below the exponent is called the **base.**

√ DEFINITION   An **exponent** is a number that tells how many times its base is to be used as a factor.

When the eponent 0 is used for any base except 0, the value of the power is defined to be 1. That is,

$$2^0 = 1$$
$$3^0 = 1$$
$$5^0 = 1$$
$$13^0 = 1$$
$$86^0 = 1$$

To understand why this definition makes sense, suppose we want to multiply $2^3 \cdot 2^4$ or $6^2 \cdot 6^3$.

$$2^3 \cdot 2^4 = \underbrace{(2 \cdot 2 \cdot 2)(2 \cdot 2 \cdot 2 \cdot 2)}_{\text{seven 2's}} = 2^7 \quad \text{or} \quad 2^3 \cdot 2^4 = 2^{3+4} = 2^7$$

$$6^2 \cdot 6^3 = \underbrace{(6 \cdot 6)(6 \cdot 6 \cdot 6)}_{\text{five 6's}} = 6^5 \quad \text{or} \quad 6^2 \cdot 6^3 = 6^{2+3} = 6^5$$

Do you see a rule for multiplying numbers with the same base? The rule, which will be discussed in detail in algebra, is to *add the exponents when multiplying numbers with the same base.*

Now suppose we want to divide $\frac{2^6}{2^2}$ or $\frac{5^4}{5^3}$.

$$\frac{2^6}{2^2} = \frac{\not{2} \cdot \not{2} \cdot 2 \cdot 2 \cdot 2 \cdot 2}{\not{2} \cdot \not{2}} = 2 \cdot 2 \cdot 2 \cdot 2 = 2^4 \quad \text{or} \quad \frac{2^6}{2^2} = 2^{6-2} = 2^4$$

$$\frac{5^4}{5^3} = \frac{\not{5} \cdot \not{5} \cdot \not{5} \cdot 5}{\not{5} \cdot \not{5} \cdot \not{5}} = 5 \quad \text{or} \quad \frac{5^4}{5^3} = 5^{4-3} = 5^1$$

The rule is to *subtract the exponents when dividing numbers with the same base.*

We want these rules to hold up even if the exponent is 0. So,

$$2^5 \cdot 2^0 = 2^{5+0} = 2^5 \quad \text{and} \quad \frac{3^4}{3^4} = 3^{4-4} = 3^0$$

But, $$2^5 \cdot 1 = 2^5 \quad \text{and} \quad \frac{3^4}{3^4} = 1$$

The point is that for the rules of exponents to make sense we must have

$$2^0 = 1 \quad \text{and} \quad 3^0 = 1$$

DEFINITION   For any nonzero whole number a, $a^0 = 1$.

EXAMPLES

               exponent

1. $6^2 = 6 \cdot 6 = 36$

       base   power

2. $8^1 = 8$	(Read: "eight to the first power is eight")
3. $7^1 = 7$	(Read: "seven to the first power is seven")
4. $7^2 = 7 \cdot 7 = 49$	(Read: "seven squared is forty-nine")
5. $5^2 = 5 \cdot 5 = 25$	(Read: "five squared is twenty-five")
6. $2^3 = 2 \cdot 2 \cdot 2 = 8$	(Read: "two cubed is eight")
7. $3^4 = 3 \cdot 3 \cdot 3 \cdot 3 = 81$	(Read: "three to the fourth power is eighty-one")
8. $2^5 = 2 \cdot 2 \cdot 2 \cdot 2 \cdot 2 = 32$	(Read: "two to the fifth power is thirty-two")

SELF QUIZ	Find each of the following powers.	ANSWERS
	1. 10^0	1. 1
	2. 3^3	2. 27
	3. 12^2	3. 144
	4. 5^3	4. 125
	5. 1^4	5. 1

EXERCISES 1.1

Odd Problems 1–59

In each of the following expressions, name (a) the exponent and (b) the base. Also, find each power.

1. 2^3	2. 2^5	3. 5^2	4. 6^2	5. 7^0
6. 11^2	7. 1^4	8. 4^3	9. 4^0	10. 3^6
11. 3^2	12. 2^4	13. 5^0	14. 1^{50}	15. 62^1
16. 12^2	17. 10^2	18. 10^3	19. 4^2	20. 2^5
21. 10^4	22. 5^3	23. 6^3	24. 10^5	25. 9^0

Find a base and exponent form for each of the following powers without using the exponent 1.

26. 4	27. 25	28. 16	29. 27	30. 32
31. 121	32. 49	33. 8	34. 9	35. 36

| **36.** 125 | **37.** 81 | **38.** 64 | **39.** 100 | **40.** 1000 |
| **41.** 10,000 | **42.** 216 | **43.** 144 | **44.** 169 | **45.** 243 |

Find as many factors as you can for each of the following numbers.

46. 20	**47.** 14	**48.** 8	**49.** 17	**50.** 51
51. 75	**52.** 100	**53.** 25	**54.** 29	**55.** 16
56. 57	**57.** 62	**58.** 24	**59.** 79	**60.** 44

1.2   THE DECIMAL SYSTEM (BASE TEN)

The term **numeral** refers to the symbols used to represent the idea of **number.** Generally, to distinguish between the two terms is a pain in the neck. However, *numeral* will be used frequently in the remainder of this chapter because we are mainly concerned with the symbols.

Consider the two numerals 70 and 700. The digit 7 has a different meaning (or value) in each of the numerals. Why? The symbol used, 7, is the same, so the difference must be in the placement of the 7.

OUR SYSTEM OF REPRESENTING NUMBERS DEPENDS ON THREE THINGS

1.   The ten digits $\{0, 1, 2, 3, 4, 5, 6, 7, 8, 9\}$.

2.   The placement of the digits.

3.   The value of each place.

Thus, in the symbol 70, the digit 7 is in the second place, and the second place has a value of ten. In the symbol 700, the digit 7 is in the third place, and the third place has a value of one hundred.

The **decimal system** (*deci* means *ten* in Latin) is a place value system that uses ten digits, and the value of each place is a power of ten (10^0, 10^1, 10^2, 10^3, 10^4, and so on). Since 10 is the base for the value of each place, the decimal system is also known as the **base ten system.**

We start, as in Figure 1.1, with a point called a **decimal point.** Each place to the left of the decimal point has ten times the value of the previous place. The value of a digit is determined by its place. The value of 5 in 25 is five units, but the value of 5 in 57 is 5 times 10, or fifty units. When we write a digit in a place, we mean that **the value of that digit is now multiplied by the value of the place.** The value of any numeral is found by adding the results of multiplication of the digits by their place values. Also, if the decimal point is not written, it is understood to be to the right of the rightmost digit. (See top of page 6 for Figure 1.1.)

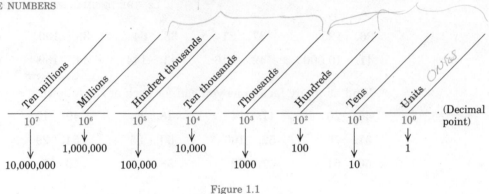

Figure 1.1

EXAMPLE   1.   Find the value of the numeral 573.

Write the value of each place under the digit in that place:

$$\frac{5}{10^2} \quad \frac{7}{10^1} \quad \frac{3}{10^0}$$

The 5 is multiplied by 10^2, or 100; the 7 is multiplied by 10^1, or 10; and the 3 is multiplied by 10^0, or 1. These products are added to find the value of the numeral:

$$
\begin{aligned}
573 &= 5(10^2) + 7(10^1) + 3(10^0) \qquad \text{(Expanded notation)}\\
&= 5(100) + 7(10) + 3(1)\\
&= 500 + 70 + 3\\
&= \text{five hundred seventy-three}
\end{aligned}
$$

Notice that 10 is not a digit but a combination of two digits, 1 and 0.

To write a numeral in expanded notation, each digit is multiplied by the value of its place, and all the products are added as in the example above. The English word equivalents can then be read from the products. Commas are placed to separate every three digits in a numeral to make it easier to read; however, in numerals of only four digits, the comma is usually omitted. Simiarly, commas are used between the words representing every three digits. Example 2 illustrates this.

EXAMPLE   2.   Write 4862 and 1,590,768 in expanded notation and in their English word equivalents.

$$
\begin{aligned}
4862 &= 4(10^3) + 8(10^2) + 6(10^1) + 2(10^0) \qquad \text{(Expanded notation)}\\
&= 4(1000) + 8(100) + 6(10) + 2(1)\\
&= 4000 + 800 + 60 + 2\\
&= \text{four thousand, eight hundred sixty-two}
\end{aligned}
$$

$$1{,}590{,}768 = 1(10^6) + 5(10^5) + 9(10^4) + 0(10^3) + 7(10^2)$$
$$+ 6(10^1) + 8(10^0) \qquad \text{(Expanded notation)}$$
$$= 1(1{,}000{,}000) + 5(100{,}000) + 9(10{,}000)$$
$$+ 0(1000) + 7(100) + 6(10) + 8(1)$$
$$= 1{,}000{,}000 + 500{,}000 + 90{,}000 + 0 + 700$$
$$+ 60 + 8$$
$$= \text{one million, five hundred ninety thousand,}$$
$$\text{seven hundred sixty-eight}$$

SELF QUIZ	Write each of the following numerals in expanded notation and in their English word equivalents.	ANSWERS
	1. 512	1. $5(10^2) + 1(10^1) + 2(10^1)$ Five hundred twelve
	2. 6394	2. $6(10^3) + 3(10^2) + 9(10^1)$ $+ 4(10^0)$ Six thousand, three hundred ninety-four
	3. Write one hundred eighty thousand, five hundred forty-three as a decimal numeral.	3. 180,543

EXERCISES 1.2

Odd Problems 1-45

Write each of the following decimal numerals in expanded notation and in their English word equivalents.

1. 37	2. 84	3. 98	4. 56
5. 122	6. 493	7. 821	8. 333
9. 828	10. 5496	11. 12,517	12. 42,100
13. 243,400	14. 891,540	15. 43,655	16. 99,999
17. 8,400,810	18. 5,663,701	19. 16,302,590	20. 71,500,000
21. 83,000,605	22. 152,403,672	23. 679,078,100	
24. 4,830,231,010	25. 8,572,003,425		

Write each of the following numerals as decimal numerals.

26. seventy-six

27. one hundred thirty-two

28. five hundred eighty

29. three thousand, eight hundred forty-two

30. two thousand, five

31. one hundred ninety-two thousand, one hundred fifty-one

32. seventy-eight thousand, nine hundred two

33. twenty-one thousand, four hundred

34. thirty-three thousand, three hundred thirty-three

35. five million, forty-five thousand

36. five million, forty-five

37. ten million, six hundred thirty-nine thousand, five hundred eighty-two

38. two hundred eighty-one million, three hundred thousand, five hundred one

39. five hundred thirty million, seven hundred

40. seven hundred fifty-eight million, three hundred fifty thousand, sixty

41. ninety million, ninety thousand, ninety

42. eighty-two million, seven hundred thousand

43. one hundred seventy-five million, two

44. thirty-six   45. seven hundred fifty-seven

1.3   ADDITION

The set of numbers $W = \{0, 1, 2, 3, 4, 5, \ldots\}$ is the set of **whole numbers.** The three dots indicate that the pattern of numbers is to continue without end. Thus, 13, 92, and 10,000,000 are all whole numbers. Zero (0) is a special whole number in that it was invented long after the counting numbers (or natural numbers) $N = \{1, 2, 3, 4, \ldots\}$ were in use in ancient number systems. Of course, the numerals used were different from those shown here. The invention of 0 actually led to the place value system.

The whole numbers and the operations of addition, subtraction, multiplication, and division with whole numbers are part of our daily lives. Do you check the addition on your bill at a restaurant? Do you count your change at the supermarket? Do you check the arithmetic on the contract when you buy a new car? Do you know how to compare prices for different amounts of the same item in a store? Do you know how to balance your checking account at the bank? All these questions involve regular use of whole numbers. Clerks and salespersons (and math teachers) can make mistakes in arithmetic, and you should be able to find any such mistakes to save yourself time and money and to avoid many other kinds of problems.

Addition with whole numbers is indicated either by writing the numbers horizontally with a plus sign (+) between them: $6 + 23 + 17$; or by writing the numbers vertically in columns with instructions to add:

$$\begin{array}{rr} \text{Add} & 6 \\ & 23 \\ & \underline{17} \end{array}$$

When writing the numbers in column form, be sure to keep the digits aligned vertically so you will be adding units digits to units digits, tens digits to tens digits, hundreds digits to hundreds digits, and so on. Neatness is a good quality to have in any discipline, but in mathematics it is a necessity.

The numbers being added are called **addends** and the result of the addition is called the **sum.**

$$\begin{array}{rl} 6 & \text{addend} \\ 23 & \text{addend} \\ \underline{17} & \text{addend} \\ 46 & \text{sum} \end{array}$$

Before you can add with speed and accuracy, you must memorize the basic addition facts, which are given in Table 1.1. If you are slow at any of the addition combinations in the table, practice those combinations until you can give the answers immediately. When adding, do not count with your lips or tap with a pencil or your fingers or any other object. Concentrate and force yourself to add mentally. Your adding speed should also increase if you learn to look quickly for combinations of digits that total ten.

TABLE 1.1   BASIC ADDITION FACTS

+	0	1	2	3	4	5	6	7	8	9
0	0	1	2	3	4	5	6	7	8	9
1	1	2	3	4	5	6	7	8	9	10
2	2	3	4	5	6	7	8	9	10	11
3	3	4	5	6	7	8	9	10	11	12
4	4	5	6	7	8	9	10	11	12	13
5	5	6	7	8	9	10	11	12	13	14
6	6	7	8	9	10	11	12	13	14	15
7	7	8	9	10	11	12	13	14	15	16
8	8	9	10	11	12	13	14	15	16	17
9	9	10	11	12	13	14	15	16	17	18

As an aid to improving your skill in addition, write all one hundred possible combinations to be added on a piece of paper in a mixed order. Perform the additions and find the ones you missed. Study these in your spare time until you have confidence that you know them.

EXAMPLE   To add the following numbers, we note the combinations that total ten and find the sums quickly.

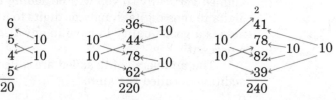

This technique can be used when the combinations are obvious to you. Of course, it would be a great advantage to know the basic addition combinations so well that you can add the digits in order with little or no hesitation.

When two numbers are added, is the order of the numbers important? That is, if we are adding 4 and 9, are we concerned about which number to write first? Do we write $4 + 9$ or $9 + 4$? Or, if we are adding figures in columns, do we write

$$\begin{matrix} 4 \\ 9 \end{matrix} \quad \text{or} \quad \begin{matrix} 9 \\ 4 \end{matrix}$$

If we check the table, we can see that $4 + 9 = 13$ and $9 + 4 = 13$. By looking at the table we can see that reversing the order of any two numbers in the table will not change their sum. This is an important fact about addition with any two whole numbers and is called the **commutative property of addition.** Using letters to represent whole numbers, we can state this property in the following way.

COMMUTATIVE   If a and b are two whole numbers, then $a + b = b + a$.
PROPERTY OF
ADDITION

Add the following three numbers mentally.

$$6 + 3 + 5$$

Which two did you add first? You did add only two and then add the third number to the sum, didn't you? Did you think $9 + 5$ or $6 + 8$? In either case, the answer is 14.

Try three more numbers.

$$8 + 4 + 7$$

Again, you added only two numbers first. Did you think $12 + 7$ or $8 + 11$? The answer, 19, is the same in either case.

Two properties of addition are illustrated here. First, addition is a **binary operation;** that is, only two numbers are added at a time. Second, two different groupings or associations of the numbers still give the same answer. We can write

$$6 + 3 + 5 = (6 + 3) + 5 = 6 + (3 + 5)$$

and $\qquad 8 + 4 + 7 = (8 + 4) + 7 = 8 + (4 + 7)$

This second property involving grouping is called the **associative property of addition.**

ASSOCIATIVE PROPERTY OF ADDITION

If a, b, and c are whole numbers, then

$$a + b + c = (a + b) + c = a + (b + c)$$

Notice that the associative property states that the *grouping,* or *association,* of the numbers together may be changed. If the *order* of two or more numbers is changed, then the commutative property is being used. The following two equations illustrate the use of these properties.

$(15 + 20) + 6 = 15 + (20 + 6)$ $\qquad$ Associative property of addition

$(15 + 20) + 6 = (20 + 15) + 6$ $\qquad$ Commutative property of addition

Another property of addition with whole numbers is the simple relationship that adding 0 gives a sum that is the same as the other addend. For example,

$$8 + 0 = 8, \qquad 0 + 13 = 13, \quad \text{and} \quad 38 + 0 = 38$$

Zero (0) is called the **additive identity,** or the **identity element for addition.**

ADDITIVE IDENTITY

If a is a whole number, then there is a unique whole number 0 with the property that $a + 0 = a$.

SELF QUIZ　　Which property of addition is illustrated?　ANSWERS

1.　$12 + 0 = 12$

2.　$15 + 3 = 3 + 15$

Find the following sums.

3.　57　4.　36
　　98　　　78
　　　　　　89

1.　Additive identity

2.　Commutative property

3.　155　　4.　203

EXERCISES 1.3

Odd Problems 1 - 49

Do the following exercises mentally and write only the answers.

1.　$6 + (3 + 7)$　　　　2.　$(4 + 5) + 6$　　　　3.　$(2 + 3) + 8$

4.　$(2 + 6) + (4 + 5)$　5.　$8 + (3 + 4) + 6$　6.　$9 + (8 + 3)$

7.　$9 + 2 + 8$　　　　　8.　$7 + (6 + 3)$　　　　9.　$(2 + 3) + 7$

10.　$8 + 7 + 2$　　　　11.　$9 + 6 + 3$　　　　12.　$4 + 3 + 6$

13.　$4 + 4 + 4$　　　　14.　$9 + 1 + 5 + 6$　　15.　$5 + 3 + 7 + 2$

16.　$3 + 6 + 2 + 8$　　17.　$5 + 4 + 6 + 4 + 8$　18.　$6 + 2 + 9 + 1$

19.　$4 + 3 + 6 + 6 + 2$　20.　$8 + 8 + 7 + 6 + 3$

Show that the following statements are true by performing the addition. State which property of addition is being illustrated.

21.　$9 + 3 = 3 + 9$　　　　　　　22.　$8 + 7 = 7 + 8$

23.　$4 + (5 + 3) = (4 + 5) + 3$　　24.　$4 + 8 = 8 + 4$

25.　$2 + (1 + 6) = (2 + 1) + 6$　　26.　$(8 + 7) + 3 = 8 + (7 + 3)$

27.　$9 + 0 = 9$　　　　　　　　　28.　$(2 + 3) + 4 = (3 + 2) + 4$

29.　$7 + (6 + 0) = 7 + 6$　　　　30.　$8 + 20 + 1 = 8 + 21$

Copy the following problems and add.

31.	65	32.	24	33.	73	34.	165	35.	876
	43		78		68		276		279
	54		95		98		394		143

36.	268	37.	981	38.	2112	39.	114	40.	1403
	93		146		147		5402		7010
	74		92		904		710		622
	192		17		1005		643		29

41.	213,116	42.	21,442	43.	438,966	44.	123,456
	116,018		32,462		1,572,486		456,123
	722,988		564,792		327,462		879,282
	24,336		801,801		181,753		617,500
	526,968		43,433		90,000		740,765

45. Mr. Jones kept the following mileage records on his car for six months: Jan., 546; Feb., 378; Mar., 496; Apr., 357; May, 503; June, 482. How many miles did he drive in the six months?

46. The Modern Products Corp. showed profits of $1,078,416 in 1974; $1,270,842 in 1975; $2,000,593 in 1976; and $1,963,472 in 1977. What were the company's total profits for the years 1974–77?

47. During six years of college, including two years of graduate school, Fred estimated his expenses each year as $2035; $2842; $2786; $3300; $4000; $3500. What were his total expenses for six years of schooling? (NOTE: He had some financial aid.)

48. Apple County has the following items budgeted: highways, $270,455; salaries, $95,479; maintenance, $127,220. What is the county's total budget for these three items?

49. The numbers of students at South Junior College enrolled in mathematics courses are: 303 in arithmetic; 476 in algebra; 293 in trigonometry; 257 in college algebra; 189 in calculus. Find the total number of students taking math.

50. In one year, the High Price Manufacturing Co. made 2476 refrigerators; 4217 gas stoves; 3947 electric stoves; 9576 tractors; 11,872 electric fans; 1742 air conditioners. What was the total number of appliances High Price produced that year?

1.4 SUBTRACTION

A long-distance runner who is 25 years old had run 17 miles of his usual 19-mile workout when a storm forced him to quit for the day. How many miles short was he of his usual daily training?

What *thinking* did you do to answer the question? You may have thought something like this: "Well, I don't need to know how old the runner is to answer the question, so his age is just extra information. Since he had already run 17 miles, I need to know what to add to 17 miles to get 19 miles. Since 17 + 2 = 19, he was 2 miles short of his usual workout."

In this problem, the sum of two addends was given, and only one of the addends was given. The other addend was the unknown quantity.

$$17 \ + \ \square \ = 19$$
addend   missing addend   sum

As you may know, this kind of addition problem is called **subtraction** and can be written:

$$19 - 17 = \quad \square$$

("Read: "19 minus 17 equals blank")

sum   addend   missing addend
(or difference)

Or
$$\begin{array}{r} 19 \\ -17 \\ \hline \square \end{array}$$

sum
addend
difference or missing addend

In other words, subtraction is a reverse addition, and the missing addend is called the **difference** between the sum and one addend.

Perform the following subtraction mentally: $17 - 8 = \square$. You should think, "8 plus what number gives 17? Since $8 + 9 = 17$, we have $17 - 8 = 9$."

EXAMPLES

1. Find the difference for
$$\begin{array}{r} 75 \\ -48 \end{array}$$

Since thinking of a number to add to 48 to get 75 is difficult, we resort to place value and write the numbers in expanded form.

EXPANDED

$$\begin{array}{rcccc} 75 = & 70 + 5 & = & 60 + 15 \\ -48 = & 40 + 8 & = & 40 + 8 \\ \hline & & & 20 + 7 & = 27 \end{array}$$

(10 is "borrowed" from 70 since no whole number can be added to 8 to get 5.)

The above procedure is commonly written as

$$\begin{array}{r} \overset{6}{\cancel{7}}{}^{1}5 \\ -4\,8 \\ \hline 2\,7 \end{array}$$

(Can you see that $\overset{6}{\cancel{7}}{}^{1}5$ is shorthand for the expanded form $60 + 15$?)

2. Find the difference for $482 - 195$ using place value.

$$\begin{array}{rccccccc} 482 = & 400 + 80 + 2 & = & 400 + 70 + 12 & = & 300 + 170 + 12 \\ -195 = & 100 + 90 + 5 & = & 100 + 90 + 5 & = & 100 + 90 + 5 \\ \hline & & & & & 200 + 80 + 7 & = 287 \end{array}$$

Notice that 10 is "borrowed" from 80, and then 100 is "borrowed" from 400. Do you know why these "borrowings" are necessary?

In shorthand,

$$\begin{array}{r} \overset{317}{\cancel{4}\,\cancel{8}{}^{1}2} \\ -1\,9\,5 \\ \hline 2\,8\,7 \end{array}$$

(You should realize that the numbers are not written in expanded form here but should be visualized that way.)

3. Find the difference for $500 - 132$.

$$500 = 490 + 10 = 400 + 90 + 10$$
$$-\,132 = 130 + 2 = 100 + 30 + 2$$
$$\ 300 + 60 + 8 = 368$$

Or
$$500 = 490 + 10$$
$$-\,132 = 130 + 2$$
$$\ 360 + 8 = 368$$

Or

$$\overset{4\ 9}{\cancel{5}\ \cancel{0}\,{}^{1}0}$$
$$-\,1\ 3\ 2$$
$$\overline{3\ 6\ 8}$$

SELF QUIZ	Find the following differences.	ANSWERS
1.	83 54	**1.** 29
2.	600 368	**2.** 232
3.	7856 6397	**3.** 1459

EXERCISES 1.4

Subtract. Do as many problems as you can mentally.

Odd Problems 1 - 53

1. $8 - 5$	**2.** $19 - 6$	**3.** $14 - 14$	**4.** $17 - 9$	**5.** $20 - 11$
6. $17 - 0$	**7.** $17 - 8$	**8.** $16 - 16$	**9.** $11 - 6$	**10.** $13 - 7$

11. 17 17	**12.** 42 31	**13.** 89 76	**14.** 53 33	**15.** 47 27
16. 96 27	**17.** 23 18	**18.** 74 29	**19.** 61 48	**20.** 52 27
21. 126 32	**22.** 174 48	**23.** 347 129	**24.** 256 118	**25.** 692 217
26. 543 167	**27.** 900 307	**28.** 603 208	**29.** 474 286	**30.** 657 179
31. 7843 6274	**32.** 6793 5827	**33.** 4376 2808	**34.** 3275 1744	**35.** 3546 3546

36. 4900	**37.** 5070	**38.** 8007	**39.** 4065	**40.** 7602
3476	4376	2136	1548	2985

41. 7,085,076
4,278,432

42. 6,543,222
2,742,663

43. 4,000,000
2,993,042

44. 8,000,000
647,561

45. 6,000,000
328,989

46. What number should be added to 978 to get a sum of 1200?

47. If the sum of two numbers is 693, and one of the numbers is 498, what is the other number?

48. A 36-year-old man and his wife both attended school for several years. Including their high-school educations, he attended for 18 years and she attended for 15 years. How many total years of education have they had?

49. Basketball Team A has twelve players and won its first three games by the following scores: 84 to 73, 97 to 78, and 101 to 63. Team B has ten players and won its first three games by 76 to 75, 83 to 70, and 94 to 84. What is the difference between the total of the differences of Team A's scores and its opponents and the total of the differences of Team B's scores and its opponents?

50. The Kingston Construction Co. made a bid of $7,043,272 to build a stretch of freeway, but the Beach City Construction Co. made a lower bid of $6,792,868. How much lower was the Beach City bid?

51. In June, Ms. White opened a checking account and deposited $1342, $238, $57, and $486. She also wrote checks for $132, $76, $25, $42, $480, $90, $17, and $327. What was her balanace at the end of June?

52. In pricing a four-door car, Mr. Kelly found that he would have to pay $5287 plus $420 for air conditioning and $80 for a radio. If he bought a two-door car, he would have to pay only $4839 plus $380 for air conditioning and $70 for a radio. What is the difference in the base prices of the cars? What would be the difference if he wanted a radio included?

53. A manufacturing company had assets of $5,027,479, which included $1,500,000 in real estate. The liabilities were $4,792,023. By how much was the company "in the black"?

54. If the sum is three million, four hundred ninety-two thousand, one hundred eighty-six, and one addend is one million, five hundred eight-three thousand, ninety, find the missing addend.

1.5 BASIC FACTS OF MULTIPLICATION

Subtraction with whole numbers is related to addition, as was discussed in Section 1.4. Multiplication is also related to addition, but in a different way. Suppose you were to use the same number as an addend four times, such as

$$7 + 7 + 7 + 7$$

This sum is not difficult to find, and you might proceed in the following manner.

$$7 + 7 + 7 + 7 = (7 + 7) + 7 + 7$$
$$= (14 + 7) + 7$$
$$= 21 + 7$$
$$= 28$$

Even if you did this type of work mentally, considerable time would be needed to add something like $279 + 279 + 279 + 279$. Multiplication is simply a shorthand method of repeated addition of the same number. Thus,

$$279 + 279 + 279 + 279 = 4 \cdot 279$$

The addend (279) and the number of times it is being used (4) are both called **factors,** and the sum is now called the **product.**

$$4 \quad \cdot \quad 279 = 279 + 279 + 279 + 279 = \quad 1116$$

factor   factor   sum   product

Several notations are useful in various situations to indicate multiplication:

(a) $4 \cdot 279$ (b) 4(279)   (c) (4)279

(d) (4)(279)   (e) 4×279 (f) 279
$$\underline{\times\ 4}$$

(g) Directions are given:

Multiply   279
$$\underline{4}$$

Generally, types (e) 4×279 and (f) 279 will be avoided, since the
$$\underline{\times\ 4}$$
times sign ($\times$) can be easily confused with the letter x.

To change a multiplication problem to a repeated addition problem every time we are to multiply two numbers would be ridiculous. For example, 48 · 137 would mean using 137 as an addend 48 times. The first step in learning the multiplication process is to *memorize* the basic multiplication facts in Table 1.2. The factors in the table are only the digits 0 through 9. Using other factors involves the place value concept, as we shall see.

TABLE 1.2   BASIC MULTIPLICATION FACTS

·	0	1	2	3	4	5	6	7	8	9
0	0	0	0	0	0	0	0	0	0	0
1	0	1	2	3	4	5	6	7	8	9
2	0	2	4	6	8	10	12	14	16	18
3	0	3	6	9	12	15	18	21	24	27
4	0	4	8	12	**16**	20	24	28	32	36
5	0	5	10	15	20	**25**	30	35	40	45
6	0	6	12	18	24	30	**36**	42	48	54
7	0	7	14	21	28	35	42	**49**	56	63
8	0	8	16	24	32	40	48	56	**64**	72
9	0	9	18	27	36	45	54	63	72	**81**

If you have difficulty with *any* of the basic facts in the table, write all the possible combinations in a mixed-up order on a sheet of paper. Write the products down as quickly as you can and then find the ones you missed. Practice these in your spare time until you are sure you know them.

Study the table closely for a few minutes. Do you notice any particular features or patterns? Draw a line diagonally across the table through the numbers 0, 1, 4, 9, 16, 25, 36, 49, 64, 81. Do you notice any pattern in the numbers on either side of this diagonal? You may have observed that the numbers at an upward angle to the right of the diagonal are the same as those at a downward angle to the left of the diagonal. For example, find the number 16 along the diagonal. Upward to the right are the numbers 15, 12, 7, 0, and downard to the left the same numbers appear in the same order — 15, 12, 7, 0:

Downward to the left of 16

$$5 \cdot 3 = 15$$
$$6 \cdot 2 = 12$$
$$7 \cdot 1 = 7$$
$$8 \cdot 0 = 0$$

Upward to the right of 16

$$3 \cdot 5 = 15$$
$$2 \cdot 6 = 12$$
$$1 \cdot 7 = 7$$
$$0 \cdot 8 = 0$$

The patterns indicate that

$$5 \cdot 3 = 3 \cdot 5, \qquad 6 \cdot 2 = 2 \cdot 6, \qquad 7 \cdot 1 = 1 \cdot 7, \qquad 8 \cdot 0 = 0 \cdot 8$$

In general, the product of any two whole numbers is the same regardless of the order of the numbers. This property is called the **commutative property of multiplication.**

COMMUTATIVE
PROPERTY OF
MULTIPLICATION

If a and b are whole numbers, then $a \cdot b = b \cdot a$.

As an illustration of the commutative property of multiplication with reference to repeated addition, consider

$$4 \cdot 7 = 7 + 7 + 7 = 7 = 28$$
and $\qquad 7 \cdot 4 = 4 + 4 + 4 + 4 + 4 + 4 + 4 = 28$

Again, reference to the table of multiplication facts shows that the first row and first column are all 0's. This indicates that multiplication by 0 gives 0. In fact, multiplication of any whole number by 0 gives a product of 0, and this result is called the **zero factor law.**

ZERO FACTOR
LAW

If a is a whole number, then $a \cdot 0 = 0$.

EXAMPLES

1. $75 \cdot 0 = 0$
2. $0 \cdot 942 = 0$

Table 1.2 also shows the second row (opposite 1) to be identical to the heading row and the second column (under 1) to be identical to the heading column. Thus, the indication is that one (1) times a number gives that number. In fact, 1 is called the **multiplicative identity.**

EXAMPLE

$$5 \cdot 1 = 1 + 1 + 1 + 1 + 1 = 5$$
$$1 \cdot 5 = 5$$

MULTIPLICATIVE
IDENTITY

If a is a whole number, then there is a unique whole number 1 with the property that $a \cdot 1 = a$.

Naming 1 the multiplicative identity corresponds with naming 0 the additive identity.

$$6 + 0 = 6 \quad \text{and} \quad 6 \cdot 1 = 6 \qquad \text{The number 6 is its sum with 0 and its product with 1.}$$

The final property of multiplication to be discussed here has to do with multiplying more than two numbers. Multiplication, as addition, is a **binary operation,** and only two numbers may be multiplied at a time. Therefore, if three (or more) factors are to be multiplied, a decision must be made as to which two factors are to be multiplied together first. For example, to find the product $2 \cdot 3 \cdot 7$, would you multiply $2 \cdot 3$ first or $3 \cdot 7$ first? Which way is correct?

$$2 \cdot 3 \cdot 7 = (2 \cdot 3) \cdot 7 = 6 \cdot 7 = 42$$
$$2 \cdot 3 \cdot 7 = 2 \cdot (3 \cdot 7) = 2 \cdot 21 = 42$$

This example shows that both ways are correct. We seem to have a choice of procedures. Try one more, $9 \cdot 2 \cdot 6$.

$$9 \cdot 2 \cdot 6 = (9 \cdot 2) \cdot 6 = 18 \cdot 6 = 108$$
$$9 \cdot 2 \cdot 6 = 9 \cdot (2 \cdot 6) = 9 \cdot 12 = 108$$

These examples illustrate the **associative property of multiplication** with whole numbers.

ASSOCIATIVE
PROPERTY OF
MULTIPLICATION

If a, b, and c are whole numbers, then $a \cdot b \cdot c = a(b \cdot c) = (a \cdot b)c$.

This property says that regardless of the grouping (or association) of the factors, the product will be the same.

SELF QUIZ	Which property of multiplication is illustrated?	ANSWERS
	1. $5 \cdot (3 \cdot 7) = (5 \cdot 3) \cdot 7$	1. Associative property
	2. $32 \cdot 0 = 0$	2. Zero factor law
	3. $6 \cdot 8 = 8 \cdot 6$	3. Commutative property

EXERCISES 1.5

Do the following problems mentally and write only the answers.

1. $8 \cdot 9$ 2. $6 \cdot 9$ 3. $8 \cdot 7$ 4. $3 \cdot 4$ 5. $0 \cdot 6$

6. $1 \cdot 7$ 7. $4 \cdot 8$ 8. $2 \cdot 7 \cdot 3$ 9. $6 \cdot 8 \cdot 2$ 10. $9 \cdot 7 \cdot 1$

11. $3(5 \cdot 1)$ 12. $(2 \cdot 6)3$ 13. $(4 \cdot 5)5$ 14. $4(5 \cdot 5)$ 15. $(2)(6)(0)(4)$

For each property listed, give two examples that illustrate the property with whole numbers.

16. Associative property of multiplication

17. Associative property of addition

18. Commutative property of addition

19. Commutative property of multiplication

20. Multiplicative identity

21. Zero factor law

22. Additive identity

23. In your own words, describe the meaning of the term *factor*.

24. Using repeated addition, show that

(a) $3 \cdot 9 = 9 \cdot 3$ (b) $2(4 \cdot 6) = (2 \cdot 4)6$

25. Fill in the missing numbers in the chart. Five is added to the given number. The sum is then doubled and ten is subtracted from the product. The answer is written in the last column. [HINT: To fill in the last two rows, you must think backwards.]

GIVEN NUMBER	ADD 5	DOUBLE	SUBTRACT 10
2	7	14	4
0	5	?	?
1	6	12	?
7	?	?	?
?	?	?	16
?	?	?	10

26. Make up a chart of your own similar to that in Problem 25. Head the columns: Given Number, Add 6, Triple, Subtract 18. Do you notice any pattern common to the results? Is there any relationship between these results and those in Problem 25?

1.6 MULTIPLICATION BY POWERS OF TEN

Some of the powers of ten are shown here for convenience.

$$10^0 = 1$$
$$10^1 = 10$$
$$10^2 = 10 \cdot 10 = 100$$
$$10^3 = 10 \cdot 10 \cdot 10 = 1000$$
$$10^4 = 10 \cdot 10 \cdot 10 \cdot 10 = 10,000$$

They are very useful in explaining multiplication with whole numbers in general.

Multiplication by 10

$10 \cdot 2 = 2 + 2 + 2 + 2 + 2 + 2 + 2 + 2 + 2 + 2 = 20$
$2 \cdot 10 = 10 + 10 = 20$

$10 \cdot 5 = 5 + 5 + 5 + 5 + 5 + 5 + 5 + 5 + 5 + 5 = 50$
$5 \cdot 10 = 10 + 10 + 10 + 10 + 10 = 50$

$10 \cdot 9 = 9 + 9 + 9 + 9 + 9 + 9 + 9 + 9 + 9 + 9 = 90$
$9 \cdot 10 = 10 + 10 + 10 + 10 + 10 + 10 + 10 + 10 + 10 = 90$

$10 \cdot 32 = 32 + 32 + 32 + 32 + 32 + 32 + 32 + 32 + 32 + 32 = 320$
$32 \cdot 10 = \underbrace{10 + 10 + 10 + \cdots + 10}_{32 \text{ 10's}} = 320$

Multiplication by 100

$100 \cdot 2 = 2 \cdot 100 = 100 + 100 = 200$
$100 \cdot 5 = 5 \cdot 100 = 100 + 100 + 100 + 100 + 100 = 500$
$100 \cdot 9 = 9 \cdot 100 = 100 + 100 + 100 + 100 + 100 + 100$
$$+ 100 + 100 + 100 = 900$$

Multiplication by 1000

$1000 \cdot 3 = 3 \cdot 1000 = 1000 + 1000 + 1000 = 3000$
$1000 \cdot 4 = 4 \cdot 1000 = 1000 + 1000 + 1000 + 1000 = 4000$
$1000 \cdot 7 = 7 \cdot 1000 = 1000 + 1000 + 1000 + 1000 + 1000$
$$+ 1000 + 1000 = 7000$$

There is a definite pattern to the products when one of the factors is a power of ten:

$$6 \cdot 1 = 6$$
$$6 \cdot 10 = 60$$
$$6 \cdot 100 = 600$$
$$6 \cdot 1000 = 6000$$

If one of two whole number factors is 1000, the product will be the other factor with three zeros (000) written to the right of it. Two zeros (00) are written to the right of the other factor when multiplying by 100, and one zero (0) is written when multiplying by 10. Will multiplication by one million (1,000,000) result in writing six zeros to the right of the other factor? The answer is yes.

Many products can be found mentally by using the properties of multiplication and the techniques of multiplying by powers of ten. The processes are written out in the following examples, but they can easily be done mentally with practice.

EXAMPLES

1. $6 \cdot 90 = 6(9 \cdot 10) = (6 \cdot 9)10 = 54 \cdot 10 = 540$

2. $3 \cdot 400 = 3(4 \cdot 100) = (3 \cdot 4)100 = 12 \cdot 100 = 1200$

3. $2 \cdot 300 = 2(3 \cdot 100) = (2 \cdot 3)100 = 6 \cdot 100 = 600$

4. $6 \cdot 700 = 6(7 \cdot 100) = (6 \cdot 7)100 = 42 \cdot 100 = 4200$

5. $40 \cdot 30 = (4 \cdot 10)(3 \cdot 10) = (4 \cdot 3)(10 \cdot 10) = 12 \cdot 100 = 1200$

6. $50 \cdot 700 = (5 \cdot 10)(7 \cdot 100) = (5 \cdot 7)(10 \cdot 100) = 35 \cdot 1000$
    $= 35,000$

7. $200 \cdot 800 = (2 \cdot 100)(8 \cdot 100) = (2 \cdot 8)(100 \cdot 100) = 16 \cdot 10,000$
    $= 160,000$

8. $7000 \cdot 9000 = (7 \cdot 1000)(9 \cdot 1000) = (7 \cdot 9)(1000 \cdot 1000)$
    $= 63 \cdot 1,000,000 = 63,000,000$

[NOTE: To find the product in each example, the nonzero digits are multiplied and the appropriate number of zeros is written.]

EXERCISES 1.6

Use the techniques of multiplying by the powers of ten to find the following products mentally.

1. $25 \cdot 10$	**2.** $76 \cdot 100$	**3.** $47 \cdot 1000$	**4.** $18 \cdot 10^0$
5. $72 \cdot 10^0$	**6.** $13 \cdot 1$	**7.** $50 \cdot 60$	**8.** $90 \cdot 80$
9. $20 \cdot 20$	**10.** $60 \cdot 60$	**11.** $30 \cdot 40$	**12.** $70 \cdot 80$
13. $90 \cdot 70$	**14.** $300 \cdot 30$	**15.** $200 \cdot 20$	**16.** $500 \cdot 70$
17. $500 \cdot 30$	**18.** $120 \cdot 30$	**19.** $130 \cdot 40$	**20.** $200 \cdot 60$
21. $200 \cdot 80$	**22.** $300 \cdot 600$	**23.** $100 \cdot 100$	**24.** $100 \cdot 50$
25. $3000 \cdot 20$	**26.** $500 \cdot 50$	**27.** $400 \cdot 30$	**28.** $50 \cdot 200$
29. $40 \cdot 6000$	**30.** $2000 \cdot 400$	**31.** $80 \cdot 600$	**32.** $3000 \cdot 5000$
33. $20,000 \cdot 30$	**34.** $4000 \cdot 4000$	**35.** $70 \cdot 9000$	**36.** $800 \cdot 4000$

1.7 MULTIPLICATION OF WHOLE NUMBERS

Consider the product $3(70 + 2)$. To find this product, you would probably add $70 + 2 = 72$ and then multiply.

$$\begin{array}{r} 72 \\ \underline{3} \\ 216 \end{array}$$

This is, of course, a correct procedure.

However, observe that

$$3(70 + 2) = 3 \cdot 70 + 3 \cdot 2$$
$$= 210 + 6$$
$$= 216$$

gives the same result. This is an example of the **distributive property of multiplication over addition.**

DISTRIBUTIVE
PROPERTY OF
MULTIPLICATION
OVER ADDITION

If a, b, and c are whole numbers, then $a(b + c) = a \cdot b + a \cdot c$.

The following examples illustrate how the distributive property is used to explain the procedure for multiplying two whole numbers.

EXAMPLES

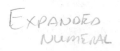

1. Multiply $68 \cdot 4$.

$$
\begin{array}{r}
60 + 8 \\
4 \\
\hline
240 + 32 = 272
\end{array}
\qquad \text{or} \qquad
\begin{array}{r}
68 \\
4 \\
\hline
32 \\
240 \\
\hline
272
\end{array}
$$

partial products   product   32, 240 } partial products   272 product

2. Find the product of the factors 25 and 6.

$$
\begin{array}{r}
20 + 5 \\
6 \\
\hline
120 + 30 = 150
\end{array}
\qquad \text{or} \qquad
\begin{array}{r}
25 \\
6 \\
\hline
30 \\
120 \\
\hline
150
\end{array}
$$

partial products   product   30, 120 } partial products   150 product

Examples 1 and 2 use the concept of place value and the distributive property. That is, $68 = 60 + 8$ and

$$4 \cdot 68 = 4(60 + 8) = 4 \cdot 60 + 4 \cdot 8 = 240 + 32 = 272$$

3. Now consider $37 \cdot 42$. The distributive property may be used three times.

(a) $37 \cdot 42 = (30 + 7)(40 + 2)$
$\qquad = (30 + 7)40 + (30 + 7)2$ (First use of distributivity)
$\qquad = 30 \cdot 40 + 7 \cdot 40 + 30 \cdot 2 + 7 \cdot 2$ (Second and third uses of distributivity)
$\qquad = 1200 + 280 + 60 + 14$ (Notice how the product $30 \cdot 40$ can be found quickly, using the techniques of Section 1.6.)
$\qquad = 1554$

(b) In another form,

42	40 + 2	The partial products are

$$
\begin{array}{r}
42 \\
37 \\
\hline
\end{array}
\qquad
\begin{array}{r}
40 + 2 \\
30 + 7 \\
\hline
280 + 14 \\
1200 + 60 \\
\hline
1200 + 340 + 14 = 1554
\end{array}
$$

The partial products are

$7 \cdot 2 = 14$
$7 \cdot 40 = 280$
$30 \cdot 2 = 60$
$30 \cdot 40 = 1200$

(c) In a third form,

$$
\begin{array}{rl}
42 & \text{factor} \\
37 & \text{factor} \\
\hline
14 & \\
280 & \\
60 & \text{partial products} \\
1200 & \\
\hline
1554 & \text{product}
\end{array}
$$

The partial products are

$7 \cdot 2 = 14$
$7 \cdot 40 = 280$
$30 \cdot 2 = 60$
$30 \cdot 40 = 1200$

These examples are given to show the underlying structure of the process of multiplication. As most of you know, the process can be greatly accelerated by noticing some key relationships.

1. When the units digit in one factor is the multiplier, the first partial product is written with its units digit directly under the units digits of the factors. For example,

$$
\begin{array}{r}
56 \\
3 \\
\hline
18 \\
150
\end{array}
$$

Notice that in the partial product 18, the 8 is directly under the 3.

2. The tens digit of the first partial product is always added to the tens digit of the second partial product.

3. The units digit of the second partial product is always zero (0), since the multiplier is a multiple of 10.

We can increase the speed of multiplication by doing the following.

(a) $\begin{array}{r} 56 \\ 3 \\ \hline 8 \end{array}$ Write the units digit (8) of the first partial product (18) directly under the units digit (3).

(b) $\begin{array}{r} 56 \\ 3 \\ \hline 8 \end{array}$ Multiply the units digit (3) by the tens digit in the other factor (5): $(3 \cdot 5 = 15)$.

(c) $\overset{1}{56}$ Realizing that the 5 in 15 should be in the tens column, mentally add
    $\underline{3}$ the tens digit (1) in the first partial product (18) to the 15:
    8 $(15 + 1 = 16)$.

(d) $\overset{1}{56}$ Since the 6 in 16 should be in the tens column, simply write it next to
    $\underline{3}$ the 8 to obtain the final result: 168.
    168

Of course, thinking in such detail is usually unnecessary. You may already be aware of this technique and be able to do it rapidly. The hope here is that, after careful discussion, you will begin to understand why the rules of multiplication work as they do.

The process illustrated in multiplying $3 \cdot 56$ can be carried over to include multiplication when the multiplier has more than one digit. For example, consider $58 \cdot 43$. We can follow the process and get the first partial product.

$$\overset{2}{43}$$
$$\underline{58}$$
$$344$$

Now, to obtain the second partial product, proceed by multiplying by the 5 just as with the 8, only be sure to write the first digit (5) in the column directly under the 5.

$$\overset{1}{43}$$
$$\underline{58}$$
$$344$$
$$215$$

Then add the partial products.

$$43$$
$$\underline{58}$$
$$344$$
$$\underline{215}$$
$$2494$$

Compare this with the method previously discussed, and you will see the difference in time saved.

43	43
58	58
24	344
320	215
150	2494
2000	
2494	

Either method is correct and may be used successfully.

Suppose you want to multiply 2000 by 423. Would you write

$$
\begin{array}{r}
2000 \\
423 \\
\hline
6000 \\
4000 \\
8000 \\
\hline
846000
\end{array}
\qquad \text{or} \qquad
\begin{array}{r}
423 \\
2000 \\
\hline
000 \\
000 \\
000 \\
846 \\
\hline
846000
\end{array}
$$

Go OVER →

Both answers are correct; neither technique is wrong. However, writing all the 0's is a waste of time, so knowledge about powers of ten is helpful. We can write

$$
\begin{array}{r}
423 \\
2000 \\
\hline
846000
\end{array}
$$

423
2 000
000

DO SELF QUIZ

We know $2000 = 2 \cdot 1000$, so we are simply multiplying $423 \cdot 2$, then the result by 1000.

EXAMPLES

1. Multiply $596 \cdot 3000$.

$$
\begin{array}{r}
596 \\
3000 \\
\hline
1,788,000
\end{array}
$$

2. Multiply $265 \cdot 15,000$.

$$
\begin{array}{r}
265 \\
15,000 \\
\hline
1\ 325\ 000 \\
2\ 65 \\
\hline
3,975,000
\end{array}
$$

SELF QUIZ	Find the following products.	ANSWERS
	1. 18 24	1. 432
	2. 300 500	2. 150,000
	3. 129 39	3. 5031

EXERCISES 1.7

Odd 1-49
And 51-54

Find the following products by writing in all the partial products.

1. 56	**2.** 27	**3.** 48	**4.** 65	**5.** 43	**6.** 72	**7.** 91
4	6	9	5	8	6	5

8. 39	**9.** 84	**10.** 95	**11.** 42	**12.** 25	**13.** 15	**14.** 29
2	3	8	56	33	22	41

15. 67	**16.** 54	**17.** 48	**18.** 93	**19.** 83	**20.** 96
36	27	20	30	85	62

4
8
12
16
20
24
28
32

Multiply each of the following.

21. 17	**22.** 28	**23.** 20	**24.** 16	**25.** 25	**26.** 93	**27.** 24
32	91	44	26	15	47	86

Go over same as pg 25

28. 72	**29.** 12	**30.** 81	**31.** 126	**32.** 232	**33.** 114	**34.** 72
65	13	36	41	76	25	106

35. 207	**36.** 420	**37.** 200	**38.** 849	**39.** 673	**40.** 192
143	104	49	205	186	467

Go back to pg 27
Go over same as pg 27

Multiply, using your knowledge of powers of ten.

41. 52 · 600	**42.** 930 · 72	**43.** 76 · 5000	**44.** 8000 · 500
45. 68 · 7300	**46.** 320 · 4700	**47.** 5300 · 41	**48.** 157 · 6000
49. 48 · 5200	**50.** 39 · 23,000		

51. Find the product of three thousand, four hundred seventy-one and one hundred sixty-eight.

52. Find the sum of seventy-seven and two hundred fourteen. Find the sum of eighty-six and thirty-four. Find the product of the sums.

53. An automobile dealer orders 213 new cars on January 1. He returns 16 of them because of damage in shipment. What does he owe the company if he pays on February 13 and pays $2315 per car?

54. An apartment owner receives $125 per month on each of six apartments and $190 per month on each of five others when they are occupied. If two of the higher priced units were vacant for three months, what would her income be for the three months?

55. If 746 is added to the product of 58 and 32, what is the sum? If 598 is subtracted from the previous sum, what is the difference? Doubling the last result is not necessary.

answer incorrect

check

1.8   DIVISION WITH WHOLE NUMBERS

Since $6 \cdot 13 = 78$, 6 and 13 are both factors of 78. They are also called **divisors** of 78. The process of **division** can be thought of as reverse multiplication. For example, why is 78 divided by 13 equal to 6, and 78 divided by 6 equal to 13? The answer is that $78 \div 13 = 6$ and $78 \div 6 = 13$ because $6 \cdot 13 = 78$.

Many times the question of division does not involve factors (or exact divisors) of a number. In these cases, division can be thought of as asking how many times is one number contained in another, and is anything left over?

Did you ever wonder why you subtract in division? The reason is that division is repeated subtraction, just as multiplication is repeated addition. Examples 1 and 2 illustrate this relationship. The relationship between division and multiplication is used to check the answers for possible mistakes.

EXAMPLES

1. Find how many times 8 is contained in 200.

$$
\begin{array}{r}
8)\overline{200} \\
-\,160 \leftarrow 20 \text{ eights} \\
\hline
40 \\
-\,40 \leftarrow \underline{5} \text{ eights} \\
\hline
0 \quad 25 \text{ eights}
\end{array}
$$

$$
\begin{array}{r}
25 \leftarrow \text{quotient} \\
8)\overline{200} \leftarrow \text{dividend} \\
\text{divisor}
\end{array}
$$

8 is called the **divisor.**
200 is called the **dividend.**
25 is called the **quotient.**

In this example, $8 \cdot 25 = 200$, and 8 and 25 are both factors of 200.

If, after subtracting the divisor as many times as possible, a number less than the divisor remains, this number is called the **remainder.** If the remainder is 0, as in Example 1, then the divisor and quotient are factors of the dividend.

2. Find the quotient and remainder for $194 \div 7$.

$$
\begin{array}{r}
7)\overline{194} \\
-\,140 \leftarrow 20 \text{ sevens} \\
\hline
54 \\
49 \leftarrow \underline{7} \text{ sevens} \\
\hline
5 \quad 27 \text{ sevens}
\end{array}
$$

remainder   quotient

GO OVER CHECK! NEXT PAGE

Check:

27	189
7	+ 5
189	194

To check division, multiply the divisor and quotient, then add the remainder. The sum should be the dividend. If it is not, a mistake has been made and the division process should be repeated.

Examples 1 and 2 have shown how division is related to repeated subtraction. The notation and placement of the quotient probably seemed strange and are not normally recommended because, as you may have guessed, there is a shorthand approach. But the understanding of division actually comes from learning the processes illustrated in Examples 1 and 2.

The shorthand method for division is known as the **division algorithm*** and is illustrated in Examples 3, 4, and 5.

3. $2076 \div 8$

2 hundreds   5 tens   9 units

(a)
```
        2
   8) 2076
    - 1600
```

(b)
```
       25
  8) 2076
   - 1600
     476
   - 400
      76
```

(c)
```
      259  quotient
 8) 2076
  - 1600
    476
  - 400
     76
  -  72
      4  remainder
```

Eliminating excess numbers, the process can be shortened as follows:

The problem
```
      259
 8) 2076
  - 1600
    476
  - 400
     76
  -  72
      4
```
becomes
```
      259
 8) 2076
  - 16
     47
   - 40
     76
   - 72
      4
```

Bring down 7 only since 8 is trial divided into 47. Now bring down the 6 to divide 8 into 76.

---

*An algorithm is a process or pattern of steps to be followed in working with numbers.

4. $746 \div 32$

(a) $32\overline{)746}$
$$\frac{2}{32\overline{)746}}$$
$$\frac{64}{10}$$

Trial divide 30 into 70 or 3 into 7 giving 2 in the tens position. Note that 10 is less than 32.

(b)
$$\frac{23\ R10}{32\overline{)746}}$$
$$\frac{64}{106}$$
$$\frac{96}{10}$$

Trial divide 30 into 100 or 3 into 10 giving 3 in the units position.

Check:

23		736
32		+ 10
46		746
69		
736		

5. $7492 \div 47$

(a)
$$\frac{1}{47\overline{)7492}}$$
$$\frac{47}{27}$$

Trial divide 40 into 70 or 4 into 7 giving 1 in the 100's position. Note that 27 is less than 47.

(b)
$$\frac{16}{47\overline{)7492}}$$
$$\frac{47}{279}$$
$$282$$

Trial divide 40 into 270 or 4 into 27. But the trial quotient 6 is too large, since $6 \cdot 47 = 282$, and 282 is larger than 279.

$$\frac{15}{47\overline{)7492}}$$
$$\frac{47}{279}$$
$$\frac{235}{44}$$

Now the trial quotient is 5. Since the 44 is smaller than 47, 5 is the desired quotient.

(c)
$$\frac{159\ R19}{47\overline{)7492}}$$
$$\frac{47}{279}$$
$$\frac{235}{442}$$
$$\frac{423}{19}$$

Trial divide 40 into 400. But this gives 10, and 10 is not a digit. We can also see that $10 \cdot 47 = 470$, which is larger than 442. So, try 9.

SELF QUIZ	Find the quotient and remainder for each of the following problems.	ANSWERS
	1. $325 \div 7$	**1.** 46 R3
	2. $16\overline{)324}$	**2.** 20 R4
	3. $41\overline{)24682}$	**3.** 602 R0

EXERCISES 1.8

Find the quotient and remainder for each of the following problems by using the method of repeated subtraction shown in Examples 1 and 2.

1. $210 \div 7$	**2.** $140 \div 14$	**3.** $168 \div 8$	**4.** $70 \div 5$
5. $132 \div 11$	**6.** $120 \div 4$	**7.** $75 \div 15$	**8.** $51 \div 3$
9. $52 \div 8$	**10.** $44 \div 6$	**11.** $600 \div 25$	**12.** $413 \div 20$
13. $161 \div 15$	**14.** $182 \div 13$	**15.** $150 \div 13$	**16.** $500 \div 14$
17. $205 \div 5$	**18.** $321 \div 7$	**19.** $1042 \div 22$	**20.** $1461 \div 12$
21. $2817 \div 12$	**22.** $5684 \div 42$	**23.** $6791 \div 32$	**24.** $4872 \div 12$
25. $4864 \div 16$			

Divide and check using the division algorithm.

26. $6\overline{)32}$	**27.** $7\overline{)17}$	**28.** $4\overline{)25}$	**29.** $5\overline{)35}$
30. $8\overline{)48}$	**31.** $6\overline{)72}$	**32.** $9\overline{)81}$	**33.** $2\overline{)76}$
34. $3\overline{)98}$	**35.** $14\overline{)52}$	**36.** $12\overline{)108}$	**37.** $11\overline{)424}$
38. $16\overline{)128}$	**39.** $20\overline{)305}$	**40.** $18\overline{)206}$	**41.** $30\overline{)847}$
42. $10\overline{)423}$	**43.** $15\overline{)750}$	**44.** $13\overline{)260}$	**45.** $17\overline{)340}$
46. $12\overline{)360}$	**47.** $19\overline{)7603}$	**48.** $16\overline{)4813}$	**49.** $11\overline{)4406}$
50. $13\overline{)3917}$	**51.** $73\overline{)148}$	**52.** $68\overline{)207}$	**53.** $49\overline{)993}$
54. $50\overline{)3065}$	**55.** $40\overline{)2163}$	**56.** $105\overline{)210}$	**57.** $116\overline{)232}$
58. $213\overline{)4760}$	**59.** $716\overline{)3056}$	**60.** $630\overline{)4768}$	**61.** $414\overline{)83276}$
62. $502\overline{)98762}$	**63.** $317\overline{)70365}$	**64.** $471\overline{)50612}$	**65.** $215\overline{)64930}$
66. $342\overline{)157904}$	**67.** $627\overline{)191235}$	**68.** $171\overline{)35226}$	**69.** $401\overline{)277893}$
70. $232\overline{)141752}$			

71. If one factor of four thousand, one hundred sixteen is forty-two, find the other factor.

ASSIGN
ODD 1-61
AND 71-74

72. What number multiplied by seventy-three gives a product of one thousand, six hundred six?

73. What is $98 \div 7$? What is $7 \div 98$? Are the results equal? What can be said about division under these circumstances?

74. One man is five years older than a second man. If the first man is 60 years old, how old is the second man?

1.9   ORDER OF OPERATIONS AND AVERAGE

What is $24 \div 6$? Well, $24 \div 6 = 4$ because $24 = 6 \cdot 4$. What is $6 \div 24$? Well,

$$
\begin{array}{r}
0 \\
24\overline{)6} \\
0 \\
\hline
6
\end{array}
$$

Thus, $6 \div 24 = 0$ R6 because $6 = 0 \cdot 24 + 6$. Therefore, $24 \div 6 \neq 6 \div 24$. And, therefore, **division is not commutative.**

Is division associative? What do you do with $36 \div 12 \div 3$? If division is associative, then

$$36 \div (12 \div 1\overset{3}{\cancel{3}}) \text{ should be the same as } (36 \div 12) \div 3$$

But, $36 \div (12 \div 3) = 36 \div 4 = 9$

and $(36 \div 12) \div 3 = 3 \div 3 = 1$

Therefore, $36 \div (12 \div 3) \neq (36 \div 12) \div 3$

and **division is not associative.**

Which answer is correct, 9 or 1? Both cannot be correct! Some agreement must be made as to the order of operations.

In fact, agreement has been reached for simplifying any numerical expression involving addition, subtraction, multiplication, division, and exponents.

RULES FOR ORDER OF OPERATIONS

1. First, simplify expressions within parentheses.

2. Second, find any powers indicated by exponents.

3. Third, moving from left to right, perform any multiplications or divisions in the order they appear.

4. Fourth, moving from left to right, perform any additions or subtractions in the order they appear.

Go to Pg 36

EXAMPLES

1. Suppose we have $14 \div 7 + 3 \cdot 2 - 5$. According to agreement,

$$14 \div 7 + 3 \cdot 2 - 5$$

$$= \quad 2 \quad + \quad 6 \quad - 5 = 8 - 5 = 3$$

2. Suppose we have $3 \cdot 6 \div 9 - 1 + 4 \cdot 7$. According to agreement,

$$3 \cdot 6 \div 9 - 1 + 4 \cdot 7$$

$$= \quad 18 \quad \div 9 - 1 + \quad 28 = 2 - 1 + 28 = 1 + 28 = 29$$

3. Simplify $(6 + 2) + (8 + 1) \div 9$.

$$(6 + 2) + (8 + 1) \div 9$$

$$= \quad 8 \quad + \quad 9 \quad \div 9 = 8 + 1 = 9$$

4. Simplify $(4 + 3)6 - 2 + 18 \div 6$.

$$(4 + 3)6 - 2 + 18 \div 6$$

$$= \quad 7 \cdot 6 - 2 + 18 \div 6$$

$$= \quad 42 \quad - 2 + \quad 3 = 43$$

5. Simplify $2 \cdot 3^2 + 18 \div 3^2$.

$$2 \cdot 3^2 + 18 \div 3^2$$

$$= 2 \cdot 9 \ + 18 \div 9$$

$$= \quad 18 \quad + \quad 2 = 20$$

6. Simplify $3 \cdot 5^2 \div 15 + 30 - 2^3 \cdot 3$.

$$3 \cdot 5^2 \div 15 + 30 - 2^3 \cdot 3$$

$$= 3 \cdot 25 \div 15 + 30 - 8 \cdot 3$$

$$= \quad 75 \quad \div 15 + 30 - \quad 24$$

$$= \quad 5 \quad + 30 - 24 = 11$$

Another topic related to addition and division and the order of these operations is **average.** In almost any newspaper or magazine the term *average* appears quite frequently. We read about the Dow Jones stock averages, the average income of American families, the batting average of a baseball player, the average IQ of five-year-old children in the local school, and so on. The average of a set of numbers is a kind of "middle number" of the set.* **The average of a set of numbers can be defined as the number found by adding the numbers in the set, then dividing this sum by the number of numbers in the set†**

EXAMPLE   Find the average of the set of numbers {32, 47, 23}.

$$
\begin{array}{r}
32 \\
47 \\
\underline{23} \\
102
\end{array}
\qquad
\begin{array}{r}
34 \quad \text{(average)} \\
3\overline{)102} \\
\underline{9} \\
12 \\
\underline{12} \\
0
\end{array}
$$

The sum, 102, is divided by 3 because there are three numbers being added.

The average of a set of whole numbers need not be a whole number. However, in this section, the problems will be set up so that the averages will be whole numbers. Other cases will be discussed later in the chapters on fractions and decimals (Chapters 3 and 4).

The average of a set of numbers can be very useful, but it can also be misleading. Judging the importance of an average is up to you, the reader of the information. For example, suppose five people had the following incomes for one year: {$8,000, $9,000, $10,000, $11,000, $12,000}. The average of these numbers is $10,000, as shown below:

$$
\begin{array}{r}
\$\ 8,000 \\
9,000 \\
10,000 \\
11,000 \\
\underline{12,000} \\
\$50,000
\end{array}
\qquad
\begin{array}{r}
\$10,000 \\
5\overline{)50,000} \\
\underline{50,000} \\
0
\end{array}
$$

Now consider the incomes {$1,000, $1,000, $1,000, $1,000, $46,000}. Averaging again gives $10,000:

$$
\begin{array}{r}
\$\ 1,000 \\
1,000 \\
1,000 \\
1,000 \\
\underline{46,000} \\
\$50,000
\end{array}
\qquad
\begin{array}{r}
\$10,000 \\
5\overline{)50,000} \\
\underline{50,000} \\
0
\end{array}
$$

---

*Such terms as the "average housewife" or "average mailman" are not related to numbers and are not so easily defined as the average of a set of numbers.
†This average is also called the arithmetic average, or mean.

In the first case, the average of $10,000 serves well as a "middle score" or "representative" of all the incomes. However, in the second case, none of the incomes is even close to $10,000. The one large income completely destroys the "representativeness" of the average. Thus, it is useful to see the numbers or at least know something about the numbers before attaching too much importance to an average.

In statistics, other measures of "representativeness" such as the median (actual middle score) and mode (most frequent score) are discussed. Either the median or the mode gives better information about the second example than the average does.

SELF QUIZ	Find the value for each of the following expressions using the rules for order of operations.	ANSWERS
	1. $15 \div 5 + 10 \cdot 2$	1. 23
	2. $3 \cdot 2^3 - 12 - 3 \cdot 2^2$	2. 0
	3. Find the average of the set of numbers $\{83, 77, 92\}$.	3. 84

EXERCISES 1.9

Find the value of each of the following expressions using the rules for order of operations.

1. $4 \div 2 + 7 - 3 \cdot 2$
2. $8 \cdot 3 \div 12 + 13$
3. $6 + 3 \cdot 2 - 10 \div 2$
4. $14 \cdot 3 \div 7 \div 2 + 6$
5. $6 \div 2 \cdot 3 - 1 + 2 \cdot 7$
6. $5 \cdot 1 \cdot 3 - 4 \div 2 + 6 \cdot 3$
7. $72 \div 4 \div 9 - 2 + 3$
8. $14 + 63 \div 3 - 35$
9. $(2 + 3 \cdot 4) \div 7 + 3$
10. $(2 + 3) \cdot 4 \div 5 + 3 \cdot 2$
11. $(7 - 3) + (2 + 5) \div 7$
12. $16(2 + 4) - 90 - 3 \cdot 2$
13. $35 \div (6 - 1) - 5 + 6 \div 2$
14. $22 - 11 \cdot 2 + 15 - 5 \cdot 3$
15. $(42 - 2 \div 2 \cdot 3) \div 13$
16. $18 + 18 \div 2 \div 3 - 3 \cdot 1$
17. $4(7 - 2) \div 10 + 5$
18. $(33 - 2 \cdot 6) \div 7 + 3 - 6$
19. $72 \div 8 + 3 \cdot 4 - 105 \div 5$
20. $6(14 - 6 \div 2 - 11)$
21. $48 \div 12 \div 4 - 1 + 6$
22. $5 - 1 \cdot 2 + 4(6 - 18 \div 3)$
23. $8 - 1 \cdot 5 + 6(13 - 39 \div 3)$
24. $(21 \div 7 - 3)42 + 6$
25. $16 - 16 \div 2 - 2 + 7 \cdot 3$
26. $(135 \div 3 + 21 \div 7) \div 12 - 4$
27. $(13 - 5) \div 4 + 12 \cdot 4 \div 3 - 72 \div 18 \cdot 2 + 16$

28. $15 \div 3 + 2 - 6 + (3)(2)(18)(0)(5)$

29. $100 \div 10 \div 10 + 1000 \div 10 \div 10 \div 10 - 2$

30. $[(85 + 5) \div 3 \cdot 2 + 15] \div 15$

31. $2 \cdot 5^2 - 4 \div 2 + 3 \cdot 7$ **32.** $16 \div 2^4 - 9 \div 3^2$

33. $(4^2 - 7) \cdot 2^3 - 8 \cdot 5 \div 10$ **34.** $4^2 - 2^4 + 5 \cdot 6^2 - 10^2$

35. $(2^5 + 1) \div 11 - 3 + 7(3^3 - 7)$ **36.** $(6 + 8^2 - 10 \div 2) \div 5 + 5 \cdot 3^2$

37.   Make up two examples that illustrate that division is not commutative.

38.   Make up two examples that illustrate that division is not associative.

39.   Show that $(35 + 15) \div 5 = 35 \div 5 + 15 \div 5$ by doing the arithmetic on both sides of the equation. Does this example indicate that division is distributive over addition? [HINT: Refer to Problem 37.]

Find the average of each of the following sets of numbers.

Go BACK TO
Pg 35

40.   {102, 113, 97, 100}   **41.**   {512, 618, 332, 478}

42.   {1000, 1000, 7000}   **43.**   {897, 182, 617, 534, 700}

44.   Three families, each with two children, had incomes of $8942. Two families, each with four children, had incomes of $10,512. Four families, each with two children, had incomes of $11,111. One family had no children and an income of $12,026. What was the average income per family? per person?

45.   Find the average of five thousand, four hundred ninety-two, and six thousand, eight hundred seventy-six. Write the answer in words.

46.   Take the average of 72 and 86 and add it to the product of 91 and 34. What number should be added to this result to get 5146?

47.   During July, Mr. Rodriquez made deposits in his checking account of $400 and $750 and wrote checks totaling $625. During August, his deposits were $632, $322, and $798, and his checks totaled $978. In September, the deposits were $520, $436, $200, and $376; the checks totaled $836. What was the average monthly difference between his deposits and his withdrawals? What was his bank balance at the end of September if he had a balance of $500 on July 1?

48.   A man bought ten shares of stock in a company at $35 per share. The next month he bought five shares in another company at $85 per share. In the third month, he bought one hundred shares in a company at $2.50 per share. How many shares did he buy all together in the three months? What was the average price (to the nearest penny) per share he paid? What was the average amount of money he spent each month on stock transactions?

49. On a math exam, two students scored 95 points, six students scored 90 points, three students scored 80 points, and one student scored 50 points. What was the class average?

50. During her first semester at school, a student bought five textbooks. One book cost $12.95, two books cost $8.75 each, and the other two cost $10.00 each. At the end of the semester, she sold the books back to the bookstore at an average price of $5.50. What was her average cost per book for the semester?

SUMMARY: CHAPTER 1

DEFINITION   An **exponent** is a number that tells how many times its base is to be used as a factor.

DEFINITION   For any nonzero whole number a, $a^0 = 1$.

> OUR SYSTEM OF REPRESENTING NUMBERS DEPENDS ON THREE THINGS
> 1. The ten digits $\{0, 1, 2, 3, 4, 5, 6, 7, 8, 9\}$.
> 2. The placement of the digits.
> 3. The value of each place.

Numbers being added are called **addends,** and the result of the addition is called the **sum.**

COMMUTATIVE PROPERTY OF ADDITION   If a and b are two whole numbers, then $a + b = b + a$.

ASSOCIATIVE PROPERTY OF ADDITION   If a, b, and c are whole numbers, then

$$a + b + c = (a + b) + c = a + (b + c)$$

ADDITIVE IDENTITY   If a is a whole number, then there is a unique whole number 0 with the property that $a + 0 = a$.

Subtraction is a reverse addition, and the missing addend is called the **difference** between the sum and one addend.

The result of multiplying two numbers is called the **product,** and the two numbers are called **factors** of the product.

COMMUTATIVE PROPERTY OF MULTIPLICATION   If a and b are whole numbers, then $a \cdot b = b \cdot a$.

ZERO FACTOR
LAW

If a is a whole number, then $a \cdot 0 = 0$.

MULTIPLICATIVE
IDENTITY

If a is a whole number, then there is a unique whole number 1 with the property that $a \cdot 1 = a$.

ASSOCIATIVE
PROPERTY OF
MULTIPLICATION

If a, b, and c are whole numbers, then

$$a \cdot b \cdot c = a(b \cdot c) = (a \cdot b)c$$

DISTRIBUTIVE
PROPERTY OF
MULTIPLICATION
OVER ADDITION

If a, b, and c are whole numbers, then

$$a(b + c) = a \cdot b + a \cdot c$$

In division:

The number dividing is called the **divisor.**

The number being divided is called the **dividend.**

The result is called the **quotient.**

The number left over is the **remainder,** and it must be less than the divisor.

RULES FOR ORDER OF OPERATIONS

1. First, simplify expressions within parentheses.

2. Second, find any powers indicated by exponents.

3. Third, moving from left to right, perform any multiplications or divisions in the order they appear.

4. Fourth, moving from left to right, perform any additions or subtractions in the order they appear.

An **average** of a set of numbers is the number found by adding the numbers in the set, then dividing this sum by the number of numbers in the set.

REVIEW QUESTIONS · CHAPTER 1

1. In the expression 5^3, 5 is called the _____ and 3 is called the _____ .

2. Since $7 \cdot 2 = 14$, 7 and 2 are called _____ of 14.

Find a base and exponent form for each of the following powers without using the exponent 1.

3. 128

4. 169

Write the following numerals in expanded form and write their values in English word equivalents.

5. 496   **6.** 7842   **7.** 8,000,570

Write each of the following numbers as decimal numbers.

8. four thousand, eight hundred fifty-six

9. fifteen million, thirty-two thousand, one hundred ninety-seven

10. six hundred seventy-two million, three hundred forty thousand, eighty-three

11. Find as many factors as you can for

(a) 102   (b) 39   (c) 76

State which property of addition or multiplication is illustrated.

12. $17 + 32 = 32 + 17$ **13.** $3(22 \cdot 5) = (3 \cdot 22)5$

14. $28 + (6 + 12) = (28 + 6) + 12$ **15.** $72 \cdot 89 = 89 \cdot 72$

16. Show that the following statement is *not true*.

$$32 \div (16 \div 2) = (32 \div 16) \div 2$$

What fact does this illustrate?

Evaluate each of the following expressions.

17. $7 + 3 \cdot 2 - 1 + 9 \div 3$ **18.** $3 \cdot 2^5 - 2 \cdot 5^2$

19. $14 \div 2 + 2 \cdot 8 + 30 \div 5 \cdot 2$ **20.** $(16 \div 2^2 + 6) \div 2 + 8$

21. $(75 - 3 \cdot 5) \div 10 - 4$ **22.** $(7^2 \cdot 2 + 2) \div 10 \div (2 + 3)$

Add.

23.
```
8445
 267
1351
 478
```

24.
```
 39
487
966
182
```

Multiply.

25. $60 \cdot 40$ **26.** $47 \cdot 0$ **27.** $5(3)(29)(0)(86)$

Subtract.

28.
```
647
139
```

29.
```
7036
4652
```

30.
```
5000
2898
```

Multiply.

31. 96
    62

32. 8973
    426

33. 4796
    3000

34. Divide using the method of repeated subtraction. $7\overline{)2046}$

Divide.

35. $529\overline{)71496}$

36. $38\overline{)26721}$

37. If the product of 17 and 51 is added to the product of 16 and 12, what is the sum?

38. Find the average of 33, 42, 25, and 40.

39. If the quotient of 546 and 6 is subtracted from 100, what is the difference?

40. Two years ago, Ms. Miller bought five shares of stock at $353 per share. One year ago, she bought another ten shares at $290 per share. Yesterday, she sold all her shares at $410 per share. What was her total profit? What was her average profit per share?

41. On a history exam, two students scored 98 points, five students scored 87 points, one student scored 81 points, and six students scored 75 points. What was the average score in the class?

42. State two identity properties, one for addition of whole numbers and one for multiplication of whole numbers.

43. What number should be added to seven hundred forty-three to get a sum of eight hundred thirteen?

44. Fill in the missing numbers in the chart according to the directions.

GIVEN NUMBER	ADD 100	DOUBLE	SUBTRACT 200
3	103	206	?
20	120	?	?
15	?	?	?
?	?	?	16

45. How many times can 35 be repeatedly subtracted from 700?

PRIME FACTORS

2.1   TESTS FOR DIVISIBILITY

This section will provide you with some tools (or gimmicks, in this case) that will prove very useful in dealing with factorizations and fractions. In certain cases, there are tests for exact divisibility *without actually dividing.* Four of these cases (and combinations) are discussed here because they occur frequently and because they are easy to apply.

For example, can you tell (without actually dividing) if 14,595 is divisible by 2? by 3? by 5? by 9? The answers are that 14,595 is divisible by 3 and 5 but not by 2 or 9. The following list of rules explains how to test for divisibility by 2, 3, 5, and 9.

TESTS FOR DIVISIBILITY BY 2, 3, 5, AND 9

1.   If the last digit (units digit) of a whole number is 0, 2, 4, 6, or 8, then the whole number is divisible by 2. In other words, even whole numbers are divisible by 2; odd whole numbers are not divisible by 2.

2.   If the sum of the digits of a whole number is divisible by 3, then the number is divisible by 3.

3.   If a whole number ends in 0 or 5, then it is divisible by 5.

4.   If the sum of the digits of a whole number is divisible by 9, then the number is divisible by 9.

EXAMPLES

1.   356 is divisible by 2 since the last digit is 6.

2.   6801 is divisible by 3 since $6 + 8 + 0 + 1 = 15$, and 15 is divisible by 3.

3.   365 is divisible by 5 since 5 is the last digit.

4.   657 is divisible by 9 since $6 + 5 + 7 = 18$ is divisible by 9.

[NOTE: In Example 2, 6801 is *not* divisible by 9 since the sum of the digits is 15, and 15 is not divisible by 9. In Example 4, however, 657 is divisible by 9 and by 3.]

If a number can be divided by 6, the same quotient can be found by dividing the number by 2 and then dividing the new quotient by 3. For example,

$$90 \div 6 = 15$$

but $90 \div 2 = 45$

and $45 \div 3 = 15$

The answer, 15, can be found either way.

> Thus, if a number is *divisible by both 2 and 3,* then it is divisible by 6.
>
> Also, if a number is *divisible by both 3 and 5,* then it is divisible by 15.
>
> If a number ends with 0, then it is divisible by 10, since the number is divisible by *both 2 and 5.*
>
> Not all numbers that are divisible by 3 are divisible by 9. However, if a number is divisible by 9, then it is also divisible by 3.

EXAMPLES

1. 5712 is divisible by 6 since it is divisible by both 2 and 3. It is even, and $5 + 7 + 1 + 2 = 15$, which is divisible by 3. [NOTE: 5712 is divisible by 3 but not by 9.]

2. 975 is divisible by both 5 and 3 since it ends in 5, and $9 + 7 + 5 = 21$, which is divisible by 3. Therefore, 975 is divisible by 15.

3. 3590 is divisible by 10 since it ends with 0.

4. 65,331 is divisible by both 3 and 9 since $6 + 5 + 3 + 3 + 1 = 18$, which is divisible by both 3 and 9.

SELF QUIZ	Using the techniques of this section, determine which numbers (if any) from the set {2, 3, 5, 6, 9, 10, 15} will divide exactly into each of the following numbers.	ANSWERS
	1. 842	1. 2
	2. 9030	2. 2, 3, 5, 6, 10, 15
	3. 4031	3. None

EXERCISES 2.1

ODD PROB 1-51

Using the techniques of this section, determine which number or numbers (if any) of the set {2, 3, 5, 6, 9, 10, 15} will divide exactly into each of the following.

1. 98   **2.** 154   **3.** 75   **4.** 333   **5.** 471

6. 370	**7.** 571	**8.** 466	**9.** 897	**10.** 695
11. 795	**12.** 777	**13.** 45,000	**14.** 885	**15.** 4422
16. 1234	**17.** 4321	**18.** 8765	**19.** 5678	**20.** 402
21. 705	**22.** 732	**23.** 441	**24.** 555	**25.** 666
26. 9000	**27.** 10,000	**28.** 576	**29.** 549	**30.** 792
31. 5700	**32.** 4391	**33.** 5476	**34.** 6930	**35.** 4380
36. 510	**37.** 8805	**38.** 7155	**39.** 8377	**40.** 2222

41. 35,622	**42.** 75,495	**43.** 12,324	**44.** 55,555
45. 632,448	**46.** 578,400	**47.** 9,737,001	**48.** 17,158,514
49. 36,762,252	**50.** 20,498,105		

51. If a number is divisible by both 2 and 9, will it be divisible by 18? Give five examples to support your answer.

52. If a number is divisible by both 3 and 9, will it be divisible by 27? Give five examples to support your answer.

2.2   PRIME NUMBERS AND COMPOSITE NUMBERS

The concept of **factor** was developed in Section 1.5. If two or more numbers are multiplied together, the result is called the **product,** and the numbers being multiplied are **factors** of the product. For example, since $18 \cdot 4 = 72$, both 18 and 4 are factors of 72. Because of the close relationship between multiplication and division, factors can also be thought of as **divisors** of a product; that is, any factor of a product will divide evenly (zero remainder) into that product. The quotient will also be a factor of the product. Since

$$
\begin{array}{r}
4 \\
18\overline{)72} \\
72 \\
\hline
0
\end{array}
$$

both 18 and 4 are divisors of 72. Thus, factors and divisors are the same. From this example, it is clear that 18 and 4 are factors, or divisors, of 72. But are they the *only* factors of 72? The answer is *no*, because $2 \cdot 36 = 72$ also; therefore 2 and 36 are factors of 72. Just how many factors does 72 have, and what are they? [NOTE: To avoid unnecessary complications with the factors of the number zero (0), the set of **natural numbers** (or counting numbers) $N = \{1, 2, 3, 4, 5, 6, \ldots\}$ will be used for this discussion. Zero is not a natural number.]

Finding all the factors of the natural number 72 might be difficult at this time. An easier problem is to find all the factors for each of the natural numbers 18, 17, 26, 12, 3. A list of these factors follows.

NATURAL NUMBER	FACTORS
18	1, 2, 3, 6, 9, 18
17	1, 17
26	1, 2, 13, 26
12	1, 2, 3, 4, 6, 12
3	1, 3

The list of factors shows that

(a) One (1) is a factor of each of the listed numbers.
(b) Each number is a factor of itself.

These two properties hold for all natural numbers. Two of the numbers on the list have only two factors, and these two numbers are special natural numbers called **prime numbers.**

DEFINITION   A **prime number** is a natural number, except 1, whose only factors (or divisors) are itself and 1.*

Note that the definition of prime number has three parts.
(a) A prime number is a natural number.
(b) 1 is not a prime number.
(c) A prime number has only two factors, itself and 1.

Except for one (1), natural numbers that are not prime numbers are called **composite numbers.** Thus, in the list discussed, 18, 26, and 12 are composite numbers, and 17 and 3 are prime numbers.

EXAMPLES   1.   Seven (7) is a *prime number* since $7 \cdot 1 = 7$, and no other natural numbers will multiply together to give 7. That is, 1 and 7 are the only factors of 7.

2.   Thirteen (13) is a *prime number* since $13 \cdot 1 = 13$, and 1 and 13 are the only divisors of 13.

3.   Twenty-one (21) is a *composite number* since not only does $21 \cdot 1 = 21$, but $7 \cdot 3 = 21$ also; that is, 21 has factors (or divisors) other than 21 and 1.

---

*Another definition for prime number is that **a prime number is a natural number with exactly two different factors.**

It would be nice to have some simple formula to use to decide whether a number is prime or not. Alas, there is no simple formula. In fact, there is no formula. So we will discuss a technique devised by a Greek mathematician named Eratosthenes, who used the concept of **multiples.** Every natural number has a set of multiples associated with it.

The multiples of 2: {2, 4, 6, 8, 10, 12, . . .}
The multiples of 3: {3, 6, 9, 12, 15, 18, . . .}
The multiples of 4: {4, 8, 12, 16, 20, 24, . . .}

To find the multiples of any natural number, just multiply every natural number {1, 2, 3, 4, 5, . . .} by the number itself. Thus, the multiples of 8 are

{8, 16, 24, 32, 40, 48, . . .}

And the multiples of 13 are

{13, 26, 39, 52, 65, 78, . . .}

Now, none of the numbers in a set of multiples can be prime except possibly the first number. Therefore, to sift out the prime numbers, the **Sieve of Eratosthenes** may be used.

1. To find the prime numbers from 1 to 50, list all the natural numbers from 1 to 50 in rows of ten.

1	2	3	4	5	6	7	8	9	10
11	12	13	14	15	16	17	18	19	20
21	22	23	24	25	26	27	28	29	30
31	32	33	34	35	36	37	38	39	40
41	42	43	44	45	46	47	48	49	50

2. Start by crossing out 1 (since 1 is not a prime number). Next, circle 2 and cross out all the other multiples of 2; that is, cross out every second number.

1	②	3	4	5	6	7	8	9	10
11	12	13	14	15	16	17	18	19	20
21	22	23	24	25	26	27	28	29	30
31	32	33	34	35	36	37	38	39	40
41	42	43	44	45	46	47	48	49	50

3. The first number after 2 not crossed out is 3. Circle 3 and cross out all multiples of 3 that are not already crossed out; that is, after 3, every third number should be crossed out.

1̸	②	③	4̸	5	6̸	7	8̸	9̸	1̸0̸
11	1̸2̸	13	1̸4̸	1̸5̸	1̸6̸	17	1̸8̸	19	2̸0̸
2̸1̸	2̸2̸	23	2̸4̸	25	2̸6̸	2̸7̸	28	29	3̸0̸
31	3̸2̸	3̸3̸	3̸4̸	35	3̸6̸	37	3̸8̸	3̸9̸	40
41	4̸2̸	43	4̸4̸	4̸5̸	4̸6̸	47	4̸8̸	49	5̸0̸

4. The next number not crossed out is 5. Circle 5 and cross out all multiples of 5 that are not already crossed out. If we proceed this way, we will have the prime numbers circled and the composite numbers crossed out. The final table is as follows:

1̸	②	③	4̸	⑤	6̸	⑦	8̸	9̸	1̸0̸
⑪	1̸2̸	⑬	1̸4̸	1̸5̸	1̸6̸	⑰	1̸8̸	⑲	2̸0̸
2̸1̸	2̸2̸	㉓	2̸4̸	2̸5̸	2̸6̸	2̸7̸	28	㉙	3̸0̸
㉛	3̸2̸	3̸3̸	3̸4̸	35	36	㊲	3̸8̸	3̸9̸	4̸0̸
㊶	4̸2̸	㊸	4̸4̸	4̸5̸	4̸6̸	㊼	4̸8̸	4̸9̸	5̸0̸

Looking at the final table (the Sieve of Eratosthenes), we see that the prime numbers less than 50 are {2, 3, 5, 7, 11, 13, 17, 19, 23, 29, 31, 37, 41, 43, 47}. You should also notice that (a) the only even prime number is 2; and (b) all other prime numbers are odd, but not all odd numbers are prime.

EXERCISES 2.2   ODD 1- 53

Find two sets of factors for each of the following composite numbers.

1. 28   **2.** 32   **3.** 16   **4.** 15   **5.** 9   **6.** 105   **7.** 35

8. 63   **9.** 36   **10.** 72   **11.** 14   **12.** 21   **13.** 10   **14.** 50

15. 24   **16.** 8   **17.** 100   **18.** 65   **19.** 51   **20.** 52

List the set of multiples for each of the following numbers.

21. 3   **22.** 7   **23.** 6   **24.** 9   **25.** 20   **26.** 16   **27.** 17

28. 12   **29.** 25   **30.** 21

31. Construct a Sieve of Eratosthenes for the numbers from 1 to 100. List the set of prime numbers from 1 to 100.

Find two factors of each of the first numbers given whose sum is the second number.

Example: (12, 8) — Two factors of 12 whose sum is 8 are 6 and 2, since $6 \cdot 2 = 12$ and $6 + 2 = 8$.

32. (24, 10) **33.** (12, 7) **34.** (16, 10) **35.** (12, 13) **36.** (14, 9)

37. (50, 27) **38.** (20, 9) **39.** (24, 11) **40.** (48, 19) **41.** (36, 15)

42. (7, 8)   **43.** (63, 24) **44.** (51, 20) **45.** (25, 10) **46.** (16, 8)

47. (60, 17) **48.** (52, 17) **49.** (27, 12) **50.** (72, 22)

51. List the set of prime numbers that are *not* odd.

52. List the set of prime numbers that *are* odd.

53. List the set of all factors of 52. Divide each factor into 52 and show that the quotient is another factor of 52.

2.3 PRIME FACTORIZATIONS

Finding the factors of a composite number is a very useful tool when working with fractions (Chapter 3). Even more useful is finding all the prime factors of a number. This is called finding the **prime factorization** of the number.

There are two techniques for finding prime factorizations of composite numbers.

---

1. To find the prime factorization of a composite number.
   (a) Factor the composite number into any two factors.

   $$60 = 6 \cdot 10$$

   (b) If either or both factors are not prime, factor each of these.

   $$6 = 2 \cdot 3 \quad \text{and} \quad 10 = 2 \cdot 5$$
   $$60 = 6 \cdot 10 = 2 \cdot 3 \cdot 2 \cdot 5$$

   (c) Continue this process until all factors are prime. The factors are customarily written in order, and exponents may be used for convenience. The prime factorization is the product of all the prime factors.

   $$60 = 2 \cdot 2 \cdot 3 \cdot 5 = 2^2 \cdot 3 \cdot 5$$

EXAMPLES

1. $70 = 7 \cdot 10 = 7 \cdot 2 \cdot 5 = 2 \cdot 5 \cdot 7$

2. $85 = 5 \cdot 17$

3. $72 = 8 \cdot 9 = 2 \cdot 4 \cdot 3 \cdot 3 = 2 \cdot 2 \cdot 2 \cdot 3 \cdot 3 = 2^3 \cdot 3^2$

4. $245 = 5 \cdot 49 = 5 \cdot 7 \cdot 7 = 5 \cdot 7^2$

Another arrangement of the factors, called a factor tree, may also be used. The following factor trees for 72 also show that no matter what two factors are used to begin with, the prime factorization is the same.

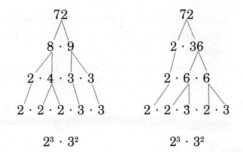

2. To find the prime factorization of a composite number:
    (a) Divide the composite number by any prime number that will divide into it.

    $$\begin{array}{r} 30 \\ 2\overline{)60} \end{array}$$

    (b) Continue to divide the *quotient* by prime numbers.

    $$\begin{array}{r} 30 \\ 2\overline{)60} \end{array} \quad \begin{array}{r} 15 \\ 2\overline{)30} \end{array} \quad \begin{array}{r} 5 \\ 3\overline{)15} \end{array}$$

    (c) The prime factorization is the product of all the prime divisors and the last prime quotient.

    $$60 = 2 \cdot 2 \cdot 3 \cdot 5 = 2^2 \cdot 3 \cdot 5$$

SHOW THIS →

THEN

SHORT CUT

NE

The division may be written in a more compact form as shown:

$$\begin{array}{l} 2\underline{)60} \\ 2\underline{)30} \\ 3\underline{)15} \\ 5 \end{array} \qquad 60 = 2^2 \cdot 3 \cdot 5$$

EXAMPLES   1.   2)70 $70 = 2 \cdot 5 \cdot 7$ 2.   5)85 $85 = 5 \cdot 17$
                  5)35   17
                     7

            3.   2)72 $72 = 2^3 \cdot 3^2$ 4.   5)245 $245 = 5 \cdot 7^2$
                  2)36   7) 49
                  2)18   7
                  3) 9
                     3

Technique 2 gives the same prime factorizations as technique 1. Either technique may be used; however, the first is usually much faster and more useful.

There is an important statement about prime factors and composite numbers called the **Fundamental Theorem of Arithmetic.**

FUNDAMENTAL
THEOREM OF
ARITHMETIC

Every composite number has a unique prime factorization.

What this theorem says is that no matter how we proceed to factor a composite number, we will always get precisely the same prime factors. Changing the order of the prime factors does not change the factorization of the composite number because multiplication is commutative.

Since one (1) is not prime, it is not included in the prime factorization of a number. If 1 were allowed to be prime, the Fundamental Theorem of Arithmetic would not be true, and a number could have several prime factorizations. For example, each of the factorizations $6 = 1 \cdot 2 \cdot 3$, $6 = 1 \cdot 1 \cdot 1 \cdot 2 \cdot 3$, and $6 = 1 \cdot 1 \cdot 1 \cdot 1 \cdot 1 \cdot 1 \cdot 1 \cdot 2 \cdot 3$ would have to be considered different. To avoid such confusion and to uphold the Fundamental Theorem, 1 is not a prime number. Remember, however, that 1 is a factor of every natural number. This fact will be of "prime" importance in the next section and in Chapter 3.

The prime factorization of a number is useful in finding *all* the factors (or divisors) of the number. Each of the prime factors is, of course, a factor. 1 is a factor. And, all possible combinations of products of the prime factors will give the remaining factors.

EXAMPLE   Find all the factors of 72.

$$72 = 8 \cdot 9 = 2^3 \cdot 3^2$$

The numbers 1, 2, and 3 are factors. Forming all possible combinations of products of prime factors gives the following.

$$2 \cdot 2 = 4 \qquad\qquad 2 \cdot 3 \cdot 3 = 18$$
$$2 \cdot 3 = 6 \qquad\qquad 2 \cdot 2 \cdot 2 \cdot 3 = 24$$
$$2 \cdot 2 \cdot 2 = 8 \qquad\qquad 2 \cdot 2 \cdot 3 \cdot 3 = 36$$
$$3 \cdot 3 = 9 \qquad 2 \cdot 2 \cdot 2 \cdot 3 \cdot 3 = 72$$
$$2 \cdot 2 \cdot 3 = 12$$

Therefore, $\{1, 2, 3, 4, 6, 8, 9, 12, 18, 24, 36, 72\}$ is the set of factors for 72.

SELF QUIZ	Find the prime factorization of each of the following numbers.	ANSWERS
	1. 44	1. $2^2 \cdot 11$
	2. 42	2. $2 \cdot 3 \cdot 7$
	3. 56	3. $2^3 \cdot 7$
	4. 230	4. $2 \cdot 5 \cdot 23$

EXERCISES 2.3   ODD 1 - 49

Find the prime factorization for each of the following numbers.

1. 24	**2.** 28	**3.** 27	**4.** 16	**5.** 36
6. 60	**7.** 72	**8.** 90	**9.** 81	**10.** 105
11. 125	**12.** 160	**13.** 75	**14.** 150	**15.** 210
16. 40	**17.** 250	**18.** 93	**19.** 168	**20.** 360
21. 126	**22.** 48	**23.** 17	**24.** 47	**25.** 51
26. 144	**27.** 121	**28.** 169	**29.** 225	**30.** 52
31. 32	**32.** 98	**33.** 108	**34.** 103	**35.** 101
36. 202	**37.** 78	**38.** 500	**39.** 10,000	**40.** 100,000

Find the set of all factors (or divisors) for each of the following numbers.

41. 12	**42.** 18	**43.** 28	**44.** 98	**45.** 121
46. 45	**47.** 105	**48.** 54	**49.** 97	**50.** 144

2.4 GREATEST COMMON DIVISOR (GCD)   GREATEST COMMON FACTOR

Consider the two numbers 12 and 18. Is there a number (or numbers) that will divide into *both* 12 and 18? To help answer this question, the divisors for 12 and 18 are listed at the top of the next page.

Set of divisors for 12: {1, 2, 3, 4, 6, 12}
Set of divisors for 18: {1, 2, 3, 6, 9, 18}

The **common divisors** for 12 and 18 are 1, 2, 3, and 6. The **greatest common divisor (GCD)** for 12 and 18 is 6; that is, of all the common divisors of 12 and 18, 6 is the largest divisor.

EXAMPLE   List the divisors of each number in the set {36, 24, 48} and find the greatest common divisor (GCD).

Set of divisors for 36: {**1**, **2**, **3**, **4**, **6**, 9, **12**, 18, 36}
Set of divisors for 24: {**1**, **2**, **3**, **4**, **6**, 8, **12**, 24}
Set of divisors for 48: {**1**, **2**, **3**, **4**, **6**, 8, **12**, 16, 24, 48}

The common divisors are **1, 2, 3, 4, 6,** and **12. GCD = 12**

DEFINITION   The **Greatest Common Divisor (GCD)*** of a set of natural numbers is the largest natural number that will divide into all the numbers in the set.

As the above example illustrates, listing all the divisors of each number before finding the GCD can be tedious and difficult. *The use of prime factorizations leads to a simple technique for finding the GCD.*

TECHNIQUE FOR FINDING THE GCD OF A SET OF NATURAL NUMBERS

1. Find the prime factorization of each number.

2. Find the prime factors common to all factorizations.

3. Form the product of these primes using each prime the number of times it is common to *all* factorizations.

4. This product is the GCD. If there are no primes common to all factorizations, the GCD is 1.

EXAMPLES   1. Find the GCD for {36, 24, 48}.

$$\left.\begin{array}{l} 36 = 2 \cdot 2 \cdot 3 \cdot 3 \\ 24 = 2 \cdot 2 \cdot 2 \cdot 3 \\ 48 = 2 \cdot 2 \cdot 2 \cdot 2 \cdot 3 \end{array}\right\} \text{GCD} = 2 \cdot 2 \cdot 3 = 12$$

---

*The largest common divisor is, of course, the largest common factor, and the GCD could be called the **greatest common factor** and be abbreviated **GCF.**

The factor 2 appears twice, and the factor 3 appears once in *all* the prime factorizations.

2. Find the GCD for $\{360, 75, 30\}$.

$$\left.\begin{array}{l} 360 = 36 \cdot 10 = 4 \cdot 9 \cdot 2 \cdot 5 = 2 \cdot 2 \cdot 2 \cdot 3 \cdot 3 \cdot 5 \\ 75 = 3 \cdot 25 = 3 \cdot 5 \cdot 5 \\ 30 = 6 \cdot 5 = 2 \cdot 3 \cdot 5 \end{array}\right\} \begin{array}{l} \text{GCD} = 3 \cdot 5 \\ \qquad = 15 \end{array}$$

Each of the factors 3 and 5 appears only once in *all* the prime factorizations.

3. Find the GCD for $\{168, 420, 504\}$.

$$\left.\begin{array}{l} 168 = 8 \cdot 21 = 2 \cdot 2 \cdot 2 \cdot 3 \cdot 7 \\ 420 = 10 \cdot 42 = 2 \cdot 5 \cdot 6 \cdot 7 \\ \qquad = 2 \cdot 2 \cdot 3 \cdot 5 \cdot 7 \\ 504 = 4 \cdot 126 = 2 \cdot 2 \cdot 6 \cdot 21 \\ \qquad = 2 \cdot 2 \cdot 2 \cdot 3 \cdot 3 \cdot 7 \end{array}\right\} \text{GCD} = 2 \cdot 2 \cdot 3 \cdot 7 = 84$$

In *all* the prime factorizations, 2 appears twice, 3 once, and 7 once.

If the GCD of two numbers is 1 (that is, they have no common prime factors), then the two numbers are said to be **relatively prime.** The numbers themselves may be prime or they may be composite.

EXAMPLES

1. Find the GCD for $\{15, 8\}$.

$$\left.\begin{array}{l} 15 = 3 \cdot 5 \\ 8 = 2 \cdot 2 \cdot 2 \end{array}\right\} \text{GCD} = 1 \qquad \text{8 and 15 are relatively prime.}$$

2. Find the GCD for $\{20, 21\}$.

$$\left.\begin{array}{l} 20 = 2 \cdot 2 \cdot 5 \\ 21 = 3 \cdot 7 \end{array}\right\} \text{GCD} = 1 \qquad \text{20 and 21 are relatively prime.}$$

SELF QUIZ	Find the GCD for each of the following sets of numbers.	ANSWERS	
	1. $\{30, 40, 50\}$	1.	GCD $= 10$
	2. $\{28, 70\}$	2.	GCD $= 14$
	3. $\{168, 140\}$	3.	GCD $= 28$

EXERCISES 2.4

ODD 1 - 35

Find the GCD for each of the following sets of numbers.

1. {12, 8}	**2.** {16, 28}	**3.** {85, 51}
4. {20, 75}	**5.** {20, 30}	**6.** {42, 48}
7. {15, 21}	**8.** {27, 18}	**9.** {18, 24}
10. {77, 66}	**11.** {182, 184}	**12.** {110, 66}
13. {8, 16, 64}	**14.** {121, 44}	**15.** {28, 52, 56}
16. {98, 147}	**17.** {60, 24, 96}	**18.** {33, 55, 77}
19. {25, 50, 75}	**20.** {30, 78, 60}	**21.** {17, 15, 21}
22. {520, 220}	**23.** {14, 55}	**24.** {210, 231, 84}
25. {140, 245, 420}		

Which of the following pairs of numbers are relatively prime?

26. {35, 24}	**27.** {11, 23}	**28.** {14, 36}	**29.** {72, 35}
30. {42, 77}	**31.** {16, 51}	**32.** {20, 21}	**33.** {8, 15}
34. {66, 22}	**35.** {10, 27}		

2.5 LEAST COMMON MULTIPLE (LCM)

The sets of multiples for the numbers 6, 8, and 12 are shown below. What numbers are common to these sets of multiples? That is, what are the common multiples of 6, 8, and 12?

Multiples of 6: {6, 12, 18, 24, 30, 36, 42, 48, . . .}
Multiples of 8: {8, 16, 24, 32, 40, 48, 56, 64, . . .}
Multiples of 12: {12, 24, 36, 48, 60, 72, 84, 96, . . .}

Studying these sets of multiples will show

{24, 48, 72, 96, 120, 144, . . .}

as the set of common multiples for 6, 8, and 12. The smallest of these common multiples, the **least common multiple (LCM)**, is 24.

DEFINITION
The **Least Common Multiple (LCM)** of a set of natural numbers is the smallest number common to all the sets of multiples of the given numbers.

Note that in the example discussed, each of the numbers 6, 8, and 12 will divide into all the common multiples. And 24, the LCM, is the smallest number possible that has 6, 8, and 12 as divisors.

Listing all the multiples of each number in a set is generally not the most efficient way of finding the LCM. Two other techniques are available for finding the LCM, one involving prime factorizations and the other involving division by prime factors.

---

1. To find the LCM of a set of natural numbers:

    (a) Find the prime factorization of each number.

    (b) Find the prime factors that appear in *any one* of the prime factorizations.

    (c) Form the product of these primes using each prime the most number of times it appears in *any one* of the prime factorizations.

EXAMPLES

1. Find the LCM for {12, 20, 18}.

$$12 = 4 \cdot 3 = 2 \cdot 2 \cdot 3 = 2^2 \cdot 3$$
$$20 = 4 \cdot 5 = 2 \cdot 2 \cdot 5 = 2^2 \cdot 5$$
$$18 = 2 \cdot 9 = 2 \cdot 3 \cdot 3 = 2 \cdot 3^2$$
$$\text{LCM} = 2^2 \cdot 3^2 \cdot 5 = 180$$

The most number of times **2** appears in *any one* of the prime factorizations is twice, **3** twice, and **5** once.

2. Find the LCM for {36, 24, 48}.

$$36 = 4 \cdot 9 = 2^2 \cdot 3^2$$
$$24 = 8 \cdot 3 = 2^3 \cdot 3$$
$$48 = 16 \cdot 3 = 2^4 \cdot 3$$
$$\text{LCM} = 2^4 \cdot 3^2 = 144$$

3. Find the LCM for {27, 30, 42}.

$$27 = 3 \cdot 9 = 3^3$$
$$30 = 6 \cdot 5 = 2 \cdot 3 \cdot 5$$
$$42 = 2 \cdot 21 = 2 \cdot 3 \cdot 7$$
$$\text{LCM} = 2 \cdot 3^3 \cdot 5 \cdot 7 = 1890$$

The second technique for finding the LCM involves division.

2. To find the LCM of a set of natural numbers:

(a) Write the numbers horizontally and find a prime number that will divide into more than one number, if possible.

(b) Divide by that prime and write the quotients beneath the dividends. Rewrite any numbers not divided beneath themselves.

(c) Continue the process until no two numbers have a common prime divisor.

(d) The LCM is the product of all the prime divisors and the last set of quotients.

EXAMPLE   Find the LCM for $\{20, 25, 18, 6\}$.

$$
\begin{array}{r|cccc}
5) & 20 & 25 & 18 & 6 \\
2) & 4 & 5 & 18 & 6 \\
3) & 2 & 5 & 9 & 3 \\
\hline
& 2 & 5 & 3 & 1
\end{array}
$$

$$
\begin{aligned}
LCM &= 5 \cdot 2 \cdot 3 \cdot 2 \cdot 5 \cdot 3 \cdot 1 \\
&= 2^2 \cdot 3^2 \cdot 5^2 \\
&= 900
\end{aligned}
$$

The LCM is the smallest number that has all the numbers in the set as factors (or divisors). Once the LCM is found, a natural question is, How many times does each number go into the LCM? Knowing the prime factorizations of each of the numbers in the set and the prime factorization of the LCM makes this question easy to answer.

EXAMPLES   1. How many times does each of the numbers in the set $\{12, 20, 18\}$ go into the LCM, 180?

$$
\left.\begin{aligned}
12 &= 2^2 \cdot 3 \\
20 &= 2^2 \cdot 5 \\
18 &= 2 \cdot 3^2
\end{aligned}\right\}
\begin{aligned}
LCM &= 2^2 \cdot 3^2 \cdot 5 \\
&= 180
\end{aligned}
$$

How many times does 12 divide into 180? Rearrange the factors of the LCM so that one group of factors are the factors of 12.

$$180 = 2^2 \cdot 3^2 \cdot 5 = (2^2 \cdot 3)(3 \cdot 5) = 12 \cdot 15$$

Answer: 12 divides 15 times into 180.

How many times does 20 divide into 180?

$$180 = 2^2 \cdot 3^2 \cdot 5 = (2^2 \cdot 5)(3^2) = 20 \cdot 9$$

Answer: 20 divides 9 times into 180.

How many times does 18 divide into 180?

$$180 = 2^2 \cdot 3^2 \cdot 5 = (2 \cdot 3^2)(2 \cdot 5) = 18 \cdot 10$$

Answer: 18 divides 10 times into 180.

Although the point of such questions may not be clear at this time, the technique for answering them will prove very useful in work with fractions.

2. Rearranging the factors for the LCM = 1890, answer the same questions for the set {27, 30, 42}.

$$\left. \begin{array}{l} 27 = 3^3 \\ 30 = 2 \cdot 3 \cdot 5 \\ 42 = 2 \cdot 3 \cdot 7 \end{array} \right\} \text{LCM} = 2 \cdot 3^3 \cdot 5 \cdot 7$$

$$1890 = 2 \cdot 3^3 \cdot 5 \cdot 7 = (3^3)(2 \cdot 5 \cdot 7) = 27 \cdot 70$$
$$1890 = 2 \cdot 3^3 \cdot 5 \cdot 7 = (2 \cdot 3 \cdot 5)(3^2 \cdot 7) = 30 \cdot 63$$
$$1890 = 2 \cdot 3^3 \cdot 5 \cdot 7 = (2 \cdot 3 \cdot 7)(3^2 \cdot 5) = 42 \cdot 45$$

Thus, 27 divides 70 times, 30 divides 63 times, and 42 divides 45 times into 1890.

SELF QUIZ	Find the LCM for each of the following sets of numbers.	ANSWERS
	1. {30, 40, 50}	1. LCM = 600
	2. {28, 70}	2. LCM = 140
	3. {168, 140}	3. LCM = 840

EXERCISES 2.5   ODD 1-37 AND 36

Find the LCM for each of the following sets of numbers.

1. {8, 12}   2. {3, 5, 7}   3. {4, 6, 9}

4. {6, 8, 27}   5. {144, 216}   6. {40, 25}

7. {40, 75}	**8.** {98, 28}	**9.** {72, 36, 54}
10. {15, 10, 35}	**11.** {13, 26, 169}	**12.** {121, 33, 66}
13. {8, 15, 13}	**14.** {125, 45, 150}	**15.** {51, 54, 34}

Find the LCM for each of the following sets of numbers and tell how many times each number will divide into the LCM. *LIST BOTH FACTORS*

16. {8, 15, 10}	**17.** {24, 15, 10}	**18.** {8, 10, 120}
19. {6, 15, 30}	**20.** {45, 18, 6, 27}	**21.** {228, 12, 95}
22. {63, 98, 45}	**23.** {56, 40, 196}	**24.** {135, 125, 225}
25. {99, 363, 143}		

Find the GCD and the LCM for each of the following sets of numbers.

26. {3, 5, 9}	**27.** {2, 5, 11}	**28.** {4, 18, 14}
29. {6, 15, 12}	**30.** {49, 25, 35}	**31.** {45, 145, 290}
32. {40, 48, 56, 24}	**33.** {81, 54, 108}	**34.** {135, 75, 45}
35. {169, 637, 845}		

36. Three people walk around a tract of homes in 9, 12, and 14 minutes, respectively. If they all start at the same time and place and continue to walk until they meet, how many minutes will it be before they meet at the starting point?

37. Two astronauts miss connections at their first rendezvous in space. If one astronaut circles the earth every 12 hours and the other every 16 hours, in how many hours will they rendezvous again? How many orbits will each astronaut make before the second rendezvous?

38. Three truck drivers have lunch together whenever all three are at the routing station at the same time. The route for the first driver takes 5 days, for the second driver 15 days, and for the third driver 6 days. How often do the three drivers have lunch together? If the first driver's route was changed to 6 days, how often would they have lunch together?

SUMMARY: CHAPTER 2   ~~Study~~

TESTS FOR DIVISIBILITY BY 2, 3, 5, AND 9

1. If the last digit (units digit) of a whole number is 0, 2, 4, 6, or 8, then the whole number is divisible by 2. In other words, even whole numbers are divisible by 2; odd whole numbers are not divisible by 2.

2. If the sum of the digits of a whole number is divisible by 3, then the number is divisible by 3.

3. If a whole number ends in 0 or 5, then it is divisible by 5.

4. If the sum of the digits of a whole number is divisible by 9, then the number is divisible by 9.

If a number is divisible by
- (a) Both 2 and 3, then it is divisible by 6.
- (b) Both 3 and 5, then it is divisible by 15.
- (c) Both 2 and 5, then it is divisible by 10.

DEFINITION   A **prime number** is a natural number, except 1, whose only factors (or divisors) are itself and 1.

Natural numbers that are not prime and not 1 are called **composite numbers.**

FUNDAMENTAL THEOREM OF ARITHMETIC   Every composite number has a unique prime factorization.

DEFINITION   The **Greatest Common Divisor (GCD)** of a set of natural numbers is the largest natural number that will divide into all the numbers in the set.

TECHNIQUE FOR FINDING THE GCD OF A SET OF NATURAL NUMBERS

1. Find the prime factorization of each number.

2. Find the prime factors common to all factorizations.

3. Form the product of these primes using each prime the number of times it is common to *all* factorizations.

4. This product is the GCD. If there are no primes common to all factorizations, the GCD is 1.

DEFINITION   The **Least Common Multiple (LCM)** of a set of natural numbers is the smallest number common to all the sets of multiples of the given numbers.

TO FIND THE LCM OF A SET OF NATURAL NUMBERS

1. (a) Find the prime factorization of each number.

   (b) Find the prime factors that appear in *any one* of the prime factorizations.

   (c) Form the product of these primes using each prime the most number of times it appears in *any one* of the prime factorizations.

2. (a) Write the numbers horizontally and find a prime number that will divide into more than one number, if possible.

   (b) Divide by that prime and write the quotients beneath the dividends. Rewrite any numbers not divided beneath themselves.

   (c) Continue the process until no two numbers have a common prime divisor.

   (d) The LCM is the product of all the prime divisors and the last set of quotients.

REVIEW QUESTIONS · CHAPTER 2

1. A prime number is any _____ number, except _____, whose only factors (or _____) are _____ and _____.

2. Except for 1, natural numbers that are not prime numbers are called _____ numbers.

3. Numbers that form a product are called _____ of the product.

4. The Fundamental Theorem of Arithmetic: Every _____ number has a unique _____.

5. Two numbers are _____ if they have no common prime factors. Their GCD is _____.

Determine which numbers, if any, from the set $\{2, 3, 5, 6, 9, 10, 15\}$ divide evenly into each of the following.

6. 45   **7.** 72   **8.** 479   **9.** 5040   **10.** 8836   **11.** 575,493

12. List the set of multiples of 14.

13. List the set of prime numbers less than 70.

14. Find two factors of 24 whose sum is 10.

15. Find two factors of 60 whose sum is 17.

16. Find the prime factorizations for each of the following numbers.
    (a) 150   (b) 65   (c) 84

17. List all the divisors of the following numbers.
    (a) 75   (b) 36   (c) 169

18. Find the GCD for each of the following sets of numbers.
    (a) {30, 50, 60}   (b) {8, 28, 52}   (c) {20, 27}

19. Find the LCM for each of the following sets of numbers.
    (a) {14, 8, 24}   (b) {8, 12, 25, 36}

20. Find the GCD and the LCM for the set {39, 18, 63}. How many times will each number divide into the LCM?

21. One racing car goes around the track every 30 seconds and the other every 35 seconds. If both cars start a race from the same point, in how many seconds will the first car be exactly one lap ahead of the second car? How many laps will each car have made when the first car is two laps ahead of the second?

RATIONAL NUMBERS

3

3.1   RATIONAL NUMBERS AND MULTIPLYING RATIONAL NUMBERS

Rational number is the technical name for the common term **fraction.** There are several important reasons for the need of rational numbers, and three of them are:

1. To indicate equal parts of a whole.

2. To indicate division.

3. To solve algebraic equations, such as $5x = 3$.

The first two reasons will be discussed in this chapter, and the third will be discussed later in Chapter 5 on ratio and proportion. Examples of rational numbers are $\dfrac{1}{2}, \dfrac{3}{4}, \dfrac{9}{10}, \dfrac{7}{100}, \dfrac{100}{3}$, and $\dfrac{0}{18}$.

DEFINITION   A **rational number** is a number that can be written in the form $\dfrac{a}{b}$, where a is a whole number and b is a natural number.

$$\frac{a}{b} \qquad \begin{array}{l}\text{numerator} \\ \text{denominator}\end{array}$$

W $\{0, 1, 2, \dots\}$

N $\{1, 2, 3, \dots\}$

[NOTE: The numerator a can be 0 since a is a whole number, but the denominator b *cannot be* 0 since b is a natural number.]

Figure 3.1 shows how a whole may be separated into equal parts in several ways. The shading indicates that the rational numbers $\dfrac{1}{2}, \dfrac{2}{4}, \dfrac{4}{8}$, and $\dfrac{6}{12}$ all represent the same amount of the whole. These numbers are **equivalent.** *Any one represents the same number the others do.* The set of rational numbers equivalent to $\dfrac{1}{2}$ is

$$\left\{\frac{1}{2}, \frac{2}{4}, \frac{3}{6}, \frac{4}{8}, \frac{5}{10}, \frac{6}{12}, \frac{7}{14}, \frac{8}{16}, \frac{9}{18}, \frac{10}{20}, \frac{11}{22} \cdots \right\}$$

The set of rational numbers equivalent to $\dfrac{2}{3}$ is

$$\left\{\frac{2}{3}, \frac{4}{6}, \frac{6}{9}, \frac{8}{12}, \frac{10}{15}, \frac{12}{18}, \frac{14}{21}, \frac{16}{24} \cdots \right\}$$

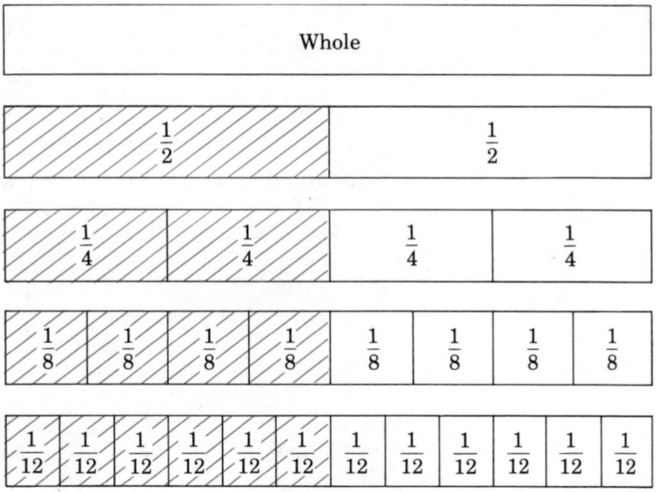

GO OVER

$\frac{1}{2}$

$\frac{2}{4}$

$\frac{4}{8}$

WORK SOME
PROBLEMS
Pg 73
PROB 1-21

$\frac{6}{12}$

Figure 3.1

We say two numbers equivalent to each other are **equal.** Thus,

WORK Some
PROBLEMS
Pg 73
Probs 22-31

$$\frac{2}{3} = \frac{4}{6} \quad \text{and} \quad \frac{9}{18} = \frac{11}{22}$$

Each whole number can also be identified with a set of equivalent rational numbers.

0 can be identified with $\left\{ \dfrac{0}{1}, \dfrac{0}{2}, \dfrac{0}{3}, \dfrac{0}{4}, \dfrac{0}{5}, \dfrac{0}{6}, \dfrac{0}{7}, \cdots \right\}$

1 can be identified with $\left\{ \dfrac{1}{1}, \dfrac{2}{2}, \dfrac{3}{3}, \dfrac{4}{4}, \dfrac{5}{5}, \dfrac{6}{6}, \dfrac{7}{7}, \cdots \right\}$

2 can be identified with $\left\{ \dfrac{2}{1}, \dfrac{4}{2}, \dfrac{6}{3}, \dfrac{8}{4}, \dfrac{10}{5}, \dfrac{12}{6}, \dfrac{14}{7}, \cdots \right\}$

3 can be identified with $\left\{ \dfrac{3}{1}, \dfrac{6}{2}, \dfrac{9}{3}, \dfrac{12}{4}, \dfrac{15}{5}, \dfrac{18}{6}, \dfrac{21}{7}, \cdots \right\}$, and so on.

Interpreting rational numbers as indicated division helps to explain how these identifications can be made. For example, if $\frac{a}{b}$ means $a \div b$, then we have the following situations.

$\dfrac{2}{2}$ means $2 \div 2$ or $2{\overline{)2}}$, with quotient 1,   so $\dfrac{2}{2} = 1$

$\dfrac{3}{3}$ means $3 \div 3$ or $3{\overline{)3}}$, with quotient 1,   so $\dfrac{3}{3} = 1$

$\dfrac{8}{4}$ means $8 \div 4$ or $4{\overline{)8}}$, with quotient 2,   so $\dfrac{8}{4} = 2$

$\dfrac{15}{5}$ means $15 \div 5$ or $5{\overline{)15}}$, with quotient 3,   so $\dfrac{15}{5} = 3$

Using the division concept in another way, $\dfrac{15}{5} = 3$ because $15 = 3 \cdot 5$. With this in mind, then $\dfrac{0}{2} = 0$ since $0 = 0 \cdot 2$. Also, $\dfrac{0}{5} = 0$ since $0 = 0 \cdot 5$. In general, if $b \neq 0$ ($\neq$ means *is not equal to*), $\dfrac{0}{b} = 0$. However, the denominator can *never* be 0.

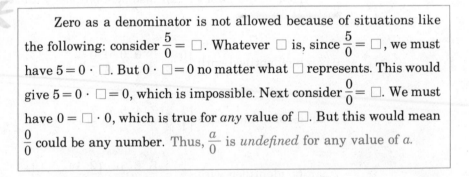

Zero as a denominator is not allowed because of situations like the following: consider $\dfrac{5}{0} = \square$. Whatever $\square$ is, since $\dfrac{5}{0} = \square$, we must have $5 = 0 \cdot \square$. But $0 \cdot \square = 0$ no matter what $\square$ represents. This would give $5 = 0 \cdot \square = 0$, which is impossible. Next consider $\dfrac{0}{0} = \square$. We must have $0 = \square \cdot 0$, which is true for *any* value of $\square$. But this would mean $\dfrac{0}{0}$ could be any number. Thus, $\dfrac{a}{0}$ is *undefined* for any value of a.

Rational numbers such as $\dfrac{41}{30}$ and $\dfrac{28}{5}$ in which the numerator is larger than the denominator are sometimes called **improper fractions.** This term is very misleading because there is nothing "improper" about $\dfrac{41}{30}$ or $\dfrac{28}{5}$. In algebra and other mathematical courses, these forms are preferred over mixed numbers such as $1\dfrac{11}{30}$ and $5\dfrac{3}{5}$. In this text, the author refers to the **fraction form** of a rational number regardless of whether the numerator or denominator is larger.

Multiplication with rational numbers can be illustrated with figures. Consider the situations shown in Figure 3.2.

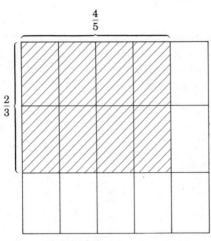

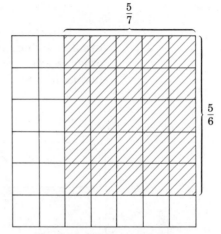

The shaded region represents $\frac{2}{3} \cdot \frac{4}{5}$. There are 15 parts and 8 are shaded. Therefore,

$$\frac{2}{3} \cdot \frac{4}{5} = \frac{8}{15}$$

(a)

The shaded region represents $\frac{5}{6} \cdot \frac{5}{7}$. There are 42 parts and 25 are shaded. Therefore,

$$\frac{5}{6} \cdot \frac{5}{7} = \frac{25}{42}$$

(b)

Figure 3.2

Diagrams such as those in Figure 3.2 are useful as aids. They suggest that to multiply two rational numbers, their numerators should be multiplied and their denominators should be multiplied. The following is a formal definition of this process.

DEFINITION   The **product** of two rational numbers $\frac{a}{b}$ and $\frac{c}{d}$ is the rational number whose numerator is the product of the numerators $(a \cdot c)$ and whose denominator is the product of the denominators $(b \cdot d)$. That is,

$$\frac{a}{b} \cdot \frac{c}{d} = \frac{a \cdot c}{b \cdot d}$$

EXAMPLES   1. $\dfrac{2}{3} \cdot \dfrac{5}{7} = \dfrac{2 \cdot 5}{3 \cdot 7} = \dfrac{10}{21}$

2. $\dfrac{1}{4} \cdot \dfrac{3}{5} = \dfrac{1 \cdot 3}{4 \cdot 5} = \dfrac{3}{20}$

3. $\dfrac{7}{5} \cdot \dfrac{2}{1} = \dfrac{7 \cdot 2}{5 \cdot 1} = \dfrac{14}{5}$

Figure 3.3 shows that, if the order of the fractional parts in Figure 3.2 is changed, the area represented by the product is the same. In other words, multiplication with rational numbers is **commutative,** just as it is with whole numbers.

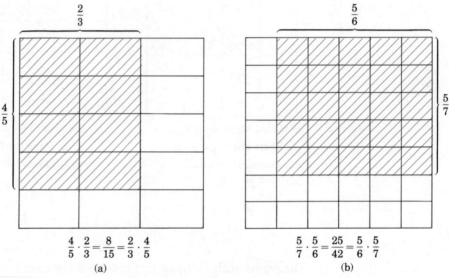

$\dfrac{4}{5} \cdot \dfrac{2}{3} = \dfrac{8}{15} = \dfrac{2}{3} \cdot \dfrac{4}{5}$

(a)

$\dfrac{5}{7} \cdot \dfrac{5}{6} = \dfrac{25}{42} = \dfrac{5}{6} \cdot \dfrac{5}{7}$

(b)

Figure 3.3

COMMUTATIVE PROPERTY OF MULTIPLICATION

ORDER

If $\dfrac{a}{b}$ and $\dfrac{c}{d}$ are rational numbers, then

$$\dfrac{a}{b} \cdot \dfrac{c}{d} = \dfrac{c}{d} \cdot \dfrac{a}{b}$$

The definition can easily be extended to multiplication with more than two rational numbers by making the numerator of the product the product of all the numerators, and the denominator of the product the product of all the denominators, as is illustrated in Examples 4 and 5.

EXAMPLES

4. $\dfrac{1}{4} \cdot \dfrac{3}{5} \cdot \dfrac{7}{2} = \dfrac{1 \cdot 3 \cdot 7}{4 \cdot 5 \cdot 2} = \dfrac{21}{40}$

5. $\dfrac{2}{3} \cdot \dfrac{11}{15} \cdot \dfrac{1}{7} = \dfrac{2 \cdot 11 \cdot 1}{3 \cdot 15 \cdot 7} = \dfrac{22}{315}$

The extension of multiplication to more than two numbers involves the property of **associativity** with rational numbers, as it does with whole numbers.

ASSOCIATIVE PROPERTY OF MULTIPLICATION

GROUPING

If $\dfrac{a}{b}, \dfrac{c}{b},$ and $\dfrac{e}{f}$ are rational numbers, then

$$\dfrac{a}{b} \cdot \dfrac{c}{d} \cdot \dfrac{e}{f} = \left(\dfrac{a}{b} \cdot \dfrac{c}{d}\right) \cdot \dfrac{e}{f} = \dfrac{a}{b} \cdot \left(\dfrac{c}{d} \cdot \dfrac{e}{f}\right)$$

Thus, in Example 4, we might have written

$$\dfrac{1}{4} \cdot \dfrac{3}{5} \cdot \dfrac{7}{2} = \left(\dfrac{1}{4} \cdot \dfrac{3}{5}\right) \cdot \dfrac{7}{2} = \left(\dfrac{3}{20}\right) \cdot \dfrac{7}{2} = \dfrac{21}{40}$$

However, this procedure does not usually clarify anything, and the technique shown in Examples 4 and 5 is quite sufficient.

The number one (1) is the **multiplicative identity** for the whole numbers because $a \cdot 1 = a$ for any whole number a. One (1) is also the **multiplicative identity** for the rational numbers since

$$\dfrac{a}{b} \cdot 1 = \dfrac{a}{b} \cdot \dfrac{1}{1} = \dfrac{a \cdot 1}{b \cdot 1} = \dfrac{a}{b}$$

MULTIPLICATIVE IDENTITY

If $\dfrac{a}{b}$ is a rational number, then

$$\dfrac{a}{b} \cdot 1 = \dfrac{a}{b}$$

Thus,

$$\dfrac{2}{3} \cdot 1 = \dfrac{2}{3}, \qquad \dfrac{5}{6} \cdot 1 = \dfrac{5}{6}, \quad \text{and} \quad \dfrac{14}{3} \cdot 1 = \dfrac{14}{3}$$

In general,

$$1 = \dfrac{k}{k}, \text{ where } k \text{ is any natural number}$$

Therefore,

$$\boxed{\dfrac{a}{b} = \dfrac{a}{b} \cdot 1 = \dfrac{a}{b} \cdot \dfrac{k}{k} = \dfrac{a \cdot k}{b \cdot k}} \qquad \text{where } k \neq 0$$

$\left(\begin{array}{c} \text{ANY} \\ \text{FRACTION} \end{array} \right) \cdot \dfrac{k}{k} =$

Equivalent rational numbers can easily be found with this property.

EXAMPLES

1. Let $k = 6$ and find a number equivalent to $\dfrac{3}{4}$.

$$\frac{3}{4} = \frac{3}{4} \cdot \frac{6}{6} = \frac{18}{24}$$

2. Let $k = 7$ and find a number equivalent to $\dfrac{8}{5}$.

$$\frac{8}{5} = \frac{8}{5} \cdot \frac{7}{7} = \frac{56}{35}$$

3. Find a number with denominator 28 equivalent to $\dfrac{3}{7}$.

$$\frac{3}{7} = \frac{3}{7} \cdot \frac{4}{4} = \frac{12}{28}$$

[NOTE: $k = 4$ was used because $7 \cdot 4 = 28$]

EXERCISES 3.1

ASSIGN
1-21 AND ODD
23-57.
WORK EXAMPLES

1. Define *rational number*.

2. Give two examples that illustrate the commutative property of multiplication with rational numbers. Do the same for the associative property of multiplication.

Write a rational number that represents each of the following numbers of equal parts of a whole.

3. 13 cards of a deck of 52

4. 16 tablets from a bottle of 100

5. 1 dollar from a pile of 25 one-dollar bills

6. 3 broken plates in a set of 12

7. one candle holder of a pair

8. one volume from a set of 20 encyclopedias

9. 2 bottles of soda from a pack of 6 bottles

10. 14 coat hangers from a rack of 36 hangers

11. 5 of the pearls on a necklace of 40 pearls

Write a rational number that represents the shaded parts in each of the following diagrams.

12.

13.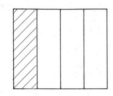

14.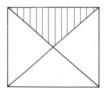

15.

16.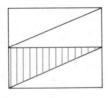

17.

18. If the rectangles in Problems 13, 14, and 16 are the same size, are the shaded portions the same shape? Are the shaded portions the same area? Draw three rectangles of the same shape and shade $\frac{3}{8}$ of each rectangle in a different way.

19. Write the set of all rational numbers equivalent to (a) $\frac{5}{6}$ and to (b) $\frac{4}{3}$.

20. Find a rational number with denominator 72 equivalent to $\frac{19}{4}$.

21. Find a rational number with numerator 18 equivalent to $\frac{2}{3}$.

State whether or not each pair of rational numbers is a pair of equivalent rational numbers.

22. $\left(\frac{2}{3}, \frac{4}{6}\right)$ 23. $\left(\frac{5}{8}, \frac{15}{24}\right)$ 24. $\left(\frac{1}{3}, \frac{2}{4}\right)$ 25. $\left(\frac{2}{5}, \frac{4}{7}\right)$

26. $\left(1, \frac{5}{5}\right)$ 27. $\left(\frac{2}{2}, \frac{10}{10}\right)$ 28. $\left(\frac{4}{3}, \frac{12}{9}\right)$ 29. $\left(\frac{3}{5}, \frac{5}{7}\right)$

30. $\left(\dfrac{7}{6}, \dfrac{35}{30}\right)$ **31.** $\left(\dfrac{4}{1}, \dfrac{12}{3}\right)$

Multiply the following rational numbers.

32. $\dfrac{1}{5} \cdot \dfrac{3}{5}$ **33.** $\dfrac{4}{1} \cdot \dfrac{2}{3}$ **34.** $\dfrac{6}{7} \cdot \dfrac{2}{5}$ **35.** $\dfrac{3}{16} \cdot \dfrac{1}{2}$

36. $\dfrac{1}{3} \cdot \dfrac{1}{3}$ **37.** $\dfrac{1}{2} \cdot \dfrac{1}{2}$ **38.** $\dfrac{1}{28} \cdot \dfrac{3}{4}$ **39.** $\dfrac{0}{4} \cdot \dfrac{5}{6}$

40. $\dfrac{22}{3} \cdot \dfrac{4}{15}$ **41.** $\dfrac{1}{5} \cdot \dfrac{2}{7} \cdot \dfrac{3}{11}$ **42.** $\dfrac{4}{39} \cdot \dfrac{16}{3}$

43. $\dfrac{4}{13} \cdot \dfrac{2}{5} \cdot \dfrac{6}{7}$ **44.** $\dfrac{7}{8} \cdot \dfrac{7}{9} \cdot \dfrac{7}{3}$ **45.** $\dfrac{1}{6} \cdot \dfrac{1}{10} \cdot \dfrac{1}{6}$

46. $\dfrac{9}{100} \cdot 1 \cdot 3$ **47.** $5 \cdot 12 \cdot 14$ **48.** $16 \cdot 3 \cdot 20$

49. $\dfrac{3}{40} \cdot \dfrac{9}{20} \cdot \dfrac{1}{5} \cdot \dfrac{11}{4}$ **50.** $\dfrac{1}{60} \cdot \dfrac{1}{20} \cdot \dfrac{1}{30} \cdot \dfrac{1}{7}$

Draw a diagram similar to those in Figure 3.2 illustrating each of the following products.

51. $\dfrac{1}{5} \cdot \dfrac{3}{4}$ **52.** $\dfrac{5}{6} \cdot \dfrac{5}{6}$ **53.** $\dfrac{7}{8} \cdot \dfrac{1}{4}$ **54.** $\dfrac{2}{3} \cdot \dfrac{4}{7}$ **55.** $\dfrac{2}{9} \cdot \dfrac{2}{3}$

56. Each of the following products has the same value. What is that value?

(a) $\dfrac{0}{4} \cdot \dfrac{5}{6}$ (b) $\dfrac{3}{10} \cdot \dfrac{0}{8}$ (c) $\dfrac{1}{100} \cdot \dfrac{0}{1000}$ (d) $\dfrac{0}{52} \cdot \dfrac{0}{13}$

57. Show by the use of two diagrams that $\dfrac{3}{5} \cdot \dfrac{3}{8} = \dfrac{3}{8} \cdot \dfrac{3}{5}$. What property has been illustrated?

58. Find the following product: $\dfrac{3}{11} \cdot \dfrac{5}{13} \cdot \dfrac{0}{8} \cdot \dfrac{17}{19}$

3.2 CHANGING TERMS

If two rational numbers are equivalent, the one with the larger denominator is said to be in **higher terms,** and the one with the smaller denominator is said to be in **lower terms.** For example, $\dfrac{3}{4} = \dfrac{9}{12}$. Then $\dfrac{9}{12}$ is in higher terms than $\dfrac{3}{4}$, and $\dfrac{3}{4}$ is in lower terms than $\dfrac{9}{12}$.

Equivalent rational numbers in higher terms are often needed when rational numbers are to be added or subtracted. To find an equivalent to $\frac{7}{12}$ with denominator 60, two thoughts are necessary:

1. What number times 12 will give 60? Or, if $12 \cdot k = 60$, what is k?

2. Using $1 = \frac{k}{k}$ and the fact that 1 is the multiplicative identity, multiply $\frac{7}{12} \cdot \frac{k}{k}$.

Thus, since $12 \cdot 5 = 60$, $\frac{7}{12} = \frac{12}{12} \cdot \frac{5}{5} = \frac{35}{60}$ (here, $k = 5$).

EXAMPLES

SOME WORK PROBLEMS 1 TO 32. Pg 79

1. Find a rational number with denominator 28 equivalent to $\frac{3}{4}$ (that is, $\frac{3}{4} = \frac{?}{28}$).

Here, $k = 7$ because $4 \cdot 7 = 28$. Therefore,

$$\frac{3}{4} = \frac{3}{4} \cdot \frac{7}{7} = \frac{21}{28}$$

2. If $\frac{9}{10} = \frac{?}{30}$, find the value for the missing numerator so that the numbers will be equivalent.

Here, $k = 3$ because $10 \cdot 3 = 30$. Therefore,

$$\frac{9}{10} = \frac{9}{10} \cdot \frac{3}{3} = \frac{27}{30}$$

Consider the number $\frac{500}{1000}$. Can you find an equivalent rational number in lower terms? Did you say $\frac{50}{100}$ or $\frac{5}{10}$? Or did you say $\frac{1}{2}$? All of these are correct, and there are many others equivalent to $\frac{500}{1000}$ in lower terms. More importantly, which rational number equivalent to $\frac{500}{1000}$ is in *lowest terms?* What is meant by **lowest terms?** Is $\frac{5}{8}$ in lowest terms? *Yes.* Is $\frac{14}{21}$ in lowest terms? *No.*

DEFINATION NEXT PAGE

DEFINITION   A rational number is in **lowest terms** if its numerator and denominator are relatively prime; that is, the numerator and denominator have only one (1) as a common factor.   *NO COMMON FACTORS*

Now 5 is prime, and $8 = 2 \cdot 2 \cdot 2$. So 5 and 8 have no common factors other than 1. Therefore, $\frac{5}{8}$ is in lowest terms. However, $14 = 7 \cdot 2$, and $21 = 7 \cdot 3$. Thus, 14 and 21 have the common factor 7; 14 and 21 are not relatively prime. Therefore, $\frac{14}{21}$ is *not* in lowest terms.

$$\frac{14}{21} = \frac{2 \cdot 7}{3 \cdot 7} = \frac{2}{3} \cdot \frac{7}{7} = \frac{2}{3} \cdot 1 = \frac{2}{3}$$

By factoring the numerator and denominator, 7 can be seen to be a common factor. And by using the multiplicative identity 1, we have $\frac{2}{3}$ equivalent to $\frac{14}{21}$ and $\frac{2}{3}$ is in lowest terms. We say that $\frac{14}{21}$ has been **reduced to lowest terms.**

EXAMPLES   1.   Reduce $\frac{15}{20}$ to lowest terms by factoring 15 and 20.

$$\frac{15}{20} = \frac{3 \cdot 5}{4 \cdot 5} = \frac{3}{4} \cdot \frac{5}{5} = \frac{3}{4} \cdot 1 = \frac{3}{4}$$

2.   Reduce $\frac{16}{20}$ to lowest terms by factoring 16 and 20.

$$\frac{16}{20} = \frac{4 \cdot 4}{5 \cdot 4} = \frac{4}{5} \cdot \frac{4}{4} = \frac{4}{5} \cdot 1 = \frac{4}{5}$$

To write continually such numbers as $\frac{4}{4}$ is a waste of time. Usually common factors in the numerator and denominator may be simply crossed out *with the understanding that they represent* 1.

$$\frac{16}{20} = \frac{4 \cdot \overset{1}{\cancel{4}}}{5 \cdot \underset{1}{\cancel{4}}} = \frac{4}{5} \quad \text{or} \quad \frac{16}{20} = \frac{4 \cdot \cancel{4}}{5 \cdot \cancel{4}} = \frac{4}{5}$$

If the numerator is a factor of the denominator, or vice versa, the factor 1 must be used. For example,

$$\frac{5}{35} = \frac{\cancel{5} \cdot 1}{\cancel{5} \cdot 7} = \frac{1}{7}$$

EXAMPLES

1. Reduce $\dfrac{90}{75}$ to lowest terms. Since the numerator and denominator are fairly large numbers, their prime factorizations should be found; otherwise some common factor may be missed.

WORK SOME
PROBLEMS
Pg 80
PROB 33-72

$$\frac{90}{75} = \frac{9 \cdot 10}{25 \cdot 3} = \frac{\cancel{3} \cdot 3 \cdot 2 \cdot \cancel{5}}{5 \cdot \cancel{5} \cdot \cancel{3}} = \frac{3 \cdot 2}{5} = \frac{6}{5}$$

2. Reduce $\dfrac{28}{15}$ to lowest terms.

$$\frac{28}{15} = \frac{4 \cdot 7}{3 \cdot 5} = \frac{2 \cdot 2 \cdot 7}{3 \cdot 5} = \frac{28}{15}$$

15 and 28 are relatively prime so $\dfrac{28}{15}$ is in lowest terms.

3. Reduce $\dfrac{8}{72}$ to lowest terms.

$$\frac{8}{72} = \frac{\cancel{8} \cdot 1}{\cancel{8} \cdot 9} = \frac{1}{9}$$

If you do not notice that 8 is a factor of 72, prime factors may be used. Be careful to include 1 as a factor in the numerator.

$$\frac{8}{72} = \frac{2 \cdot 2 \cdot 2}{6 \cdot 12} = \frac{\cancel{2} \cdot \cancel{2} \cdot \cancel{2} \cdot 1}{\cancel{2} \cdot 3 \cdot \cancel{2} \cdot \cancel{2} \cdot 3} = \frac{1}{3 \cdot 3} = \frac{1}{9}$$

Prime factors are very useful when multiplying rational numbers with common factors in numerators and denominators. **Factor and reduce before multiplying.**

EXAMPLES

1. $\dfrac{15}{28} \cdot \dfrac{7}{9} = \dfrac{15 \cdot 7}{28 \cdot 9} = \dfrac{\cancel{3} \cdot 5 \cdot \cancel{7}}{4 \cdot \cancel{7} \cdot \cancel{3} \cdot 3} = \dfrac{5}{4 \cdot 3} = \dfrac{5}{12}$

WORK SOME
PROBLEMS
Pg 80
PROB 73-84

2. $\dfrac{9}{10} \cdot \dfrac{25}{32} \cdot \dfrac{44}{33} = \dfrac{9 \cdot 25 \cdot 44}{10 \cdot 32 \cdot 33} = \dfrac{3 \cdot \cancel{3} \cdot 5 \cdot \cancel{5} \cdot \cancel{4} \cdot \cancel{11}}{2 \cdot \cancel{5} \cdot \cancel{4} \cdot 8 \cdot \cancel{3} \cdot \cancel{11}} = \dfrac{3 \cdot 5}{2 \cdot 8} = \dfrac{15}{16}$

3. $\dfrac{36}{49} \cdot \dfrac{14}{75} \cdot \dfrac{15}{18} = \dfrac{36 \cdot 14 \cdot 15}{49 \cdot 75 \cdot 18}$

$= \dfrac{\cancel{2} \cdot 2 \cdot \cancel{3} \cdot \cancel{3} \cdot 2 \cdot \cancel{7} \cdot \cancel{3} \cdot \cancel{5}}{\cancel{7} \cdot 7 \cdot \cancel{3} \cdot 5 \cdot \cancel{5} \cdot \cancel{2} \cdot \cancel{3} \cdot \cancel{3}} = \dfrac{2 \cdot 2}{7 \cdot 5} = \dfrac{4}{35}$

4. $\dfrac{55}{26} \cdot \dfrac{8}{44} \cdot \dfrac{91}{35} = \dfrac{55 \cdot 8 \cdot 91}{26 \cdot 44 \cdot 35}$

$= \dfrac{\cancel{5} \cdot \cancel{11} \cdot \cancel{2} \cdot \cancel{2} \cdot 2 \cdot \cancel{7} \cdot \cancel{13}}{\cancel{2} \cdot \cancel{13} \cdot \cancel{2} \cdot \cancel{2} \cdot \cancel{11} \cdot \cancel{7} \cdot \cancel{5}} = \dfrac{1}{1} = 1$

The use of prime factors in multiplying and reducing rational numbers is a sound technique and particularly thorough for students who have difficulty with fractions. It also gives the student a consistent approach to multiplying with rational numbers.

When common factors are easily seen, even though they are not prime numbers, they can be divided into both the numerator and denominator, and the quotients can be written. This technique is commonly used, but the student must be careful and organized. The last four examples are shown again.

EXAMPLES

1'. $\dfrac{\overset{5}{\cancel{15}}}{\underset{4}{\cancel{28}}} \cdot \dfrac{\overset{1}{\cancel{7}}}{\underset{3}{\cancel{9}}} = \dfrac{5}{12}$

3 is divided into both 15 and 9.
7 is divided into both 7 and 28.

2'. $\dfrac{\overset{3}{\cancel{9}}}{\underset{2}{\cancel{10}}} \cdot \dfrac{\overset{5}{\cancel{25}}}{\underset{8}{\cancel{32}}} \cdot \dfrac{\overset{1}{\overset{\cancel{4}}{\cancel{44}}}}{\underset{1}{\underset{\cancel{3}}{\cancel{33}}}} = \dfrac{15}{16}$

11 is divided into both 44 and 33.
5 is divided into both 25 and 10.
4 is divided into both 4 and 32.
3 is divided into both 3 and 9.

3'. $\dfrac{\overset{2}{\cancel{36}}}{\underset{7}{\cancel{49}}} \cdot \dfrac{\overset{2}{\cancel{14}}}{\underset{5}{\cancel{75}}} \cdot \dfrac{\overset{1}{\cancel{15}}}{\underset{1}{\cancel{18}}} = \dfrac{4}{35}$

18 is divided into both 18 and 36.
7 is divided into both 14 and 49.
15 is divided into both 15 and 75.

4'. $\dfrac{\overset{1}{\overset{\cancel{5}}{\cancel{55}}}}{\underset{1}{\underset{\cancel{2}}{\cancel{26}}}} \cdot \dfrac{\overset{1}{\cancel{8}}}{\underset{\cancel{44}}{\cancel{44}}} \cdot \dfrac{\overset{1}{\overset{\cancel{7}}{\cancel{91}}}}{\underset{1}{\underset{\cancel{7}}{\cancel{35}}}} = \dfrac{1}{1}$

11 is divided into both 55 and 44.
13 is divided into both 26 and 91.
5 is divided into both 5 and 35.
7 is divided into both 7 and 7.
2 is divided into both 2 and 8.
4 is divided into both 4 and 4.

Choose the technique that best suits your own needs and abilities.

SELF QUIZ	Reduce the following rational numbers to lowest terms.	ANSWERS
	1. $\dfrac{33}{66}$	1. $\dfrac{1}{2}$
	2. $\dfrac{25}{55}$	2. $\dfrac{5}{11}$
	3. $\dfrac{28}{51}$	3. $\dfrac{28}{51}$
	4. $\dfrac{96}{64}$	4. $\dfrac{3}{2}$
	5. Multiply $\dfrac{17}{100} \cdot \dfrac{27}{34} \cdot \dfrac{25}{9}$	5. $\dfrac{3}{8}$

EXERCISES 3.2

Supply the missing numerator or denominator so that the two rational numbers in each exercise will be equivalent.

ASSIGN ODD PROBS 1 – 83

Example: $\dfrac{2}{3} = \dfrac{}{12}$ $\qquad$ $\dfrac{2}{3} = \dfrac{2 \cdot 4}{3 \cdot 4} = \dfrac{8}{12}$

1. $\dfrac{7}{8} = \dfrac{}{24}$ $\qquad$ 2. $\dfrac{1}{16} = \dfrac{}{64}$ $\qquad$ 3. $\dfrac{2}{5} = \dfrac{}{25}$ $\qquad$ 4. $\dfrac{6}{7} = \dfrac{}{49}$

5. $\dfrac{1}{9} = \dfrac{5}{}$ $\qquad$ 6. $\dfrac{3}{4} = \dfrac{15}{}$ $\qquad$ 7. $\dfrac{5}{8} = \dfrac{10}{}$ $\qquad$ 8. $\dfrac{6}{5} = \dfrac{}{45}$

9. $\dfrac{14}{3} = \dfrac{}{9}$ $\qquad$ 10. $\dfrac{5}{8} = \dfrac{}{96}$ $\qquad$ 11. $\dfrac{9}{16} = \dfrac{}{96}$ $\qquad$ 12. $\dfrac{7}{2} = \dfrac{}{20}$

13. $\dfrac{10}{11} = \dfrac{}{44}$ $\qquad$ 14. $\dfrac{3}{16} = \dfrac{}{80}$ $\qquad$ 15. $\dfrac{11}{12} = \dfrac{}{48}$ $\qquad$ 16. $\dfrac{3}{7} = \dfrac{}{105}$

17. $\dfrac{5}{21} = \dfrac{}{42}$ $\qquad$ 18. $\dfrac{2}{3} = \dfrac{}{48}$ $\qquad$ 19. $\dfrac{5}{12} = \dfrac{}{108}$ $\qquad$ 20. $\dfrac{1}{13} = \dfrac{}{39}$

21. $\dfrac{5}{3} = \dfrac{}{48}$ $\qquad$ 22. $\dfrac{3}{40} = \dfrac{18}{}$ $\qquad$ 23. $\dfrac{6}{25} = \dfrac{}{75}$ $\qquad$ 24. $\dfrac{25}{27} = \dfrac{100}{}$

25. $\dfrac{1}{36} = \dfrac{}{72}$ $\qquad$ 26. $\dfrac{7}{24} = \dfrac{}{96}$ $\qquad$ 27. $\dfrac{7}{12} = \dfrac{}{60}$ $\qquad$ 28. $\dfrac{3}{7} = \dfrac{}{91}$

29. $\dfrac{3}{100} = \dfrac{}{1000}$ $\qquad$ 30. $\dfrac{7}{10} = \dfrac{}{1000}$

31. $\dfrac{19}{10,000} = \dfrac{}{100,000}$ $\qquad$ 32. $\dfrac{43}{10} = \dfrac{}{100}$

Reduce the following rational numbers to lowest terms.

33. $\dfrac{3}{9}$ **34.** $\dfrac{16}{24}$ **35.** $\dfrac{9}{12}$ **36.** $\dfrac{6}{20}$ **37.** $\dfrac{16}{40}$ **38.** $\dfrac{24}{30}$

39. $\dfrac{14}{36}$ **40.** $\dfrac{5}{11}$ **41.** $\dfrac{0}{25}$ **42.** $\dfrac{75}{100}$ **43.** $\dfrac{22}{55}$ **44.** $\dfrac{60}{75}$

45. $\dfrac{30}{36}$ **46.** $\dfrac{7}{28}$ **47.** $\dfrac{26}{39}$ **48.** $\dfrac{27}{56}$ **49.** $\dfrac{34}{51}$ **50.** $\dfrac{36}{48}$

51. $\dfrac{24}{100}$ **52.** $\dfrac{16}{32}$ **53.** $\dfrac{30}{45}$ **54.** $\dfrac{28}{42}$ **55.** $\dfrac{12}{35}$ **56.** $\dfrac{66}{84}$

57. $\dfrac{14}{63}$ **58.** $\dfrac{30}{70}$ **59.** $\dfrac{25}{76}$ **60.** $\dfrac{70}{84}$ **61.** $\dfrac{50}{100}$ **62.** $\dfrac{48}{12}$

63. $\dfrac{54}{9}$ **64.** $\dfrac{51}{6}$ **65.** $\dfrac{6}{51}$ **66.** $\dfrac{27}{72}$ **67.** $\dfrac{18}{40}$ **68.** $\dfrac{144}{156}$

69. $\dfrac{150}{135}$ **70.** $\dfrac{121}{165}$ **71.** $\dfrac{140}{112}$ **72.** $\dfrac{96}{108}$

Multiply. [HINT: Factor before multiplying.]

73. $\dfrac{23}{36} \cdot \dfrac{20}{46}$ **74.** $\dfrac{7}{8} \cdot \dfrac{4}{21}$ **75.** $\dfrac{5}{15} \cdot \dfrac{18}{24}$

76. $\dfrac{20}{32} \cdot \dfrac{9}{13} \cdot \dfrac{26}{7}$ **77.** $\dfrac{69}{15} \cdot \dfrac{30}{8} \cdot \dfrac{14}{46}$ **78.** $\dfrac{42}{52} \cdot \dfrac{27}{22} \cdot \dfrac{33}{9}$

79. $\dfrac{3}{4} \cdot 18 \cdot \dfrac{7}{2} \cdot \dfrac{22}{54}$ **80.** $\dfrac{9}{10} \cdot \dfrac{35}{40} \cdot \dfrac{65}{15}$ **81.** $\dfrac{66}{84} \cdot \dfrac{12}{5} \cdot \dfrac{28}{33}$

82. $\dfrac{24}{100} \cdot \dfrac{36}{48} \cdot \dfrac{15}{9}$ **83.** $\dfrac{17}{10} \cdot \dfrac{5}{42} \cdot \dfrac{18}{51} \cdot 4$ **84.** $\dfrac{75}{8} \cdot \dfrac{16}{36} \cdot 9 \cdot \dfrac{7}{25}$

3.3 ADDING RATIONAL NUMBERS

Consider the two rational numbers $\dfrac{3}{7}$ and $\dfrac{1}{7}$ illustrated by the shaded portions in the diagrams in Figure 3.4.

Go over

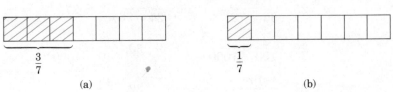

(a)   (b)

Figure 3.4

If these two shaded portions are combined, that is, put into one diagram, the result is the *sum* of the two numbers $\frac{3}{7} + \frac{1}{7}$, as in Figure 3.5.

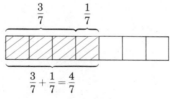

$$\frac{3}{7} + \frac{1}{7} = \frac{4}{7}$$

Figure 3.5

Similar diagrams can be used to illustrate the sum of any two (or more) rational numbers with the same denominator. More formally, addition with rational numbers with the same denominator can be defined as follows:

DEFINITION   The **sum** of two rational numbers $\frac{a}{b}$ and $\frac{c}{b}$ with common denominator b ($b \neq 0$) is the rational number whose numerator is the sum of the numerators, $a + c$, and whose denominator is b. That is,

$$\frac{a}{b} + \frac{c}{b} = \frac{a + c}{b}$$

The diagrams in Figures 3.4 and 3.5 are useful for understanding addition, and the following discussion may be helpful to some students. Do not be discouraged if you do not understand it completely.

We need to know the following two properties of rational numbers.

1. Numbers such as $\frac{5}{16}$ can be written $\frac{1}{16} \cdot 5$. This is reasonable because $5 = \frac{5}{1}$, and, thus, $\frac{1}{16} \cdot 5 = \frac{1}{16} \cdot \frac{5}{1} = \frac{5}{16}$. In general,

$$\boxed{\frac{a}{b} = \frac{1}{b} \cdot a}$$

2. The distributive property (see Section 1.7) holds for rational numbers. Thus, $\frac{1}{16} \cdot 5 + \frac{1}{16} \cdot 7 = \frac{1}{16}(5 + 7)$. In general,

$$\boxed{\frac{1}{b} \cdot a + \frac{1}{b} \cdot c = \frac{1}{b}(a + c)}$$

Putting these properties together, we get

$$\frac{5}{16} + \frac{7}{16} = \frac{1}{16} \cdot 5 + \frac{1}{16} \cdot 7 = \frac{1}{16}(5 + 7) = \frac{1}{16} \cdot 12 = \frac{12}{16}$$

In general,

$$\frac{a}{b} + \frac{c}{b} = \frac{1}{b} \cdot a + \frac{1}{b} \cdot c = \frac{1}{b}(a + c) = \frac{a + c}{b}$$

For the remainder of this chapter, we will follow the customary procedure of reducing all answers to lowest terms, whether adding, subtracting, multiplying, or dividing with rational numbers.

EXAMPLES

1. $\dfrac{1}{3} + \dfrac{1}{3} = \dfrac{1 + 1}{3} = \dfrac{2}{3}$

WORK SOME PROBLEMS
Pg 87
PROB 6 - 20

2. $\dfrac{4}{15} + \dfrac{6}{15} = \dfrac{4 + 6}{15} = \dfrac{10}{15} = \dfrac{2 \cdot 5}{3 \cdot 5} = \dfrac{2}{3}$

Addition can be easily extended to include more than two numbers by simply adding all the numerators and using the same denominator.

3. $\dfrac{1}{4} + \dfrac{2}{4} + \dfrac{1}{4} = \dfrac{1 + 2 + 1}{4} = \dfrac{4}{4} = 1$

4. $\dfrac{2}{7} + \dfrac{3}{7} + \dfrac{1}{7} + \dfrac{6}{7} = \dfrac{2 + 3 + 1 + 6}{7} = \dfrac{12}{7}$

Of course, numbers to be added will not always have the same denominator. In such a case, find numbers with a common denominator that are equivalent to each number to be added. For example, to add $\dfrac{1}{6}$ and $\dfrac{3}{10}$, list the numbers equivalent to $\dfrac{1}{6}$ and those equivalent to $\dfrac{3}{10}$, then find two with a common denominator.

LIST MULTIPLES

$$\left\{ \frac{1}{6}, \frac{2}{12}, \frac{3}{18}, \frac{4}{24}, \frac{5}{30}, \frac{6}{36}, \frac{7}{42}, \frac{8}{48}, \frac{9}{54}, \frac{10}{60}, \frac{11}{66}, \cdots \right\}$$

$$\left\{ \frac{3}{10}, \frac{6}{20}, \frac{9}{30}, \frac{12}{40}, \frac{15}{50}, \frac{18}{60}, \frac{21}{70}, \cdots \right\}$$

Thus, we could write

$$\frac{1}{6} + \frac{3}{10} = \frac{5}{30} + \frac{9}{30} \quad \text{or} \quad \frac{1}{6} + \frac{3}{10} = \frac{10}{60} + \frac{18}{60}$$

Other possible combinations can be found by looking further at the sets of equivalent rational numbers. The *preferred choice* is the pair with the smallest common denominator because the numbers are easier to work with. So

$$\frac{1}{6} + \frac{3}{10} = \frac{5}{30} + \frac{9}{30} = \frac{14}{30} = \frac{\cancel{2} \cdot 7}{\cancel{2} \cdot 15} = \frac{7}{15}$$

This technique is not very efficient, especially if there are several numbers to be added or the denominators are large. A better method is to **find the least common multiple (LCM) of the denominators and then change the terms of each number so that it has the LCM as denominator.**

EXAMPLES   1. $\dfrac{1}{4} + \dfrac{3}{8} + \dfrac{7}{10}$

Find the LCM of $\{4, 8, 10\}$.

$$\left.\begin{array}{l} 4 = 2 \cdot 2 \\ 8 = 2 \cdot 2 \cdot 2 \\ 10 = 2 \cdot 5 \end{array}\right\} \text{LCM} = 2 \cdot 2 \cdot 2 \cdot 5 = 40$$

Now find numbers equivalent to $\dfrac{1}{4}, \dfrac{3}{8},$ and $\dfrac{7}{10}$ with denominator 40.

$$\frac{1}{4} = \frac{10}{10} \cdot \frac{1}{4} = \frac{10}{40} \qquad \frac{3}{8} = \frac{5}{5} \cdot \frac{3}{8} = \frac{15}{40} \qquad \frac{7}{10} = \frac{4}{4} \cdot \frac{7}{10} = \frac{28}{40}$$

$\uparrow$ $\uparrow$ $\uparrow$

$\frac{10}{10}$ is used since $\frac{5}{5}$ is used since $\frac{4}{4}$ is used since

$10 \cdot 4 = 40$ $5 \cdot 8 = 40$ $4 \cdot 10 = 40$

Thus, replacing each number with an equivalent,

$$\frac{1}{4} + \frac{3}{8} + \frac{7}{10} = \frac{10}{40} + \frac{15}{40} + \frac{28}{40} = \frac{10 + 15 + 28}{40} = \frac{53}{40}$$

We could also write

$$\frac{1}{4} + \frac{3}{8} + \frac{7}{10} = \frac{10}{10} \cdot \frac{1}{4} + \frac{5}{5} \cdot \frac{3}{8} + \frac{4}{4} \cdot \frac{7}{10} = \frac{10}{40} + \frac{15}{40} + \frac{28}{40} = \frac{53}{40}$$

2. $\dfrac{1}{4} + \dfrac{1}{8}$

The LCM of $\{4, 8\}$ is obviously 8, and no formal work is necessary.

$$\frac{1}{4} + \frac{1}{8} = \frac{2}{2} \cdot \frac{1}{4} + \frac{1}{8} = \frac{2}{8} + \frac{1}{8} = \frac{3}{8}$$

The numbers can also be written vertically, one under the other.

WORK SOME
PROBLEMS
Pg 87
Prob 21-60

$$\frac{1}{4} = \frac{2}{2} \cdot \frac{1}{4} = \frac{2}{8}$$

$$\frac{\dfrac{1}{8}}{\quad} = \frac{\dfrac{1}{8}}{\quad} = \frac{\dfrac{1}{8}}{\dfrac{3}{8}}$$

3. $\dfrac{5}{21} + \dfrac{5}{28}$

Find the LCM of $\{21, 28\}$.

$$\left.\begin{array}{l} 21 = 3 \cdot 7 \\ 28 = 2 \cdot 2 \cdot 7 \end{array}\right\} \text{LCM} = 2 \cdot 2 \cdot 3 \cdot 7 = 84$$

$$84 = (2 \cdot 2) \cdot (3 \cdot 7) = 4 \cdot 21$$
$$84 = (2 \cdot 2 \cdot 7) \cdot 3 = 3 \cdot 28$$

$$\frac{5}{21} + \frac{5}{28} = \frac{4}{4} \cdot \frac{5}{21} + \frac{3}{3} \cdot \frac{5}{28} = \frac{20}{84} + \frac{15}{84} = \frac{35}{84} = \frac{\not{7} \cdot 5}{2 \cdot 2 \; 3 \cdot \not{7}} = \frac{5}{12}$$

Or, writing the numbers vertically,

$$\frac{5}{21} = \frac{4}{4} \cdot \frac{5}{21} = \frac{20}{84}$$

$$\frac{\dfrac{5}{28}}{\quad} = \frac{3}{3} \cdot \frac{5}{28} = \frac{15}{84}$$

$$\frac{35}{84} = \frac{\not{7} \cdot 5}{2 \cdot 2 \cdot 3 \cdot \not{7}} = \frac{5}{12}$$

Even if the numbers are written vertically, you still find the LCM of the denominator using prime factorization.

An alternate method is illustrated as

$$\frac{a}{b} \bowtie \frac{c}{d} = \frac{ad + bc}{bd}$$

This method is used *only* when adding *two* rational numbers and does *not* always give the smallest common denominator. Its proof is as follows:

$$\frac{a}{b} + \frac{c}{d} = \frac{a}{b} \cdot \frac{d}{d} + \frac{c}{d} \cdot \frac{b}{b} = \frac{ad}{bd} + \frac{cb}{db} = \frac{ad + bc}{bd}$$

EXAMPLE

4. $\dfrac{1}{4} + \dfrac{1}{8} = \dfrac{1 \cdot 8 + 4 \cdot 1}{4 \cdot 8} = \dfrac{8 + 4}{32} = \dfrac{12}{32} = \dfrac{\cancel{4} \cdot 3}{\cancel{4} \cdot 8} = \dfrac{3}{8}$

Comparison of Example 4 with Example 2 shows the differences in the two methods. Also, the method of Example 4 cannot be used with Example 1. The method of Example 4 is particularly convenient when a calculator is available and large numbers are not a problem.

What if the order of addition of two rational numbers is changed? Is the result changed? That is, does $\dfrac{1}{4} + \dfrac{5}{8} = \dfrac{5}{8} + \dfrac{1}{4}$? If three or more rational numbers are added, does the sum change if the grouping of the addends changes? That is, does $\dfrac{5}{7} + \left(\dfrac{3}{7} + \dfrac{4}{7}\right) = \left(\dfrac{5}{7} + \dfrac{3}{7}\right) + \dfrac{4}{7}$? In other words, is addition with rational numbers **commutative** and **associative?** The answer is *yes,* because addition involves the addition of the numerators, which are whole numbers, and whole numbers are commutative and associative under addition.

COMMUTATIVE PROPERTY OF ADDITION

ORDER

If $\dfrac{a}{b}$ and $\dfrac{c}{d}$ are rational numbers, then

$$\frac{a}{b} + \frac{c}{d} = \frac{c}{d} + \frac{a}{b}$$

ASSOCIATIVE PROPERTY OF ADDITION

GROUPING

If $\dfrac{a}{b}, \dfrac{c}{d},$ and $\dfrac{e}{f}$ are rational numbers, then

$$\frac{a}{b} + \frac{c}{d} + \frac{e}{f} = \left(\frac{a}{b} + \frac{c}{d}\right) + \frac{e}{f} = \frac{a}{b} + \left(\frac{c}{d} + \frac{e}{f}\right)$$

EXAMPLES

1. Show that $\dfrac{1}{5} + \dfrac{2}{3} = \dfrac{2}{3} + \dfrac{1}{5}$.

$$\frac{1}{5} + \frac{2}{3} = \frac{3}{3} \cdot \frac{1}{5} + \frac{5}{5} \cdot \frac{2}{3} = \frac{3}{15} + \frac{10}{15} = \frac{13}{15}$$

$$\frac{2}{3} + \frac{1}{5} = \frac{10}{15} + \frac{3}{15} = \frac{13}{15}$$

2. Show that $\frac{1}{2} + \left(\frac{1}{6} + \frac{1}{8}\right) = \left(\frac{1}{2} + \frac{1}{6}\right) + \frac{1}{8}$.

$$\frac{1}{2} + \left(\frac{1}{6} + \frac{1}{8}\right) = \frac{12}{12} \cdot \frac{1}{2} + \left(\frac{4}{4} \cdot \frac{1}{6} + \frac{3}{3} \cdot \frac{1}{8}\right) = \frac{12}{24} + \left(\frac{4}{24} + \frac{3}{24}\right)$$

$$= \frac{12}{24} + \frac{7}{24} = \frac{19}{24}$$

$$\left(\frac{1}{2} + \frac{1}{6}\right) + \frac{1}{8} = \left(\frac{12}{24} + \frac{4}{24}\right) + \frac{3}{24} = \frac{16}{24} + \frac{3}{24} = \frac{19}{24}$$

What happens if you add 0 to $\frac{3}{8}$?

$$\frac{3}{8} + 0 = \frac{3}{8} + \frac{0}{8} = \frac{3+0}{8} = \frac{3}{8}$$

In general, adding 0 (or $\frac{0}{b}$ where $b \neq 0$) to a rational number gives the same rational number. Thus, 0 is the **additive identity** for the rational numbers.

ADDITIVE IDENTITY

If $\frac{a}{b}$ is a rational number, then

$$\frac{a}{b} + 0 = \frac{a}{b}$$

EXAMPLE

Show that $\frac{9}{16} + 0 = \frac{9}{16}$.

$$\frac{9}{16} + 0 = \frac{9}{16} + \frac{0}{16} = \frac{9+0}{16} = \frac{9}{16}$$

SELF QUIZ	Find the following sums. Reduce all answers.	ANSWERS
1. $\frac{1}{8} + \frac{3}{8} + \frac{7}{8}$		1. $\frac{11}{8}$
2. $\frac{2}{3} + \frac{5}{8} + \frac{1}{6}$		2. $\frac{35}{24}$
3. $\frac{7}{10} + \frac{1}{100} + \frac{5}{1000}$		3. $\frac{715}{1000} = \frac{143}{200}$

EXERCISES 3.3

1. Draw diagrams similar to Figure 3.5 illustrating the sums.

(a) $\dfrac{2}{5} + \dfrac{2}{5}$ (b) $\dfrac{3}{10} + \dfrac{7}{10}$ (c) $\dfrac{3}{7} + \dfrac{6}{7}$

2. Give two examples illustrating the commutative property of addition with rational numbers.

3. Show that the associative property of addition is true for the numbers $\dfrac{3}{100}, \dfrac{7}{100},$ and $\dfrac{1}{10}.$

4. Show that the associative property of multiplication holds for the same three numbers in Exercise 3.

5. Explain in your own words why zero (0) is the additive identity for the rational numbers.

Add the following rational numbers and reduce all answers.

6. $\dfrac{6}{10} + \dfrac{4}{10}$ 7. $\dfrac{3}{14} + \dfrac{2}{14}$ 8. $\dfrac{1}{20} + \dfrac{3}{20}$ 9. $\dfrac{3}{4} + \dfrac{3}{4}$

10. $\dfrac{5}{6} + \dfrac{4}{6}$ 11. $\dfrac{7}{5} + \dfrac{3}{5}$ 12. $\dfrac{11}{15} + \dfrac{7}{15}$ 13. $\dfrac{7}{9} + \dfrac{8}{9}$

14. $\dfrac{3}{25} + \dfrac{12}{25}$ 15. $\dfrac{7}{90} + \dfrac{37}{90} + \dfrac{21}{90}$ 16. $\dfrac{11}{75} + \dfrac{12}{75} + \dfrac{62}{75}$

17. $\dfrac{14}{32} + \dfrac{7}{32} + \dfrac{1}{32}$ 18. $\dfrac{4}{100} + \dfrac{35}{100} + \dfrac{76}{100}$ 19. $\dfrac{21}{95} + \dfrac{33}{95} + \dfrac{3}{95}$

20. $\dfrac{1}{200} + \dfrac{17}{200} + \dfrac{25}{200}$ 21. $\dfrac{1}{12} + \dfrac{2}{3} + \dfrac{1}{4}$ 22. $\dfrac{3}{8} + \dfrac{5}{16}$

23. $\dfrac{2}{5} + \dfrac{3}{10} + \dfrac{3}{20}$ 24. $\dfrac{3}{4} + \dfrac{1}{16} + \dfrac{6}{32}$ 25. $\dfrac{2}{7} + \dfrac{4}{21} + \dfrac{1}{3}$

26. $\dfrac{1}{6} + \dfrac{1}{4} + \dfrac{1}{3}$ 27. $\dfrac{2}{39} + \dfrac{1}{3} + \dfrac{4}{13}$ 28. $\dfrac{1}{2} + \dfrac{3}{10} + \dfrac{4}{5}$

29. $\dfrac{1}{27} + \dfrac{4}{18} + \dfrac{1}{6}$ 30. $\dfrac{2}{7} + \dfrac{3}{20} + \dfrac{9}{14}$ 31. $\dfrac{1}{8} + \dfrac{1}{12} + \dfrac{1}{9}$

32. $\dfrac{2}{5} + \dfrac{4}{7} + \dfrac{3}{8}$ 33. $\dfrac{1}{63} + \dfrac{2}{27} + \dfrac{1}{45}$ 34. $\dfrac{5}{8} + \dfrac{4}{27} + \dfrac{1}{48}$

35. $\dfrac{3}{16} + \dfrac{5}{48} + \dfrac{1}{32}$ 36. $\dfrac{18}{34} + \dfrac{4}{9} + \dfrac{27}{85}$ 37. $\dfrac{72}{105} + \dfrac{2}{45} + \dfrac{15}{21}$

38. $\dfrac{5}{16} + \dfrac{19}{72} + \dfrac{17}{18}$ 39. $\dfrac{46}{75} + \dfrac{31}{60} + \dfrac{21}{48}$ 40. $\dfrac{7}{10} + \dfrac{3}{25} + \dfrac{3}{4}$

41. $\dfrac{1}{10} + \dfrac{2}{5} + \dfrac{17}{35} + \dfrac{3}{7}$

42. $\dfrac{4}{21} + \dfrac{3}{4} + \dfrac{1}{9} + \dfrac{13}{16}$

43. $\dfrac{15}{56} + \dfrac{13}{42} + \dfrac{1}{49} + \dfrac{3}{8}$

44. $\dfrac{16}{30} + \dfrac{7}{12} + \dfrac{15}{36} + \dfrac{5}{72}$

45. $\dfrac{0}{27} + \dfrac{0}{16} + \dfrac{1}{5}$

46. $\dfrac{5}{6} + \dfrac{0}{100} + \dfrac{0}{70} + \dfrac{1}{3}$

47. $\dfrac{3}{10} + \dfrac{1}{100} + \dfrac{7}{1000}$

48. $\dfrac{11}{100} + \dfrac{15}{10} + \dfrac{1}{10}$

49. $\dfrac{17}{1000} + \dfrac{1}{100} + \dfrac{1}{10,000}$

50. $6 + \dfrac{1}{100} + \dfrac{3}{10}$

51. $8 + \dfrac{1}{10} + \dfrac{9}{100} + \dfrac{1}{1000}$

52. $\dfrac{1}{10} + \dfrac{3}{10} + \dfrac{9}{100}$

53. $\dfrac{7}{10} + \dfrac{5}{100} + \dfrac{3}{1000}$

54. $\dfrac{1}{2} + \dfrac{3}{4} + \dfrac{1}{100}$

55. $\dfrac{1}{4} + \dfrac{1}{8} + \dfrac{7}{100}$

56. $\dfrac{9}{1000} + \dfrac{7}{1000} + \dfrac{21}{10,000}$

57. $\dfrac{11}{100} + \dfrac{1}{2} + \dfrac{3}{1000}$

58. $\dfrac{3}{4} + \dfrac{17}{1000} + \dfrac{13}{10,000} + 2$

59. $5 + \dfrac{1}{10} + \dfrac{3}{100} + \dfrac{4}{1000}$

60. $\dfrac{13}{10,000} + \dfrac{1}{100,000} + \dfrac{21}{1,000,000}$

3.4   SUBTRACTING RATIONAL NUMBERS

As with whole numbers, subtraction with rational numbers is a reverse addition; that is, one of the addends is subtracted from the sum, and the missing addend is the difference.

$$\dfrac{9}{17} - \dfrac{4}{17} = \square$$

sum   addend   missing addend, or difference

$$\dfrac{9}{17} - \dfrac{4}{17} = \dfrac{5}{17} \quad \text{since} \quad \dfrac{9}{17} = \dfrac{4}{17} + \dfrac{5}{17}$$

The following discussion, however, develops a more direct method for finding differences.

Diagrams can be used to illustrate the difference between two rational numbers, such as $\dfrac{5}{8} - \dfrac{3}{8}$ as shown in Figure 3.6.

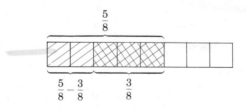

Figure 3.6

For the purposes of this chapter, the smaller number must be subtracted from the larger number. Therefore, the relative sizes of the numbers are important. Which is smaller, $\dfrac{2}{3}$ or $\dfrac{5}{7}$? The question can be answered by finding equivalent rational numbers with a common denominator and then comparing numerators.

$$\frac{2}{3} = \frac{7}{7} \cdot \frac{2}{3} = \frac{14}{21} \quad \text{and} \quad \frac{5}{7} = \frac{3}{3} \cdot \frac{5}{7} = \frac{15}{21}$$

Since 14 is smaller than 15, $\dfrac{2}{3}$ is smaller than $\dfrac{5}{7}$.

EXAMPLES

1. Which is smaller, $\dfrac{5}{6}$ or $\dfrac{7}{8}$?

$$\frac{5}{6} = \frac{4}{4} \cdot \frac{5}{6} = \frac{20}{24} \qquad \frac{7}{8} = \frac{3}{3} \cdot \frac{7}{8} = \frac{21}{24}$$

Since 20 is smaller than 21, $\dfrac{5}{6}$ is smaller than $\dfrac{7}{8}$.

2. Arrange the numbers $\dfrac{2}{3}, \dfrac{7}{10},$ and $\dfrac{9}{15}$ in order, smallest to largest.

$$\frac{7}{10} = \frac{3}{3} \cdot \frac{7}{10} = \frac{21}{30} \qquad \frac{2}{3} = \frac{10}{10} \cdot \frac{2}{3} = \frac{20}{30} \qquad \frac{9}{15} = \frac{2}{2} \cdot \frac{9}{15} = \frac{18}{30}$$

In order, $\dfrac{9}{15}, \dfrac{2}{3}, \dfrac{7}{10}.$

DEFINITION

The **difference** of two rational numbers $\frac{a}{b}$ and $\frac{c}{b}$ with a greater than or equal to c is a rational number whose numerator is the difference of the numerators and whose denominator is the common denominator b. In symbols,

$$\frac{a}{b} - \frac{c}{b} = \frac{a-c}{b}$$

EXAMPLES

1. $\dfrac{5}{6} - \dfrac{1}{6} = \dfrac{5-1}{6} = \dfrac{4}{6} = \dfrac{2 \cdot 2}{2 \cdot 3} = \dfrac{2}{3}$

2. $\dfrac{9}{10} - \dfrac{7}{10} = \dfrac{9-7}{10} = \dfrac{2}{10} = \dfrac{2 \cdot 1}{2 \cdot 5} = \dfrac{1}{5}$

As with addition, the numbers need not have the same denominator. In this case, find equivalent numbers with a common denominator. Again, the smallest common denominator is preferred.

3. $\dfrac{9}{10} - \dfrac{2}{15} = \dfrac{3}{3} \cdot \dfrac{9}{10} - \dfrac{2}{2} \cdot \dfrac{2}{15} = \dfrac{27}{30} - \dfrac{4}{30} = \dfrac{27-4}{30} = \dfrac{23}{30}$

4. $5 - \dfrac{1}{7} = \dfrac{7}{7} \cdot \dfrac{5}{1} - \dfrac{1}{7} = \dfrac{35}{7} - \dfrac{1}{7} = \dfrac{35-1}{7} = \dfrac{34}{7}$

SELF QUIZ	Find the following differences. Reduce all answers.	ANSWERS
	1. $\dfrac{5}{9} - \dfrac{1}{9}$	1. $\dfrac{4}{9}$
	2. $\dfrac{7}{6} - \dfrac{2}{3}$	2. $\dfrac{1}{2}$
	3. $\dfrac{7}{10} - \dfrac{7}{15}$	3. $\dfrac{7}{30}$
	4. $\dfrac{1}{45} - \dfrac{1}{72}$	4. $\dfrac{1}{120}$

ODD PROB 1- 49

EXERCISES 3.4

1. Using figures similar to that in Figure 3.6, illustrate the differences.

(a) $\dfrac{2}{3} - \dfrac{1}{3}$ (b) $\dfrac{3}{4} - \dfrac{5}{12}$ (c) $2 - \dfrac{5}{4}$

Find the larger of the following pairs of numbers and tell how much larger it is.

2. $\dfrac{2}{3}, \dfrac{3}{4}$ 3. $\dfrac{5}{6}, \dfrac{7}{8}$ 4. $\dfrac{4}{5}, \dfrac{17}{20}$ 5. $\dfrac{4}{10}, \dfrac{3}{8}$

6. $\dfrac{13}{20}, \dfrac{5}{8}$ 7. $\dfrac{13}{16}, \dfrac{21}{25}$ 8. $\dfrac{14}{35}, \dfrac{12}{30}$ 9. $\dfrac{10}{36}, \dfrac{7}{24}$

10. $\dfrac{17}{80}, \dfrac{11}{48}$ 11. $\dfrac{37}{100}, \dfrac{24}{75}$

Arrange the following numbers in order, smallest to largest.

12. $\dfrac{2}{3}, \dfrac{3}{5}, \dfrac{7}{10}$ 13. $\dfrac{8}{9}, \dfrac{9}{10}, \dfrac{11}{12}$ 14. $\dfrac{7}{6}, \dfrac{11}{12}, \dfrac{19}{20}$

15. $\dfrac{17}{12}, \dfrac{40}{36}, \dfrac{31}{24}$ 16. $\dfrac{1}{3}, \dfrac{5}{42}, \dfrac{3}{7}$ 17. $\dfrac{7}{8}, \dfrac{31}{36}, \dfrac{13}{18}$

18. $\dfrac{1}{100}, \dfrac{3}{1000}, \dfrac{20}{10,000}$ 19. $\dfrac{32}{100}, \dfrac{298}{1000}, \dfrac{3333}{10,000}, \dfrac{3}{10}$

20. $\dfrac{72}{120}, \dfrac{80}{150}, \dfrac{35}{60}, \dfrac{15}{24}$

Find the difference in each of the following exercises.

21. $\dfrac{4}{7} - \dfrac{1}{7}$ 22. $\dfrac{9}{10} - \dfrac{3}{10}$ 23. $\dfrac{5}{8} - \dfrac{1}{8}$ 24. $\dfrac{11}{12} - \dfrac{7}{12}$

25. $\dfrac{13}{15} - \dfrac{4}{15}$ 26. $\dfrac{5}{6} - \dfrac{1}{3}$ 27. $\dfrac{11}{15} - \dfrac{3}{10}$ 28. $\dfrac{3}{4} - \dfrac{2}{3}$

29. $\dfrac{15}{16} - \dfrac{21}{32}$ 30. $\dfrac{5}{8} - \dfrac{3}{5}$ 31. $\dfrac{14}{27} - \dfrac{7}{18}$ 32. $\dfrac{8}{45} - \dfrac{11}{72}$

33. $\dfrac{46}{55} - \dfrac{10}{33}$ 34. $\dfrac{5}{36} - \dfrac{1}{30}$ 35. $\dfrac{5}{36} - \dfrac{1}{32}$ 36. $\dfrac{15}{16} - \dfrac{60}{72}$

37. $\dfrac{25}{60} - \dfrac{6}{24}$ 38. $\dfrac{26}{56} - \dfrac{5}{42}$ 39. $\dfrac{21}{45} - \dfrac{34}{105}$ 40. $\dfrac{137}{280} - \dfrac{33}{100}$

41. $\dfrac{9}{10} - \dfrac{3}{100}$ 42. $\dfrac{159}{1000} - \dfrac{1}{10}$ 43. $\dfrac{76}{100} - \dfrac{82}{10,000}$ 44. $\dfrac{999}{1000} - \dfrac{99}{100}$

45. Subtract the sum of $\dfrac{1}{4}, \dfrac{1}{8}$, and $\dfrac{3}{16}$ from the sum of $\dfrac{3}{10}, \dfrac{9}{8}$, and $\dfrac{1}{6}$.

46. Find the sum of the difference between $\dfrac{5}{9}$ and $\dfrac{2}{3}$ and the difference between $\dfrac{7}{8}$ and $\dfrac{5}{6}$.

47. $\dfrac{11}{12}$ is a nice number. If you were to add $\dfrac{7}{16}$ and $\dfrac{5}{32}$ and multiply the sum by $\dfrac{3}{19}$, what would the product be?

48. If the product of $\dfrac{9}{10}$ and $\dfrac{9}{10}$ is added to the difference between $\dfrac{3}{4}$ and $\dfrac{1}{25}$, what is the sum? Is the sum bigger or smaller than 1? How much bigger or smaller?

49. A length of pipe is cut into three equal parts. One of these parts is cut into six equal parts, and one of the six is cut in half. If one piece of each size that was cut is chosen, what total fractional part of the original length is chosen?

3.5 MIXED NUMBERS

We have already discussed the fact that whole numbers can be represented as rational numbers. Quite frequently, whole numbers and other rational numbers that are not whole numbers are added together. For example,

$$5 + \frac{2}{3} \quad \text{or} \quad 7 + \frac{1}{4}$$

These can be added by finding a common denominator and adding, just as in the previous sections.

$$5 + \frac{2}{3} = \frac{5}{1} + \frac{2}{3} = \frac{3}{3} \cdot \frac{5}{1} + \frac{2}{3} = \frac{15}{3} + \frac{2}{3} = \frac{15 + 2}{3} = \frac{17}{3}$$

$$7 + \frac{1}{4} = \frac{7}{1} + \frac{1}{4} = \frac{4}{4} \cdot \frac{7}{1} + \frac{1}{4} = \frac{28}{4} + \frac{1}{4} = \frac{28 + 1}{4} = \frac{29}{4}$$

There is, however, a very convenient shorthand method of writing these sums. Simply write the whole number and other rational number next to each other.

$$5 + \frac{2}{3} = 5\frac{2}{3} \quad \text{(read: "five and two thirds")}$$

$$7 + \frac{1}{4} = 7\frac{1}{4} \quad \text{(read: "seven and one fourth")}$$

These numbers are called **mixed numbers.** A mixed number is a rational number in another form.

DEFINITION   A **mixed number** is the sum of *a whole number* and *a rational number not a whole number,* indicated by writing the numbers next to each other.

We shall call the rational number that is not a whole number the *fraction part of the mixed number.* Note carefully that a mixed number indicates *addition,* not multiplication. For multiplication, either put the numbers in parentheses or use a dot. Thus,

$$5 \cdot \frac{2}{3} = 5\left(\frac{2}{3}\right) = (5)\frac{2}{3}$$

$$\left[\text{REMEMBER: } 5\left(\frac{2}{3}\right) = \frac{5}{1}\left(\frac{2}{3}\right) = \frac{5 \cdot 2}{1 \cdot 3} = \frac{10}{3} \quad \text{but} \quad 5\frac{2}{3} = 5 + \frac{2}{3} = \frac{17}{3}\right]$$

A rational number with numerator larger than denominator can be written as a mixed number (or as a whole number if the denominator is a factor of the numerator). Since a rational number can indicate division, divide the numerator by the denominator. If there is a remainder, indicate the continuing division by a rational number added to the whole number quotient.

EXAMPLES   1. Change $\frac{7}{3}$ to a mixed number:

$$3\overline{)7} \atop \begin{array}{r} 2 \\ \hline 6 \\ \hline 1 \end{array}$$

$$\frac{7}{3} = 2 + \frac{1}{3} = 2\frac{1}{3}$$

2. Change $\frac{29}{4}$ to a mixed number:

$$4\overline{)29} \atop \begin{array}{r} 7 \\ \hline 28 \\ \hline 1 \end{array}$$

$$\frac{29}{4} = 7 + \frac{1}{4} = 7\frac{1}{4}$$

3. Change $\frac{59}{3}$ to a mixed number:

$$3\overline{)59} \atop \begin{array}{r} 19 \\ \hline 3 \\ \hline 29 \\ 27 \\ \hline 2 \end{array}$$

$$\frac{59}{3} = 19 + \frac{2}{3} = 19\frac{2}{3}$$

Changing a rational number to a mixed number is not reducing it. *Reducing* involves finding common factors in the numerator and denominator, while *changing to a mixed number* has nothing to do with common factors.

EXAMPLE   Reduce $\dfrac{14}{10}$. $\dfrac{14}{10} = \dfrac{2 \cdot 7}{2 \cdot 5} = \dfrac{7}{5}$

Change $\dfrac{14}{10}$ to a mixed number. $10\overline{)14}$ $\dfrac{14}{10} = 1 + \dfrac{4}{10} = 1\dfrac{4}{10}$
$$\begin{array}{r} 1 \\ 10\overline{)14} \\ 10 \\ \hline 4 \end{array}$$

TO CHANGE RATIONAL NUMBERS TO MIXED NUMBERS

1. Reduce first, then change to mixed numbers.
2. Or change to mixed numbers first, then reduce the fraction part.

EXAMPLE   Change $\dfrac{14}{10}$ to a mixed number and reduce the answer.

(a) $\dfrac{14}{10} = \dfrac{\cancel{2} \cdot 7}{\cancel{2} \cdot 5} = \dfrac{7}{5}$ $5\overline{)7}$ $\dfrac{7}{5} = 1 + \dfrac{2}{5} = 1\dfrac{2}{5}$
$$\begin{array}{r} 1 \\ 5\overline{)7} \\ 5 \\ \hline 2 \end{array}$$

(b) $\dfrac{14}{10}$ $10\overline{)14}$ $\dfrac{14}{10} = 1 + \dfrac{4}{10} = 1\dfrac{4}{10}$
$$\begin{array}{r} 1 \\ 10\overline{)14} \\ 10 \\ \hline 4 \end{array}$$

$\dfrac{4}{10} = \dfrac{\cancel{2} \cdot 2}{\cancel{2} \cdot 5} = \dfrac{2}{5}$ $1\dfrac{4}{10} = 1\dfrac{2}{5}$

Mixed numbers are rational numbers and can be changed to fraction form. Remembering that a mixed number indicates addition, we can add the whole number and the fraction part by finding a common denominator. Of course, since the denominator of the whole number is 1, the common denominator will be the denominator of the fraction part.

As mentioned before, the term *improper fraction* is sometimes used for rational numbers such as $\dfrac{17}{3}$ and $\dfrac{29}{4}$. This usage is acceptable only if the student understands there is nothing "improper" about these numbers. The mixed number form and the improper fraction form are both useful and correct.

EXAMPLES   1.   Change $7\frac{2}{5}$ to fraction form.

$$7\frac{2}{5} = 7 + \frac{2}{5} = \frac{7}{1} + \frac{2}{5} = \frac{5}{5} \cdot \frac{7}{1} + \frac{2}{5} = \frac{35}{5} + \frac{2}{5} = \frac{35 + 2}{5} = \frac{37}{5}$$

2.   Change $6\frac{2}{9}$ to fraction form.

$$6\frac{2}{9} = 6 + \frac{2}{9} = \frac{6}{1} + \frac{2}{9} = \frac{9}{9} \cdot \frac{6}{1} + \frac{2}{9} = \frac{54}{9} + \frac{2}{9} = \frac{54 + 2}{9} = \frac{56}{9}$$

3.   Change $10\frac{1}{3}$ to fraction form.

$$10\frac{1}{3} = 10 + \frac{1}{3} = \frac{10}{1} + \frac{1}{3} = \frac{3}{3} \cdot \frac{10}{1} + \frac{1}{3} = \frac{30}{3} + \frac{1}{3} = \frac{30 + 1}{3} = \frac{31}{3}$$

There is a very convenient shortcut for changing mixed numbers to fraction form. Note in the examples that the whole number is always multiplied by the denominator of the fraction, and the numerator of the fraction is added to this product to get the new numerator. We have already noted that the denominator is the denominator of the fraction. These multiplications and additions can be done in your head, without writing anything down, in the following manner.

EXAMPLES   1'.   Change $7\frac{2}{5}$ to fraction form. Multiply $5 \cdot 7 = 35$ and add 2 giving $35 + 2 = 37$. Thus,

$$7\frac{2}{5} = \frac{37}{5}$$

Prob 31-50

2'.   Change $6\frac{2}{9}$ to fraction form. Multiply $9 \cdot 6 = 54$ and add 2 giving $54 + 2 = 56$. Thus,

$$6\frac{2}{9} = \frac{56}{9}$$

3'.   Change $10\frac{1}{3}$ to fraction form. Multiply $3 \cdot 10 = 30$ and add 1 giving $30 + 1 = 31$. Thus,

$$10\frac{1}{3} = \frac{31}{3}$$

This shortcut should be used only if you remember that a mixed number indicates addition and not multiplication.

SELF QUIZ	ANSWERS
1. Reduce $\frac{18}{16}$ to lowest terms.	1. $\frac{9}{8}$
2. Change $\frac{51}{34}$ to a mixed number.	2. $1\frac{1}{2}$
3. Change $6\frac{2}{3}$ to fraction form.	3. $\frac{20}{3}$

OPP 1- 43

EXERCISES 3.5

1. Define *mixed number*.

Reduce the following rational numbers to lowest terms.

2. $\frac{25}{10}$ 3. $\frac{16}{12}$ 4. $\frac{10}{8}$ 5. $\frac{39}{26}$ 6. $\frac{48}{32}$ 7. $\frac{35}{25}$ 8. $\frac{18}{16}$

9. $\frac{80}{64}$ 10. $\frac{75}{60}$ 11. $\frac{100}{24}$

Change the following rational numbers to mixed numbers.

12. $\frac{25}{10}$ 13. $\frac{16}{12}$ 14. $\frac{10}{8}$ 15. $\frac{39}{26}$ 16. $\frac{42}{8}$ 17. $\frac{43}{7}$

18. $\frac{34}{16}$ 19. $\frac{45}{6}$ 20. $\frac{75}{12}$ 21. $\frac{56}{18}$ 22. $\frac{31}{15}$ 23. $\frac{36}{12}$

24. $\frac{48}{16}$ 25. $\frac{72}{16}$ 26. $\frac{70}{34}$ 27. $\frac{45}{15}$ 28. $\frac{60}{36}$ 29. $\frac{35}{20}$

30. $\frac{185}{100}$

Change the following mixed numbers to fraction form.

31. $4\frac{5}{8}$ 32. $3\frac{3}{4}$ 33. $5\frac{1}{15}$ 34. $1\frac{3}{5}$ 35. $4\frac{2}{11}$

36. $2\dfrac{11}{44}$ **37.** $2\dfrac{9}{27}$ **38.** $4\dfrac{6}{7}$ **39.** $10\dfrac{8}{12}$ **40.** $11\dfrac{3}{8}$

41. $6\dfrac{8}{10}$ **42.** $14\dfrac{1}{5}$ **43.** $16\dfrac{2}{3}$ **44.** $12\dfrac{4}{8}$ **45.** $20\dfrac{3}{15}$

46. $9\dfrac{4}{10}$ **47.** $13\dfrac{1}{7}$ **48.** $49\dfrac{0}{12}$ **49.** $17\dfrac{0}{3}$ **50.** $3\dfrac{1}{50}$

3.6   ADDING MIXED NUMBERS

When adding mixed numbers, again it is important to remember that a mixed number itself indicates addition. With this in mind, we can simply add the whole numbers and the fraction parts separately and combine the answers as another mixed number.

EXAMPLES

1. $4\dfrac{1}{3} + 7\dfrac{1}{5} = 4 + \dfrac{1}{3} + 7 + \dfrac{1}{5} = (4 + 7) + \left(\dfrac{1}{3} + \dfrac{1}{5}\right)$

$$= 11 + \left(\dfrac{5}{5} \cdot \dfrac{1}{3} + \dfrac{3}{3} \cdot \dfrac{1}{5}\right) = 11 + \left(\dfrac{5}{15} + \dfrac{3}{15}\right) = 11 + \dfrac{8}{15} = 11\dfrac{8}{15}$$

2. $16\dfrac{2}{7} + 10\dfrac{4}{21} = 16 + \dfrac{2}{7} + 10 + \dfrac{4}{21} = (16 + 10) + \left(\dfrac{2}{7} + \dfrac{4}{21}\right)$

$$= 26 + \left(\dfrac{3}{3} \cdot \dfrac{2}{7} + \dfrac{4}{21}\right) = 26 + \left(\dfrac{6}{21} + \dfrac{4}{21}\right)$$

$$= 26 + \dfrac{10}{21} = 26\dfrac{10}{21}$$

3. $6\dfrac{3}{5} + 2\dfrac{6}{7} = 6 + \dfrac{3}{5} + 2 + \dfrac{6}{7} = (6 + 2) + \left(\dfrac{3}{5} + \dfrac{6}{7}\right) = 8 + \left(\dfrac{7}{7} \cdot \dfrac{3}{5} + \dfrac{5}{5} \cdot \dfrac{6}{7}\right)$

$$= 8 + \left(\dfrac{21}{35} + \dfrac{30}{35}\right) = 8 + \dfrac{51}{35} = 8 + 1 + \dfrac{16}{35} = 9 + \dfrac{16}{35} = 9\dfrac{16}{35}$$

Sometimes the numbers are written one under the other, but the procedure is the same: add the whole numbers and the fraction parts separately.

4. $4\dfrac{1}{3} = 4\dfrac{5}{15}$ $\dfrac{1}{3} = \dfrac{5}{5} \cdot \dfrac{1}{3} = \dfrac{5}{15}$

$\underline{7\dfrac{1}{5} = \ 7\dfrac{3}{15}}$ $\dfrac{1}{5} = \dfrac{3}{3} \cdot \dfrac{1}{5} = \dfrac{3}{15}$

$\qquad\quad 11\dfrac{8}{15}$

5. $6\dfrac{3}{5}=6\dfrac{21}{35}$ $\dfrac{3}{5}=\dfrac{7}{7}\cdot\dfrac{3}{5}=\dfrac{21}{35}$

    $2\dfrac{6}{7}=2\dfrac{30}{35}$ $\dfrac{6}{7}=\dfrac{5}{5}\cdot\dfrac{6}{7}=\dfrac{30}{35}$

    $\overline{8\dfrac{51}{35}}=9\dfrac{16}{35}$

Problems involving the sum of mixed numbers can be applied in geometry. For example, in any triangle, the sum of any two of the three sides must be greater than the third side. (This is one way of saying that the shortest distance between two points is a straight line.) Thus, the numbers $4\dfrac{1}{2}$ inches, 9 inches, and $3\dfrac{1}{2}$ inches cannot be the sides of a triangle because $4\dfrac{1}{2}+3\dfrac{1}{2}=8$, and 8 is less than 9. (See Figure 3.7.) No triangle exists with sides $4\dfrac{1}{2}$ inches, $3\dfrac{1}{2}$ inches, and 9 inches.

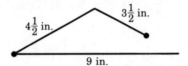

$4\dfrac{1}{2}$ inches + $3\dfrac{1}{2}$ inches will not reach to the point 9 inches away

Figure 3.7

EXERCISES 3.6   *ODD 1-33*

Find the sums as indicated.

1. $4\frac{1}{2} + 3\frac{1}{6}$
2. $3\frac{1}{4} + 7\frac{1}{8}$
3. $25\frac{1}{10} + 17\frac{1}{4}$
4. $5\frac{1}{7} + 3\frac{1}{3}$

5. $6\frac{5}{12} + 4\frac{1}{3}$
6. $5\frac{3}{10} + 2\frac{1}{14}$
7. $8\frac{2}{9} + 4\frac{1}{27}$
8. $11\frac{3}{4} + 2\frac{5}{16}$

9. $6\frac{4}{9} + 12\frac{1}{15}$
10. $4\frac{1}{6} + 13\frac{9}{10}$
11. $21\frac{3}{4} + 6\frac{3}{4}$
12. $3\frac{5}{8} + 3\frac{5}{8}$

13. $7\frac{3}{5} + 2\frac{1}{8}$
14. $9\frac{1}{8} + 3\frac{7}{12}$
15. $3\frac{1}{3} + 4\frac{1}{4} + 5\frac{1}{5}$

16. $\frac{3}{7} + 2\frac{1}{14} + 2\frac{1}{6}$
17. $20\frac{5}{8} + 42\frac{5}{6}$
18. $25\frac{2}{3} + 1\frac{1}{16}$

19. $32\frac{1}{64} + 4\frac{1}{24} + 17\frac{3}{8}$
20. $3\frac{1}{20} + 7\frac{1}{15} + 2\frac{3}{10}$

21. $46\frac{3}{8}$
    $25\frac{1}{10}$

22. $28\frac{3}{7}$
    $10\frac{4}{21}$

23. $56\frac{5}{18}$
    $29\frac{4}{27}$

24. $33\frac{5}{6}$
    $27\frac{3}{10}$

25. $13\frac{1}{24}$
    $14\frac{1}{36}$

26. $4\frac{3}{5}$
    $20\frac{1}{7}$

27. $15\frac{1}{25}$
    $95\frac{2}{10}$

28. $102\frac{3}{4}$
    $86\frac{1}{14}$

29. $35\frac{4}{39}$
    $27\frac{1}{26}$
    $74\frac{2}{65}$

30. $75\frac{3}{10}$
    $45\frac{16}{100}$
    $93\frac{476}{1000}$

31. A construction company in England built three sections of highway. One section was $20\frac{7}{8}$ kilometers, a second was $3\frac{3}{10}$ kilometers, and a third section was $11\frac{1}{10}$ kilometers. What was the total length?

32. A triangle has sides of $42\frac{3}{4}$ meters, $23\frac{1}{2}$ meters, and $22\frac{9}{10}$ meters. What is the perimeter (distance around) of the triangle?

33. In each case, decide whether there could be a triangle with the given measurements as the lengths of its sides. (See page 100.)

(a) $15\frac{1}{2}$ centimeters, $16\frac{9}{10}$ centimeters, $1\frac{3}{10}$ centimeters

(b) $9\frac{3}{4}$ inches, $5\frac{3}{8}$ inches, $14\frac{9}{16}$ inches

(c) $13\frac{1}{10}$ meters, $4\frac{7}{10}$ meters, $1\frac{1}{2}$ meters

34.   A certain bus trip consists of three parts. The first part takes $2\frac{1}{3}$ hours, the second part takes $3\frac{3}{4}$ hours, and the third part takes $2\frac{14}{15}$ hours. How many hours does the trip take? If an hour has 60 minutes, how many minutes are there in $\frac{14}{15}$ hours?

3.7   SUBTRACTING MIXED NUMBERS

As with addition of mixed numbers, the whole numbers and fraction parts can be handled separately in subtraction.

EXAMPLES

1. $4\frac{3}{5} - 1\frac{2}{5} = (4-1) + \left(\frac{3}{5} - \frac{2}{5}\right) = 3 + \frac{1}{5} = 3\frac{1}{5}$

2. $5\frac{6}{7} - 3\frac{2}{7} = (5-3) + \left(\frac{6}{7} - \frac{2}{7}\right) = 2 + \frac{4}{7} + 2\frac{4}{7}$

3. $10\frac{3}{5} - 6\frac{3}{20} = (10-6) + \left(\frac{3}{5} - \frac{3}{20}\right) = 4 + \left(\frac{12}{20} - \frac{3}{20}\right) = 4 + \frac{9}{20} = 4\frac{9}{20}$

Or, writing the numbers one under the other,

$$
\begin{array}{ccc}
4\frac{3}{5} & 5\frac{6}{7} & 10\frac{3}{5} = 10\frac{12}{20} \\
-1\frac{2}{5} & -3\frac{2}{7} & -\ 6\frac{3}{20} = 6\frac{3}{20} \\
\hline
3\frac{1}{5} & 2\frac{4}{7} & 4\frac{9}{20}
\end{array}
$$

A second technique, which was not discussed with addition but could be used there, is to change the mixed numbers to fraction form before subtracting. The numbers may be large, but the technique is helpful to some people.

EXAMPLES

$1'. \quad 4\dfrac{3}{5} - 1\dfrac{2}{5} = \dfrac{23}{5} - \dfrac{7}{5} = \dfrac{16}{5} = 3\dfrac{1}{5}$

$2'. \quad 5\dfrac{6}{7} - 3\dfrac{2}{7} = \dfrac{41}{7} - \dfrac{23}{7} = \dfrac{18}{7} = 2\dfrac{4}{7}$

$3'. \quad 10\dfrac{3}{5} - 6\dfrac{3}{20} = \dfrac{53}{5} - \dfrac{123}{20} = \dfrac{212}{20} - \dfrac{123}{20} = \dfrac{89}{20} = 4\dfrac{9}{20}$

(handwritten margin note: CHANGE TO FRACTION WORK $4\tfrac{3}{5} - 3\tfrac{1}{3}$ $8\tfrac{1}{7} - 2\tfrac{3}{5}$ *)*

Now consider the difference $4\dfrac{1}{5} - 2\dfrac{3}{5}$. Certainly $4\dfrac{1}{5}$ is bigger than $2\dfrac{3}{5}$. But, $\dfrac{3}{5}$ is bigger than $\dfrac{1}{5}$, so the fraction parts cannot be subtracted. Rewrite $4\dfrac{1}{5}$ by "borrowing" 1 from the 4; that is, write

$$4\dfrac{1}{5} = 3 + 1 + \dfrac{1}{5} = 3\dfrac{6}{5}$$

Therefore,

$$4\dfrac{1}{5} - 2\dfrac{3}{5} = 3\dfrac{6}{5} - 2\dfrac{3}{5} = (3 - 2) + \left(\dfrac{6}{5} - \dfrac{3}{5}\right) = 1 + \dfrac{3}{5} = 1\dfrac{3}{5}$$

Or,

$$
\begin{aligned}
4\dfrac{1}{5} &= 3\dfrac{6}{5} \\
-\,2\dfrac{3}{5} &= 2\dfrac{3}{5} \\
\hline
&\;1\dfrac{3}{5}
\end{aligned}
$$

EXAMPLES

Find the difference $14\dfrac{2}{3} - 2\dfrac{3}{4}$.

(handwritten margin note: WORK PROBLEMS FROM PROBLEM SET *)*

$$
\begin{aligned}
14\dfrac{2}{3} &= 14\dfrac{8}{12} = 13\dfrac{20}{12} \\
-\;2\dfrac{3}{4} &= 2\dfrac{9}{12} = 2\dfrac{9}{12} \\
\hline
&11\dfrac{11}{12}
\end{aligned}
$$

Since $\dfrac{9}{12}$ is bigger than $\dfrac{8}{12}$, 1 $\left(\text{or } \dfrac{12}{12}\right)$ is "borrowed" from 14.

The same problem can be done by changing to fraction form.

$$14\dfrac{2}{3} - 2\dfrac{3}{4} = \dfrac{44}{3} - \dfrac{11}{4} = \dfrac{4}{4} \cdot \dfrac{44}{3} - \dfrac{3}{3} \cdot \dfrac{11}{4} = \dfrac{176}{12} - \dfrac{33}{12} = \dfrac{143}{12} = 11\dfrac{11}{12}$$

SELF QUIZ	Find the following differences. Reduce all answers.	ANSWERS
	1. $17\frac{7}{8} - 4\frac{3}{8}$	1. $13\frac{1}{2}$
	2. $5 - 1\frac{3}{4}$	2. $3\frac{1}{4}$
	3. $14\frac{5}{36} - 6\frac{17}{24}$	3. $7\frac{31}{72}$

EXERCISES 3.7   *Odd 1-39*

Find the differences as indicated.

1. $5\frac{3}{4} - 2\frac{1}{4}$

2. $7\frac{9}{10} - 3\frac{3}{10}$

3. $4\frac{7}{8} - 1\frac{1}{4}$

4. $9\frac{5}{6} - 2\frac{1}{4}$

5. $15\frac{5}{8} - 11\frac{3}{4}$

6. $14\frac{6}{10} - 3\frac{4}{5}$

7. $8\frac{3}{32} - 4\frac{3}{16}$

8. $12\frac{3}{4} - 7\frac{1}{6}$

9. $8\frac{11}{12} - 5\frac{9}{10}$

10. $4\frac{7}{16} - 3$

11. $5\frac{9}{10} - 2$

12. $7 - 6\frac{2}{3}$

13. $12 - 4\frac{1}{5}$

14. $2 - 1\frac{3}{8}$

15. $75 - 17\frac{5}{6}$

16. $4\frac{9}{16} - 2\frac{7}{8}$

17. $3\frac{7}{10} - 2\frac{5}{6}$

18. $15\frac{11}{16} - 13\frac{7}{8}$

19. $20\frac{3}{6} - 3\frac{4}{8}$

20. $17\frac{3}{12} - 12\frac{2}{8}$

21. $10\frac{2}{15} - 5\frac{7}{12}$

22. $9\frac{7}{24} - 3\frac{13}{16}$

23. $3 - 1\frac{1}{6}$

24. $8 - 5\frac{1}{7}$

25. $29\frac{2}{3} - 17\frac{7}{12}$

26. $18\frac{3}{5} - 2\frac{1}{10}$

27. $7\frac{1}{9} - 3\frac{4}{27}$

28. $23\frac{1}{25} - 3\frac{1}{15}$

29. $76\frac{5}{36} - 62\frac{17}{24}$

30. $59\frac{2}{51} - 29\frac{3}{34}$

31. Sara can paint a room in $3\frac{3}{5}$ hours, and Emily can paint the same size room in $4\frac{1}{5}$ hours. How many hours are saved by having Sara paint the room? How many minutes are saved?

32. A teacher graded two sets of test papers. The first set took $3\frac{3}{4}$ hours to grade, and the second set took $2\frac{3}{5}$ hours. How much faster did the teacher grade the second set?

33. A certain box is $4\frac{5}{8}$ inches long, $3\frac{3}{16}$ inches wide, and $2\frac{3}{8}$ inches deep. The dimensions of another box are $7\frac{1}{2}$ inches, $5\frac{3}{10}$ inches, and $4\frac{4}{5}$ inches. What is the difference between the total dimensions of the two boxes?

34. Find the difference between the sum of $28\frac{1}{10}$ and $25\frac{3}{4}$ and the sum of $16\frac{3}{100}$ and $39\frac{1}{10}$.

35. A quadrilateral (four-sided figure) has sides of $13\frac{1}{2}$ centimeters, $14\frac{3}{10}$ centimeters, $16\frac{1}{2}$ centimeters, and $11\frac{9}{10}$ centimeters. How much longer is the longest side than the shortest side? What is the difference in length between the sum of the longest and shortest sides and the total of the other two sides? What is the length of the perimeter of the quadrilateral?

36. Mike takes $1\frac{1}{2}$ hours to clean the swimming pool, and Tom takes $2\frac{1}{3}$ hours to clean the same pool. How much longer does Tom take to clean the pool in hours? in minutes?

37. Find the sum of the difference between $14\frac{3}{4}$ and $5\frac{9}{10}$ and the difference between $12\frac{10}{11}$ and $8\frac{1}{2}$.

38. If a triangle has sides $42\frac{1}{2}$ inches, $35\frac{5}{8}$ inches, and $21\frac{9}{16}$ inches, find how much longer the sum of any two sides of the triangle is than the third side. (There are three answers.)

39. During each week of six weeks of dieting, Mr. Johnson, who weighed 240 pounds, lost $5\frac{1}{2}$ pounds, $2\frac{3}{4}$ pounds, $4\frac{5}{16}$ pounds, $1\frac{3}{4}$ pounds, $2\frac{5}{8}$ pounds, and $3\frac{1}{4}$ pounds. What did he weigh at the end of the six weeks if he was 35 years old?

40. Two books are side by side on a shelf, and the thickness of each front and back cover is $\frac{1}{4}$ inch. Without the covers, the two books are $2\frac{1}{2}$ inches thick and $1\frac{3}{4}$ inches thick. What is the distance between the last page of the book on the right and the first page of the book on the left?

3.8   MULTIPLYING MIXED NUMBERS

Probably the simplest way to multiply mixed numbers is to change them to fraction form and then multiply. The answers may be changed back to mixed numbers for convenience or they may be left in fraction form. As the following examples illustrate, numerators and denominators should be factored and numbers reduced before multiplication.

EXAMPLES

1. $\left(1\frac{2}{3}\right)\left(2\frac{3}{4}\right) = \left(\frac{5}{3}\right)\left(\frac{11}{4}\right) = \frac{5 \cdot 11}{3 \cdot 4} = \frac{55}{12}$ or $4\frac{7}{12}$

Only THIS WAY

2. $\left(5\frac{2}{3}\right)\left(2\frac{1}{4}\right) = \left(\frac{17}{3}\right)\left(\frac{9}{4}\right) = \frac{17 \cdot 3 \cdot 3}{3 \cdot 4} = \frac{51}{4}$ or $12\frac{3}{4}$

3. $\left(2\frac{1}{3}\right)\left(4\frac{1}{5}\right) = \left(\frac{7}{3}\right)\left(\frac{21}{5}\right) = \frac{7 \cdot 3 \cdot 7}{3 \cdot 5} = \frac{49}{5}$ or $9\frac{4}{5}$

4. $\frac{5}{6} \cdot 3\frac{3}{10} = \left(\frac{5}{6}\right)\left(\frac{33}{10}\right) = \frac{5 \cdot 33}{6 \cdot 10} = \frac{5 \cdot 3 \cdot 11}{2 \cdot 3 \cdot 2 \cdot 5} = \frac{11}{4}$ or $2\frac{3}{4}$

5. $4\frac{1}{2} \cdot 1\frac{1}{6} \cdot 3\frac{1}{3} = \frac{9}{2} \cdot \frac{7}{6} \cdot \frac{10}{3} = \frac{9 \cdot 7 \cdot 10}{2 \cdot 6 \cdot 3} = \frac{3 \cdot 3 \cdot 7 \cdot 2 \cdot 5}{2 \cdot 2 \cdot 3 \cdot 3} = \frac{35}{2}$ or $17\frac{1}{2}$

The following procedure may be used to multiply mixed numbers, particularly when the numbers are large. It involves several applications of the distributive property.

TALK ABOUT THIS

WORK PROBLEMS

6. $24\frac{3}{8} \cdot 45\frac{1}{4} = \left(24 + \frac{3}{8}\right)\left(45 + \frac{1}{4}\right)$

$\qquad = \left(24 + \frac{3}{8}\right) \cdot 45 + \left(24 + \frac{3}{8}\right) \cdot \frac{1}{4}$

$\qquad = 24 \cdot 45 + \frac{3}{8} \cdot 45 + 24 \cdot \frac{1}{4} + \frac{3}{8} \cdot \frac{1}{4}$

$$24\frac{3}{8} \cdot 45\frac{1}{4} = 1080 + \frac{135}{8} + 6 + \frac{3}{32}$$

$$= 1086 + \frac{540}{32} + \frac{3}{32} = 1086 + \frac{543}{32}$$

$$= 1086 + 16\frac{31}{32} = 1102\frac{31}{32}$$

Writing one number under the other gives the following form of the process given in Example 6.

---

1. Write the factors one under the other. Multiply the fraction in the bottom factor separately by the whole number and fraction in the top factor.

2. Multiply the whole number in the bottom factor separately by the whole number and fraction in the top factor.

3. Add all four partial products to get the product.

6′. $24\frac{3}{8} \cdot 45\frac{1}{4}$

(a) $24\frac{3}{8}$

$45\frac{1}{4}$

$\dfrac{3}{32}$

6

$\dfrac{1}{4} \cdot \dfrac{3}{8} = \dfrac{1 \cdot 3}{4 \cdot 8} = \dfrac{3}{32}$

$\dfrac{1}{4} \cdot 24 = \dfrac{1}{4} \cdot \dfrac{24}{1} = \dfrac{\cancel{4} \cdot 6}{\cancel{4} \cdot 1} = 6$

(b) $24\frac{3}{8}$

$45\frac{1}{4}$

$\dfrac{3}{32}$

6

$16\frac{7}{8}$

120

$\underline{960}$

$45 \cdot \dfrac{3}{8} = \dfrac{45}{1} \cdot \dfrac{3}{8} = \dfrac{45 \cdot 3}{1 \cdot 8} = \dfrac{135}{8} = 16\frac{7}{8}$

Multiply the whole numbers without rewriting them.

(c)

$$24\frac{3}{8}$$

$$45\frac{1}{4}$$

$$\frac{3}{32}$$

$$6$$

$$16\frac{7}{8}$$

$$120$$

$$960$$

$$\overline{1102\frac{31}{32}}$$

$$\frac{3}{32}+\frac{7}{8}=\frac{3}{32}+\frac{4}{4}\cdot\frac{7}{8}=\frac{3}{32}+\frac{28}{32}=\frac{3+28}{32}=\frac{31}{32}$$

6″. The same product may be found by changing the mixed numbers to fraction form. Generally, this is a simpler procedure, even though the numbers will be large.

$$24\frac{3}{8}\cdot45\frac{1}{4}=\frac{195}{8}\cdot\frac{181}{4}=\frac{35,295}{32}=1102\frac{31}{32}$$

SELF QUIZ   Find the following products. Reduce all answers.   ANSWERS

1. $\left(3\frac{1}{2}\right)\left(5\frac{1}{6}\right)$

2. $4\frac{1}{3}\cdot\frac{2}{13}$

3. $25\frac{1}{3}\cdot16\frac{1}{2}$

1. $\frac{217}{12}$ or $18\frac{1}{12}$

2. $\frac{2}{3}$

3. 418

EXERCISES 3.8   ODD 1-51

Find the indicated products.

1. $\left(2\frac{1}{3}\right)\left(3\frac{1}{4}\right)$

2. $\left(1\frac{1}{5}\right)\left(1\frac{1}{7}\right)$

3. $4\frac{1}{2}\left(2\frac{1}{3}\right)$

4. $3\frac{1}{3}\left(2\frac{1}{5}\right)$

5. $6\frac{1}{4}\left(3\frac{3}{5}\right)$

6. $5\frac{1}{3}\left(2\frac{1}{4}\right)$

7. $\left(8\frac{1}{2}\right)\left(3\frac{2}{3}\right)$

8. $\left(9\frac{1}{3}\right)2\frac{1}{7}$

9. $\left(6\frac{2}{7}\right)1\frac{3}{11}$

10. $\left(11\frac{1}{4}\right)1\frac{1}{15}$

11. $6\frac{2}{3}\cdot4\frac{1}{2}$

12. $4\frac{3}{8}\cdot2\frac{2}{7}$

13. $9\frac{3}{4} \cdot 2\frac{6}{26}$ 14. $7\frac{1}{2} \cdot \frac{2}{15}$ 15. $\frac{3}{4} \cdot 1\frac{1}{3}$

16. $3\frac{4}{5} \cdot 2\frac{1}{7}$ 17. $12\frac{1}{2} \cdot 2\frac{1}{5}$ 18. $9\frac{3}{5} \cdot 1\frac{1}{16}$

19. $6\frac{1}{8} \cdot 3\frac{1}{7}$ 20. $5\frac{1}{4} \cdot 11\frac{1}{3}$ 21. $\frac{1}{4} \cdot \frac{2}{3} \cdot \frac{6}{7}$

22. $\frac{7}{8} \cdot \frac{24}{25} \cdot \frac{5}{21}$ 23. $\frac{3}{16} \cdot \frac{8}{9} \cdot \frac{3}{5}$ 24. $\frac{2}{5} \cdot \frac{1}{5} \cdot \frac{4}{7}$

25. $\frac{6}{7} \cdot \frac{2}{11} \cdot \frac{3}{5}$ 26. $\left(3\frac{1}{2}\right)\left(2\frac{1}{7}\right)\left(5\frac{1}{4}\right)$ 27. $\left(4\frac{3}{8}\right)\left(2\frac{1}{5}\right)\left(1\frac{1}{7}\right)$

28. $\left(6\frac{3}{16}\right)\left(2\frac{1}{11}\right)\left(5\frac{3}{5}\right)$ 29. $7\frac{1}{3} \cdot 5\frac{1}{4} \cdot 6\frac{2}{7}$ 30. $2\frac{5}{8} \cdot 3\frac{2}{5} \cdot 1\frac{3}{4}$

31. $2\frac{1}{16} \cdot 4\frac{1}{3} \cdot 1\frac{3}{11}$ 32. $5\frac{1}{10} \cdot 3\frac{1}{7} \cdot 2\frac{1}{17}$ 33. $2\frac{1}{4} \cdot 6\frac{3}{8} \cdot 1\frac{5}{27}$

34. $1\frac{3}{32} \cdot 1\frac{1}{7} \cdot 1\frac{1}{25}$ 35. $1\frac{5}{16} \cdot 1\frac{1}{3} \cdot 1\frac{1}{5}$ 36. $24\frac{1}{5} \cdot 35\frac{1}{6}$

37. $72\frac{3}{5} \cdot 25\frac{1}{6}$ 38. $42\frac{5}{6} \cdot 30\frac{1}{7}$ 39. $75\frac{1}{3} \cdot 40\frac{1}{25}$

40. $36\frac{3}{4} \cdot 17\frac{5}{12}$ 41. $45\frac{2}{3} \cdot 7\frac{2}{15}$ 42. $26\frac{5}{8} \cdot 22\frac{1}{4}$

43. $70\frac{3}{10} \cdot 40\frac{7}{10}$ 44. $28\frac{4}{15} \cdot 33\frac{1}{6}$ 45. $30\frac{1}{24} \cdot 40\frac{1}{12}$

46. The total distance around a square (its perimeter) is found by multiplying the length of one side by 4. Find the perimeter of a square if the length of one side is $5\frac{1}{16}$ inches.

47. A man driving to work drives $17\frac{7}{10}$ miles one way five days a week. How many miles does he drive each week going to and from work?

48. A length of pipe is $27\frac{5}{8}$ feet. What would be the total length if $36\frac{1}{2}$ of these pipes were laid end to end?

49. In constructing a small building, $\frac{5}{16}$ of a cement piling must be underground and $\frac{11}{16}$ above ground. If a certain piling is $32\frac{1}{5}$ feet long, how much of it is above ground and how much below?

50. The product of $\dfrac{19}{100}$ and $\dfrac{10}{57}$ is added to the product of $\dfrac{1}{10}$ and $\dfrac{1}{3}$. What is the difference between this sum and $\dfrac{29}{60}$?

51. Three towns are located on the same highway. Two towns are $53\dfrac{1}{2}$ kilometers apart. If one of these towns is $46\dfrac{9}{10}$ kilometers from the third town, how far is the other town from the third town? (Be careful.)

52. Using a yardstick, a student measures the length and width of her rectangular desk top to be $28\dfrac{1}{2}$ inches and $22\dfrac{3}{4}$ inches, respectively. What is the area of the desk top? (Area is length times width.) What is the perimeter of the desk top? The desk cost $32.50.

3.9   DIVIDING RATIONAL NUMBERS

Multiply $\dfrac{2}{3}$ by $\dfrac{3}{2}$ and multiply $\dfrac{6}{5}$ by $\dfrac{5}{6}$.

$$\frac{2}{3} \cdot \frac{3}{2} = 1 \quad \text{and} \quad \frac{6}{5} \cdot \frac{5}{6} = 1$$

In both cases, the product is 1. If the product of two nonzero numbers is 1, then the numbers are called **reciprocals** of each other.

DEFINITION   The **reciprocal** of a rational number $\dfrac{a}{b}$ where $a \neq 0$ and $b \neq 0$ is $\dfrac{b}{a}$ because

IS A NUMBER WHOSE PRODUCT IS EQUAL TO 1

$$\frac{a}{b} \cdot \frac{b}{a} = 1$$

[NOTE: If $a = 0$, then $\dfrac{0}{b}$ has no reciprocal since $\dfrac{b}{0}$ is undefined.]

EXAMPLES   **1.** The reciprocal of $\dfrac{10}{17}$ is $\dfrac{17}{10}$ since

WORK SEVERAL THEN GO TO NEXT PAGE

$$\frac{10}{17} \cdot \frac{17}{10} = \frac{10 \cdot 17}{17 \cdot 10} = \frac{170}{170} = 1$$

2. The reciprocal of $\frac{1}{4}$ is $\frac{4}{1}$ since

$$\frac{1}{4} \cdot \frac{4}{1} = \frac{1 \cdot 4}{4 \cdot 1} = \frac{4}{4} = 1$$

In division with whole numbers, $6 \div 3 = 2$ because $6 = 2 \cdot 3$. The same is true for rational numbers. That is, if

$$\frac{3}{4} \div \frac{1}{5} = \square \quad \text{then} \quad \frac{3}{4} = \square \cdot \frac{1}{5}$$

What goes in the box? Try $\frac{3}{4} \cdot \frac{5}{1}$.

Does $\frac{3}{4} = \boxed{\frac{3}{4} \cdot \frac{5}{1}} \cdot \frac{1}{5}$? $\qquad \boxed{\frac{3}{4} \cdot \frac{5}{1}} \cdot \frac{1}{5} = \frac{3}{4} \cdot \left(\frac{5}{1} \cdot \frac{1}{5}\right) = \frac{3}{4} \cdot 1 = \frac{3}{4}$

Therefore, $\frac{3}{4} \cdot \frac{5}{1}$ is the answer. That is,

$$\frac{3}{4} \div \frac{1}{5} = \boxed{\frac{3}{4} \cdot \frac{5}{1}}$$

Can this procedure be justified for rational numbers in general? Does $\frac{5}{8} \div \frac{9}{10} = \frac{5}{8} \cdot \frac{10}{9}$? Does $\frac{16}{7} \div \frac{3}{4} = \frac{16}{7} \cdot \frac{4}{3}$? The answers are all *yes*, as the following development shows.

Note that

$$\frac{\frac{d}{c}}{\frac{d}{c}} = 1 \quad \text{where } d \neq 0 \text{ and } c \neq 0$$

$$\frac{a}{b} \div \frac{c}{d} = \frac{\frac{a}{b}}{\frac{c}{d}} = \frac{\frac{a}{b}}{\frac{c}{d}} \cdot 1 = \frac{\frac{a}{b} \cdot \frac{d}{c}}{\frac{c}{d} \cdot \frac{d}{c}} = \frac{\frac{a}{b} \cdot \frac{d}{c}}{1} = \frac{a}{b} \cdot \frac{d}{c}$$

This proves the statement

$$\frac{a}{b} \div \frac{c}{d} = \frac{a}{b} \cdot \frac{d}{c} \quad \text{where } b, c, d \neq 0$$

In words:
To divide by any number (except 0), multiply by its reciprocal.

EXAMPLES

1. $\dfrac{3}{4} \div \dfrac{2}{3} = \dfrac{3}{4} \cdot \dfrac{3}{2} = \dfrac{3 \cdot 3}{4 \cdot 2} = \dfrac{9}{8}$ or $1\dfrac{1}{8}$

Note that the divisor is $\frac{2}{3}$, and we multiply by its reciprocal, $\frac{3}{2}$.

2. $\dfrac{7}{16} \div 7 = \dfrac{7}{16} \cdot \dfrac{1}{7} = \dfrac{7 \cdot 1}{16 \cdot 7} = \dfrac{7}{7} \cdot \dfrac{1}{16} = \dfrac{1}{16}$

Note that the divisor is 7, and we multiply by its reciprocal, $\frac{1}{7}$.

3. $\dfrac{16}{27} \div \dfrac{4}{9} = \dfrac{16}{27} \cdot \dfrac{9}{4} = \dfrac{4 \cdot \cancel{4} \cdot \cancel{9}}{\cancel{9} \cdot 3 \cdot \cancel{4}}$

   $= \dfrac{4}{3}$ or $1\dfrac{1}{3}$

Note that the divisor is $\frac{4}{9}$, and we multiply by its reciprocal, $\frac{9}{4}$.

4. $3\dfrac{1}{4} \div 7\dfrac{4}{5} = \dfrac{13}{4} \div \dfrac{39}{5} = \dfrac{13}{4} \cdot \dfrac{5}{39} = \dfrac{\cancel{13} \cdot 5}{4 \cdot \cancel{13} \cdot 3}$

   $= \dfrac{5}{12}$

Note that the divisor is $\frac{39}{5}$, and we multiply by its reciprocal, $\frac{5}{39}$.

5. $7\dfrac{4}{5} \div 3\dfrac{1}{4} = \dfrac{39}{5} \div \dfrac{13}{4} = \dfrac{39}{5} \cdot \dfrac{4}{13} = \dfrac{3 \cdot \cancel{13} \cdot 4}{5 \cdot \cancel{13}}$

   $= \dfrac{12}{5}$ or $2\dfrac{2}{5}$

Note that the divisor is $\frac{13}{4}$, and we multiply by its reciprocal, $\frac{4}{13}$.

SELF QUIZ

Find the following quotients. Reduce all answers.

ANSWERS

1. $\dfrac{2}{5} \div \dfrac{5}{2}$

1. $\dfrac{4}{25}$

2. $\dfrac{12}{27} \div \dfrac{16}{18}$

2. $\dfrac{1}{2}$

3. Evaluate the expression

   $\dfrac{1}{4} \cdot \dfrac{1}{3} + \dfrac{5}{6} \div \dfrac{15}{2}$

3. $\dfrac{7}{36}$

EXERCISES 3.9   ODD 1-59

Find the quotients as indicated.

1. $\dfrac{2}{3} \div \dfrac{3}{4}$ 2. $\dfrac{1}{5} \div \dfrac{3}{4}$ 3. $\dfrac{3}{7} \div \dfrac{3}{5}$ 4. $\dfrac{2}{11} \div \dfrac{2}{3}$

5. $\dfrac{3}{5} \div \dfrac{3}{7}$ 6. $\dfrac{2}{3} \div \dfrac{2}{11}$ 7. $\dfrac{5}{16} \div \dfrac{15}{16}$ 8. $\dfrac{7}{18} \div \dfrac{3}{9}$

9. $\dfrac{3}{14} \div \dfrac{2}{7}$ 10. $\dfrac{13}{40} \div \dfrac{26}{35}$ 11. $\dfrac{5}{12} \div \dfrac{15}{16}$ 12. $\dfrac{12}{27} \div \dfrac{10}{18}$

13. $\dfrac{17}{48} \div \dfrac{51}{90}$ 14. $\dfrac{3}{5} \div \dfrac{7}{8}$ 15. $\dfrac{13}{16} \div \dfrac{2}{3}$ 16. $\dfrac{5}{6} \div \dfrac{3}{4}$

17. $\dfrac{3}{4} \div \dfrac{5}{6}$ 18. $\dfrac{14}{15} \div \dfrac{21}{25}$ 19. $\dfrac{3}{7} \div \dfrac{3}{7}$ 20. $\dfrac{6}{13} \div \dfrac{6}{13}$

21. $\dfrac{16}{27} \div \dfrac{7}{18}$ 22. $\dfrac{20}{21} \div \dfrac{15}{42}$ 23. $\dfrac{25}{36} \div \dfrac{5}{24}$ 24. $\dfrac{17}{20} \div \dfrac{3}{14}$

25. $\dfrac{26}{35} \div \dfrac{39}{40}$ 26. $\dfrac{5}{6} \div 3\dfrac{1}{4}$ 27. $\dfrac{7}{8} \div 7\dfrac{1}{2}$ 28. $\dfrac{29}{50} \div 3\dfrac{1}{10}$

29. $4\dfrac{1}{5} \div 3\dfrac{1}{3}$ 30. $2\dfrac{1}{17} \div 1\dfrac{1}{4}$ 31. $5\dfrac{1}{6} \div 3\dfrac{1}{4}$ 32. $2\dfrac{2}{49} \div 3\dfrac{1}{14}$

33. $6\dfrac{5}{6} \div 2$ 34. $4\dfrac{1}{5} \div 3$ 35. $6\dfrac{5}{6} \div \dfrac{1}{2}$ 36. $4\dfrac{5}{8} \div 4$

37. $4\dfrac{5}{8} \div \dfrac{1}{4}$ 38. $1\dfrac{1}{32} \div 3\dfrac{2}{3}$ 39. $7\dfrac{5}{11} \div 4\dfrac{1}{10}$ 40. $13\dfrac{1}{7} \div 4\dfrac{2}{11}$

Evaluate the following expressions using the rules of multiplication and division first, then addition and subtraction from left to right.

41. $\dfrac{1}{2} \div \dfrac{7}{8} + \dfrac{1}{7} \cdot \dfrac{2}{3}$ 42. $\dfrac{3}{5} \cdot \dfrac{1}{6} + \dfrac{1}{5} \div 2$ 43. $\dfrac{1}{2} \div \dfrac{1}{2} + 1 - \dfrac{2}{3} \cdot 3$

44. $\dfrac{3}{4} + 4\dfrac{1}{2} \cdot \dfrac{1}{3} \div \dfrac{5}{6}$ 45. $\dfrac{2}{15} \cdot \dfrac{1}{4} \div \dfrac{3}{5} + \dfrac{1}{25}$ 46. $3\dfrac{1}{2} \cdot 5\dfrac{1}{3} + \dfrac{5}{12} \div \dfrac{15}{16}$

47. $2\dfrac{1}{4} + 1\dfrac{1}{5} + 2 \div \dfrac{20}{21}$ 48. $\dfrac{5}{8} - \dfrac{1}{3} \cdot \dfrac{2}{5} + 6\dfrac{1}{10}$

49. $1\dfrac{1}{6} \cdot 1\dfrac{2}{19} \div \dfrac{7}{8} + \dfrac{1}{38}$ 50. $\dfrac{3}{10} + \dfrac{5}{6} \div \dfrac{1}{4} \cdot \dfrac{1}{8} - \dfrac{7}{60}$

51. Find the average of the numbers $\dfrac{5}{6}, \dfrac{7}{15}, \dfrac{8}{21}$.

52. Find the average of the numbers $\dfrac{7}{8}, \dfrac{9}{10}, 2\dfrac{1}{2}, 1\dfrac{3}{4}$.

53. The product of $\frac{9}{10}$ with another number is $\frac{5}{3}$. What is the other number?

54. The result of multiplying two numbers is $10\frac{1}{3}$. If one of the numbers is $7\frac{1}{6}$, what is the other one?

55. The product of $7\frac{2}{3}$ with some other rational number is $4\frac{1}{2}$. What is the other number?

56. An airplane is carrying 150 passengers. This is only $\frac{5}{7}$ of its capacity. How many passengers can the plane carry?

57. The sale price of a coat is $36. This is $\frac{1}{4}$ off the original price. What was the original price? (Be careful.)

58. An estate of $180,000 was to be shared by two nephews and a son. Each nephew received $\frac{1}{8}$ of the estate. How much did the son receive?

59. One hundred shares of a stock were bought at $7\frac{1}{4}$ and another hundred were bought at $8\frac{1}{2}$. What was the profit if all shares were sold at $9\frac{3}{4}$ per share? What was the average cost per share? What was the difference between the selling price per share and the average price per share? Multiply the difference by 200. What does this answer represent?

60. If a woman makes $18,000 in one year and has deductions of $5000, how much income tax does she pay if she pays $2830 plus $\frac{36}{100}$ of the excess over $12,000 after deductions?

3.10 COMPLEX FRACTIONS

If the numerator of a fraction or the denominator of a fraction or both contain rational numbers that are not whole numbers, then the fraction is called a **complex fraction.** Since addition, subtraction, multiplication, and division (nonzero division) with rational numbers always give rational numbers, a complex fraction is a rational number. The problem discussed in this section is how to write a complex fraction in the form $\frac{a}{b}$ where a and b are whole numbers.

Generally speaking, there are two ways to approach such a problem.

TECHNIQUES FOR SIMPLIFYING COMPLEX FRACTIONS

1. Treat the numerator and denominator as separate problems and simplify each of them first, then perform the division of the numerator by the denominator.

2. Find the least common multiple (LCM) of all the denominators of the fractions that are in the numerator and denominator, then multiply the numerator and denominator by this LCM using the distributive principle.

EXAMPLES

1. Simplify $\dfrac{\frac{3}{4}+\frac{1}{2}}{1-\frac{1}{3}}$ using the first technique.

Simplifying the numerator, we have $\dfrac{3}{4}+\dfrac{1}{2}=\dfrac{3}{4}+\dfrac{2}{4}=\dfrac{5}{4}$.

Simplifying the denominator, we have $1-\dfrac{1}{3}=\dfrac{3}{3}-\dfrac{1}{3}=\dfrac{2}{3}$.

Therefore, $\dfrac{\frac{3}{4}+\frac{1}{2}}{1-\frac{1}{3}}=\dfrac{\frac{5}{4}}{\frac{2}{3}}=\dfrac{5}{4}\cdot\dfrac{3}{2}=\dfrac{15}{8}$.

2. Simplify $\dfrac{\frac{3}{4}+\frac{1}{2}}{1-\frac{1}{3}}$ using the second technique.

The denominators are 4, 2, and 3. The LCM of $\{4, 2, 3\}$ is 12.

Therefore, $\dfrac{\frac{3}{4}+\frac{1}{2}}{1-\frac{1}{3}}=\dfrac{\left(\frac{3}{4}+\frac{1}{2}\right)}{\left(1-\frac{1}{3}\right)}\cdot\dfrac{12}{12}=\dfrac{\left(\frac{3}{4}+\frac{1}{2}\right)12}{\left(1-\frac{1}{3}\right)12}$

$=\dfrac{\frac{3}{4}\cdot 12+\frac{1}{2}\cdot 12}{1\cdot 12-\frac{1}{3}\cdot 12}=\dfrac{9+6}{12-4}=\dfrac{15}{8}$

By using the distributive principle and multiplying by the LCM of the denominators, we are assured that the new equivalent fraction will have only whole numbers in its numerator and denominator. This technique should be used only if you are confident in using the distributive principle of multiplication.

EXAMPLE   Simplify $\dfrac{2\frac{1}{3}}{\frac{1}{4}+\frac{1}{3}}$

First simplification:

$$2\frac{1}{3}=\frac{7}{3}\quad\text{and}\quad\frac{1}{4}+\frac{1}{3}=\frac{1}{4}\cdot\frac{3}{3}+\frac{1}{3}\cdot\frac{4}{4}=\frac{3}{12}+\frac{4}{12}=\frac{7}{12}$$

[NOTE: The mixed number is changed to fraction form.]

$$\frac{2\frac{1}{3}}{\frac{1}{4}+\frac{1}{3}}=\frac{\frac{7}{3}}{\frac{7}{12}}=\frac{7}{3}\cdot\frac{12}{7}=\frac{7\cdot 3\cdot 4}{3\cdot 7\cdot 1}=\frac{4}{1}=4$$

Second simplification:
The LCM of {4, 3} is 12. Apply the distributive principle and multiply the numerator and denominator by 12.

$$\frac{2\frac{1}{3}}{\frac{1}{4}+\frac{1}{3}}=\frac{\left(\frac{7}{3}\right)}{\left(\frac{1}{4}+\frac{1}{3}\right)}\cdot\frac{12}{12}=\frac{\frac{7}{3}\cdot 12}{\frac{1}{4}\cdot 12+\frac{1}{3}\cdot 12}=\frac{28}{3+4}=\frac{28}{7}=4$$

EXERCISES 3.10   O𝒹𝒹 1 – 33

Rewrite each complex fraction in the form $\dfrac{a}{b}$ where a and b are whole numbers.

1. $\dfrac{1+\frac{3}{7}}{\frac{2}{3}}$ 2. $\dfrac{2+\frac{1}{5}}{1+\frac{1}{4}}$ 3. $\dfrac{\frac{1}{5}+\frac{1}{6}}{2\frac{1}{3}}$ 4. $\dfrac{\frac{2}{3}+\frac{1}{5}}{4\frac{1}{2}}$

5. $\dfrac{\frac{5}{6}-\frac{1}{3}}{\frac{1}{2}+\frac{1}{5}}$ 6. $\dfrac{5\frac{1}{7}}{2+1}$ 7. $\dfrac{3\frac{4}{5}}{11+8}$ 8. $\dfrac{4+\frac{1}{3}}{6+\frac{1}{4}}$

9. $\dfrac{7 + \dfrac{2}{5}}{2 + \dfrac{1}{15}}$ **10.** $\dfrac{2 - \dfrac{1}{3}}{1 - \dfrac{1}{3}}$ **11.** $\dfrac{\dfrac{2}{3} - \dfrac{1}{4}}{\dfrac{3}{5} - \dfrac{1}{4}}$ **12.** $\dfrac{\dfrac{5}{6} - \dfrac{2}{3}}{\dfrac{5}{8} - \dfrac{1}{16}}$

13. $\dfrac{\dfrac{7}{8} - \dfrac{3}{16}}{\dfrac{1}{3} - \dfrac{1}{4}}$ **14.** $\dfrac{\dfrac{3}{5} + \dfrac{4}{7}}{\dfrac{3}{8} + \dfrac{1}{10}}$ **15.** $\dfrac{\dfrac{4}{15} + \dfrac{6}{25}}{\dfrac{3}{5} + \dfrac{3}{10}}$ **16.** $\dfrac{3\dfrac{1}{4} + 2\dfrac{1}{2}}{5\dfrac{1}{8} + 1\dfrac{5}{8}}$

17. $\dfrac{7\dfrac{1}{3} + 2\dfrac{1}{5}}{6\dfrac{1}{9} + 2}$ **18.** $\dfrac{5\dfrac{2}{3} - 1\dfrac{1}{6}}{3\dfrac{1}{2} + 3\dfrac{1}{6}}$ **19.** $\dfrac{2\dfrac{4}{9} + 1\dfrac{1}{18}}{1\dfrac{2}{9} - \dfrac{1}{6}}$ **20.** $\dfrac{7\dfrac{1}{2} + 3\dfrac{5}{11}}{2\dfrac{3}{4} + 1\dfrac{7}{8}}$

21. $\dfrac{8\dfrac{1}{10} + \dfrac{3}{10}}{\dfrac{6}{100}}$ **22.** $\dfrac{6\dfrac{1}{100} + 5\dfrac{3}{100}}{2\dfrac{1}{2} + 3\dfrac{1}{10}}$ **23.** $\dfrac{4\dfrac{7}{10} - 2\dfrac{9}{10}}{5\dfrac{1}{100}}$

24. $\dfrac{7\dfrac{1}{2} + 3\dfrac{1}{1000}}{\dfrac{99}{1000} - \dfrac{9}{100}}$ **25.** $\dfrac{\dfrac{16}{10,000} + \dfrac{3}{10}}{\dfrac{3}{10} - \dfrac{16}{10,000}}$ **26.** $\dfrac{17\dfrac{1}{10} - 16\dfrac{3}{10}}{\dfrac{4}{5}}$

27. $\dfrac{21\dfrac{1}{2} - 21\dfrac{50}{100}}{16\dfrac{1}{10} - 5\dfrac{1}{10}}$ **28.** $\dfrac{5\dfrac{1}{10} + 13\dfrac{1}{2}}{2\dfrac{9}{10} - 1\dfrac{3}{10}}$ **29.** $\dfrac{46\dfrac{83}{100} - 45\dfrac{7}{10}}{32\dfrac{57}{100} - 31\dfrac{11}{25}}$

30. $\dfrac{1\dfrac{1}{100} + 1\dfrac{1}{10}}{6\dfrac{1}{100} + 6\dfrac{1}{10}}$

31. If the sum of $2\dfrac{1}{2}$ and $3\dfrac{5}{8}$ is divided by the difference of $7\dfrac{1}{10}$ and $2\dfrac{1}{2}$, find the quotient.

32. The average of $3\dfrac{1}{4}$, $5\dfrac{1}{2}$, and $6\dfrac{1}{12}$ is divided by the average of $4\dfrac{5}{6}$ and $2\dfrac{1}{8}$. What is the quotient? Which average is larger? How much larger?

33. Find the quotient if the sum of $5\dfrac{1}{10}$ and $4\dfrac{1}{2}$ is divided by the average of $\dfrac{3}{4}$, $\dfrac{7}{10}$, and $\dfrac{1}{5}$.

34. If the product of two numbers is $10\frac{3}{4}$, and one of the numbers is the difference between $5\frac{9}{10}$ and $2\frac{3}{4}$, what is the other number?

SUMMARY: CHAPTER 3

DEFINITION

A **rational number** is a number that can be written in the form $\frac{a}{b}$ where a is a whole number and b is a natural number.

$$\frac{a}{b} \qquad \begin{matrix} \text{numerator} \\ \text{denominator} \end{matrix}$$

[NOTE: The numerator a can be 0 since a is a whole number, but the denominator b *cannot be* 0 since b is a natural number.]

> Zero as a denominator is not allowed because of situations like the following: consider $\frac{5}{0} = \square$. Whatever $\square$ is, since $\frac{5}{0} = \square$, we must have $5 = 0 \cdot \square$. But $0 \cdot \square = 0$ no matter what $\square$ represents. This would give $5 = 0 \cdot \square = 0$, which is impossible. Next consider $\frac{0}{0} = \square$. We must have $0 = \square \cdot 0$, which is true for *any* value of $\square$. But this would mean $\frac{0}{0}$ could be any number. Thus, $\frac{a}{0}$ is *undefined* for any value of a.

DEFINITION

The **product** of two rational numbers $\frac{a}{b}$ and $\frac{c}{d}$ is the rational number whose numerator is the product of the numerators ($a \cdot c$) and whose denominator is the product of the denominators ($b \cdot d$). That is,

$$\frac{a}{b} \cdot \frac{c}{d} = \frac{a \cdot c}{b \cdot d}$$

COMMUTATIVE
PROPERTY OF
MULTIPLICATION

If $\frac{a}{b}$ and $\frac{c}{d}$ are rational numbers, then

$$\frac{a}{b} \cdot \frac{c}{d} = \frac{c}{d} \cdot \frac{a}{b}$$

ASSOCIATIVE
PROPERTY OF
MULTIPLICATION

If $\frac{a}{b}, \frac{c}{d}$, and $\frac{e}{f}$ are rational numbers, then

$$\frac{a}{b} \cdot \frac{c}{d} \cdot \frac{e}{f} = \left(\frac{a}{b} \cdot \frac{c}{d} \right) \cdot \frac{e}{f} = \frac{a}{b} \cdot \left(\frac{c}{d} \cdot \frac{e}{f} \right)$$

MULTIPLICATIVE IDENTITY If $\dfrac{a}{b}$ is a rational number, then

$$\frac{a}{b} \cdot 1 = \frac{a}{b}$$

$$\frac{a}{b} = \frac{a}{b} \cdot 1 = \frac{a}{b} \cdot \frac{k}{k} = \frac{a \cdot k}{b \cdot k} \qquad \text{where } k \neq 0$$

DEFINITION A rational number is in **lowest terms** if its numerator and denominator are relatively prime; that is, the numerator and denominator have only one (1) as a common factor.

DEFINITION The **sum** of two rational numbers $\dfrac{a}{b}$ and $\dfrac{c}{d}$ with common denominator b ($b \neq 0$) is the rational number whose numerator is the sum of the numerators, $a + c$, and whose denominator is b. That is,

$$\frac{a}{b} + \frac{c}{b} = \frac{a + c}{b}$$

COMMUTATIVE PROPERTY OF ADDITION If $\dfrac{a}{b}$ and $\dfrac{c}{d}$ are rational numbers, then

$$\frac{a}{b} + \frac{c}{d} = \frac{c}{d} + \frac{a}{b}$$

ASSOCIATIVE PROPERTY OF ADDITION If $\dfrac{a}{b}, \dfrac{c}{d}$, and $\dfrac{e}{f}$ are rational numbers, then

$$\frac{a}{b} + \frac{c}{d} + \frac{e}{f} = \left(\frac{a}{b} + \frac{c}{d}\right) + \frac{e}{f} = \frac{a}{b} + \left(\frac{c}{d} + \frac{e}{f}\right)$$

ADDITIVE IDENTITY If $\dfrac{a}{b}$ is a rational number, then

$$\frac{a}{b} + 0 = \frac{a}{b}$$

DEFINITION The **difference** of two rational numbers $\dfrac{a}{b}$ and $\dfrac{c}{b}$ with a greater than or equal to c is a rational number whose numerator is the difference of the numerators and whose denominator is the common denominator b. In symbols,

$$\frac{a}{b} - \frac{c}{b} = \frac{a - c}{b}$$

DEFINITION   A **mixed number** is the sum of *a whole number* and *a rational number not a whole number,* indicated by writing the numbers next to each other.

> TO CHANGE RATIONAL NUMBERS TO MIXED NUMBERS
>
> 1. Reduce first, then change to mixed numbers.
>
> 2. Or change to mixed numbers first, then reduce the fraction part.

DEFINITION   The **reciprocal** of a rational number $\frac{a}{b}$ where $a \neq 0$ and $b \neq 0$ is $\frac{b}{a}$ because

$$\frac{a}{b} \cdot \frac{b}{a} = 1$$

[NOTE: If $a = 0$, then $\frac{0}{b}$ has no reciprocal since $\frac{b}{0}$ is undefined.]

> $$\frac{a}{b} \div \frac{c}{d} = \frac{a}{b} \cdot \frac{d}{c} \quad \text{where } b, c, d \neq 0$$
>
> In words:
> To divide by any number (except 0), multiply by its reciprocal.

A **complex fraction** is a fraction in which the numerator or denominator or both contain rational numbers that are not whole numbers.

REVIEW QUESTIONS · CHAPTER 3

1. A rational number is a number that can be written in the form _____, where _____ is a whole number and _____ is a natural number.

2. $0 \div 7 =$ _____ , while $7 \div 0 =$ _____ , and $0 \div 0 =$ _____ .

3. The multiplicative identity for the rational numbers is _____ .

4. When a numerator and denominator are relatively prime, the fraction is in _____ _____ .

5. If two whole numbers are relatively prime, their only common factor is _____ .

6. Which property of addition or multiplication with rational numbers is illustrated by each of the following statements?

(a) $\dfrac{1}{3} + \left(\dfrac{5}{6} + \dfrac{1}{2}\right) = \left(\dfrac{1}{3} + \dfrac{5}{6}\right) + \dfrac{1}{2}$ (b) $\dfrac{9}{10} \cdot \dfrac{3}{10} = \dfrac{3}{10} \cdot \dfrac{9}{10}$

7. Draw a diagram illustrating the product $\dfrac{2}{3} \cdot \dfrac{5}{6}$.

8. There are 100 centimeters in a meter. Write a rational number that represents 25 of the one hundred equal parts of one meter.

9. Find a rational number with denominator 72 equivalent to $\dfrac{7}{9}$.

10. Draw a diagram illustrating the difference $\dfrac{1}{2} - \dfrac{1}{3}$. [HINT: Equivalent rational numbers may be helpful.]

Multiply and reduce all answers.

11. $\dfrac{1}{3} \cdot \dfrac{1}{2} \cdot \dfrac{1}{5}$ **12.** $\dfrac{1}{7} \cdot \dfrac{3}{7}$ **13.** $\dfrac{35}{56} \cdot \dfrac{4}{15} \cdot \dfrac{5}{10}$

Fill in the missing terms so that each equation is true.

14. $\dfrac{1}{6} = \dfrac{}{12}$ **15.** $\dfrac{9}{10} = \dfrac{54}{}$ **16.** $\dfrac{15}{11} = \dfrac{75}{}$ **17.** $\dfrac{0}{9} = \dfrac{}{27}$

Reduce each of the following to lowest terms.

18. $\dfrac{15}{20}$ **19.** $\dfrac{150}{120}$ **20.** $\dfrac{99}{88}$ **21.** $\dfrac{0}{4}$ **22.** $\dfrac{6}{11}$

Add or subtract as indicated and reduce all answers.

23. $\dfrac{3}{7} + \dfrac{2}{7}$ **24.** $\dfrac{5}{6} - \dfrac{1}{6}$ **25.** $\dfrac{3}{90} + \dfrac{6}{90}$

26. $\dfrac{1}{12} + \dfrac{5}{36} + \dfrac{11}{24}$ **27.** $\dfrac{5}{8} - \dfrac{3}{8}$ **28.** $\dfrac{13}{22} - \dfrac{5}{33}$

29. $\dfrac{9}{10} - \dfrac{0}{32}$ **30.** $\dfrac{71}{56} - \dfrac{96}{84}$ **31.** $\dfrac{5}{27} + \dfrac{0}{960} + \dfrac{5}{18} + \dfrac{4}{5}$

Change to mixed numbers.

32. $\dfrac{47}{6}$ **33.** $\dfrac{25}{4}$ **34.** $\dfrac{342}{100}$ **35.** $\dfrac{174}{35}$

Change to fraction form.

36. $5\frac{1}{10}$ **37.** $36\frac{2}{5}$ **38.** $7\frac{1}{4}$ **39.** $13\frac{2}{3}$

Add or subtract as indicated and reduce all answers.

40. $27\frac{1}{4} + 3\frac{1}{2}$ **41.** $12\frac{5}{6} - 6\frac{1}{4}$ **42.** $35\frac{1}{12} + 8\frac{1}{10}$

43. $15 - \frac{9}{10}$ **44.** $4\frac{5}{8} + 2\frac{3}{14}$ **45.** $7\frac{3}{4} - 2\frac{11}{12}$

Multiply or divide as indicated and reduce all answers.

46. $\frac{7}{8} \cdot \frac{16}{25} \cdot \frac{5}{14}$ **47.** $7\frac{1}{11} \left(2\frac{3}{4}\right)\left(5\frac{1}{3}\right)$ **48.** $\frac{5}{6} \div 3\frac{3}{4}$

49. $4\frac{3}{8} \div 7\frac{1}{2}$ **50.** $16\frac{2}{3} \div 22\frac{2}{9}$

51. Find the average of $\frac{3}{4}, \frac{5}{8}, \frac{9}{10}$ and $\frac{2}{3}$.

52. Evaluate: $\frac{5}{8} \cdot \frac{3}{10} + \frac{1}{14} \div 2$.

Simplify.

53. $\dfrac{\frac{7}{8} - \frac{3}{16}}{\frac{1}{3} - \frac{1}{4}}$ **54.** $\dfrac{6\frac{1}{2} + 2\frac{3}{11}}{3\frac{1}{9} - 2\frac{5}{6}}$

55. If a telephone pole is $42\frac{1}{2}$ feet long and $\frac{1}{4}$ of the pole is below the ground level, how many feet are above ground?

56. A baseball team played 55 games. A pitcher won 6 games, and the leading hitter hit safely in 40 games. What rational number represents the number of games won by the pitcher divided by the number of games in the season? If he pitched in only 11 games, what rational number represents the number of games won divided by the number of games he played in?

57. If gas to run a water heater costs $\frac{1}{5}$ cent per hour, what is the cost of running a water heater for one week?

58. The product of $10\frac{1}{3}$ with $\frac{6}{7}$ is added to the quotient of $10\frac{1}{2}$ divided by $8\frac{1}{6}$. What is the sum?

59. If one of two rational number factors of $6\frac{1}{2}$ is $20\frac{2}{3}$, what is the other rational number factor?

60. One paperboy can deliver his entire route in $2\frac{3}{4}$ hours, but it takes a new paperboy $3\frac{1}{2}$ hours to deliver the same route. If the first boy is 16 years old and the second is 14 years old, what is the difference in their delivery times in hours? in minutes?

DECIMAL NUMBERS

4

4.1   DECIMAL NUMBERS

The numbers

$$\frac{7}{8}, \frac{21}{32}, \frac{93}{500}, \frac{17}{2}, \frac{16}{73}, \text{ and } \frac{56}{3}$$

are *not* decimal numbers, but the numbers

$$\frac{3}{10}, \frac{41}{100}, \frac{29}{1000}, \frac{776}{10}, \frac{894}{100}, \frac{1432}{10,000}, \text{ and } \frac{5}{1}$$

are decimal numbers. What distinguishes the second group of numbers from the first? Both are groups of rational numbers, but the denominators in the second group are all powers of ten, that is, $10^0 = 1$, $10^1 = 10$, $10^2 = 100$, $10^3 = 1000$, and so on.

DEFINITION   A **decimal number** is a rational number that has a power of ten as its denominator.

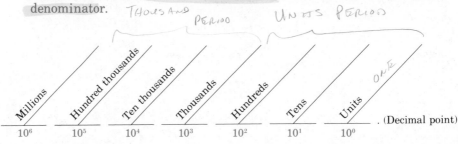

Figure 4.1

The usual notation for decimal numbers is not the fraction form. The notation used is an extension of the base ten place-value system (Figure 4.1). In the decimal system, each place is 10 times the value of the place immediately to the right. Therefore, the *ones* (or units) place is 10 times the value of the place to the right of the decimal point. What is the value of the place immediately to the right of the decimal point?

$$1 = 10 \cdot \square \qquad \square = \frac{1}{10}$$

Thus, the first place to the right is the $\frac{1}{10}$ths place (or the *tenths* place). Using the same reasoning, the value of the second place can be found.

$$\frac{1}{10} = 10 \cdot \square \qquad \square = \frac{1}{100}$$

Figure 4.2 indicates the results of such a procedure and gives the names for the values of each place. Note that the values of the places to

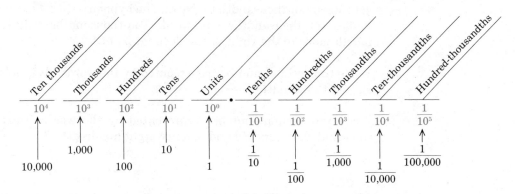

Figure 4.2

the right of the decimal point are the reciprocals of the values of the places to the left, except for 1. Thus, the ones place is a sort of place of symmetry, or the "middle" place, in the decimal number system.

In reading and writing decimal numbers, the word **and** indicates the decimal point. For example, 742.3 is read "seven hundred forty-two *and* three tenths." Be careful *not* to read 742 as "seven hundred and forty-two" because there is no decimal point between 7 and 4.

EXAMPLES

1. 4693. is read "four thousand, six hundred ninety-three."

2. 28,551. is read "twenty-eight thousand, five hundred fifty-one."

3. 476,476,476. is read "four hundred seventy-six million, four hundred seventy-six thousand, four hundred seventy-six."

    [NOTE: The numbers are read in groups of three, starting from the decimal point, and the value of the last digit in each group is read.]

If the decimal number (or decimal) has a fraction part, the whole number part is read as before; the word *and* indicates the decimal point; and the fraction part is read just as a whole number with the name of the place of the last digit.

EXAMPLES

1. 37.56 is read "thirty-seven *and* fifty-six hundredths." [NOTE: The digit 6 is in the hundredths position.]

2. 5.398 is read "five *and* three hundred ninety-eight thousandths."

3. 900.4217 is read "nine hundred *and* four thousand two hundred seventeen ten-thousandths." [NOTE: The hyphen (-) in *ten-thousandths* makes ten-thousandths one word. Ten-thousandths indicates the fourth place to the right of the decimal point.]

4. Write the decimal number represented by the words "five and thirty-six ten-thousandths." Answer: 5.0036.

5. Write the decimal number represented by the words "thirty-nine thousand, four hundred and seventy-eight hundredths." Answer: 39,400.78.

SPECIAL NOTES

A. The *th* at the end of a word indicates a fraction part (a part to the right of the decimal point).
    eight hundred = 800
    eight hundred*ths* = .08

B. The hyphen (-) indicates one word.
    eight hundred thousand = 800,000
    eight hundred-thousandths = .00008

Decimal numbers can be written in expanded form in a manner similar to writing whole numbers in expanded form. This technique may be helpful in understanding how to read and write decimal numbers. The fraction part of a decimal is the same as the fraction part of a mixed number and is sometimes called a **decimal fraction**.

EXAMPLES   1. $24.327 = 2(10^1) + 4(10^0) + 3\left(\dfrac{1}{10}\right) + 2\left(\dfrac{1}{100}\right) + 7\left(\dfrac{1}{1000}\right)$

$$= 20 + 4 + \frac{3}{10} + \frac{2}{100} + \frac{7}{1000}$$

$$= 24 + \frac{300}{1000} + \frac{20}{1000} + \frac{7}{1000}$$

$$= 24 + \frac{327}{1000}$$

$$= 24\frac{327}{1000} \quad \text{(read: "twenty-four and three hundred twenty-seven thousandths.")}$$

2. $4.006 = 4(1) + 0\left(\dfrac{1}{10}\right) + 0\left(\dfrac{1}{100}\right) + 6\left(\dfrac{1}{1000}\right) = 4 + 0 + 0 + \dfrac{6}{1000}$

$= 4 + \dfrac{6}{1000} = 4\dfrac{6}{1000}$ (read: "four and six thousandths.")

Writing checks points out the need for writing numbers in word form. Because of the problem of spelling, poor penmanship, and so on, we are required to write the amount of the check both in words and in numerals as a safety measure (Figure 4.3).

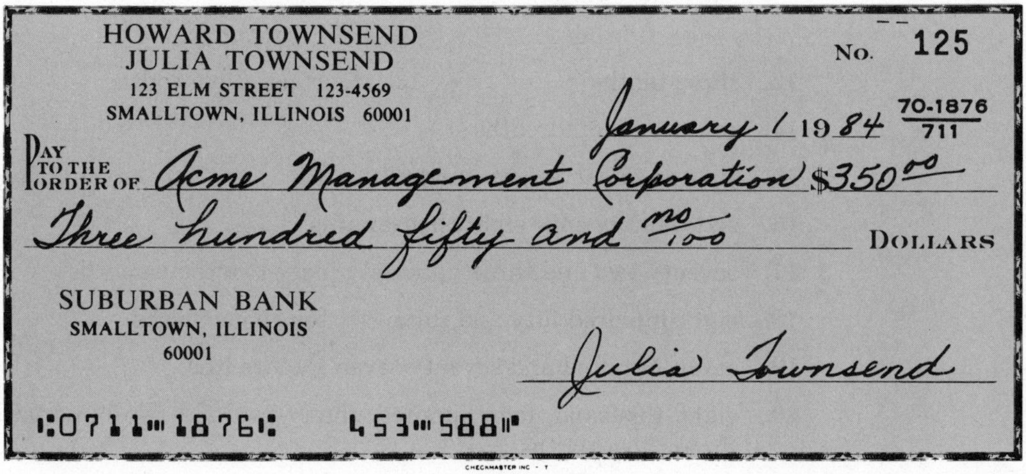

Figure 4.3

SELF QUIZ	Write the following numbers in decimal notation.	ANSWERS
	1. five thousandths	1. .005
	2. seventy-two and forty-three ten-thousandths	2. 72.0043
	Write the following decimal numbers in words.	
	3. 52.003	3. fifty-two and three thousandths
	4. 100.12	4. one hundred and twelve hundredths

EXERCISES 4.1

1. Define *decimal number*.

Write the following numbers in expanded notation using the powers of ten and their reciprocals.

Example: $49.376 = 4(10^1) + 9(10^0) + 3\left(\dfrac{1}{10}\right) + 7\left(\dfrac{1}{10^2}\right) + 6\left(\dfrac{1}{10^3}\right)$

2.	37.498	3.	562.3	4.	76.387	5.	946.346
6.	7862.3571	7.	894.6771	8.	4378.6666	9.	200.00793
10.	40.0005	11.	2936.78103				

Write the following numbers in decimal notation.

12. three tenths　　　　13. fourteen thousandths

14. seventeen hundredths

15. six and twenty-eight hundredths

16. sixty and twenty-eight thousandths

17. seventy-two and three hundred ninety-two thousandths

18. eight hundred fifty and thirty-six ten-thousandths

19. seven hundred and seventy-seven hundredths

20. eight thousand, four hundred ninety-two and two hundred sixty-three thousandths

21. six hundred thousand, five hundred and four hundred two thousandths

22. three hundred twenty-two million, four hundred seventy-six thousand and eighty-four thousandths

23. seven hundred five and four thousand, two hundred ninety-three ten-thousandths

24. five thousand, sixty-three and five hundred-thousandths

25. one million and thirty-seven ten-thousandths

Write the following decimal numbers in words.

26.	.5	27.	.93	28.	5.06	29.	32.58
30.	71.06	31.	35.078	32.	7.003	33.	18.102
34.	50.008	35.	607.607	36.	593.86	37.	593.860
38.	4700.617	39.	5000.005	40.	603.0065	41.	900.4638

42. 5005.505   **43.** 4.05671   **44.** 17.46685   **45.** 879.58932

46. Duplicate the check form in the text and write sample checks for

(a) $372.58   (b) $577.50   (c) $2405.37   (d) $1,476,324.75

4.2 ROUNDING OFF

Measuring instruments such as yardsticks, tape measures, micrometers (see Figure 4.4), or meter sticks are all man-made and can never give **exact** measurements. Any number found as the result of using a measuring instrument (not a counting instrument) is an **approximate** number.

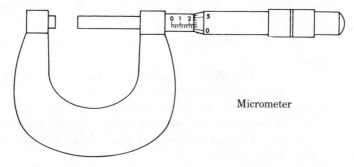

Micrometer

Figure 4.4

As an example, suppose the rod in Figure 4.5 is measured with a meter stick with markings for millimeters. Is the rod 52 or 53 millimeters long? Obviously, neither one is exactly the length of the rod, but since a length is to be reported, the number closer to the length is chosen, and this number, 53 millimeters, is an approximate number.

The operations with approximate numbers require that the work be done with the accuracy of the least accurate number. Such operations with approximate numbers are the subject of math courses in shopwork, engineering, surveying, and so on, and are not part of the discussion here. They do point out a use for the idea of rounding off measurements and in particular rounding off decimal numbers.

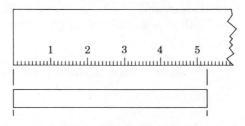

Figure 4.5

To **round off** a number means to find another number close to the original number. How close? The answer depends on the accuracy desired. The statement "round off 872" makes no sense unless you are told more. 872 is between 800 and 900, as shown in Figure 4.6.

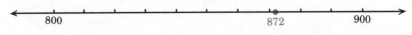

Figure 4.6

Is 872 closer to 800 or 900? Obviously, as shown in Figure 4.6, 872 is closer to 900. So, *to the nearest hundred,* 872 rounds off to 900. 872 is also between 870 and 880, as shown in Figure 4.7.

Figure 4.7

But 872 is closer to 870 than it is to 880. This means that *to the nearest ten,* 872 rounds off to 870.

Which answer is correct, 900 or 870? Each is correct for the accuracy it indicates. The *closer hundred* is 900, and the *closer ten* is 870.

Careful analysis of the "closer to" rounding off idea leads to the following rule.

RULE FOR ROUNDING OFF DECIMAL NUMBERS

1. Look at the single digit just to the right of the digit that is in the place of desired accuracy.

2. If this digit is 5 or greater, make the digit in the desired place of accuracy one larger and replace all digits to the right with zeros.

3. If this digit is less than 5, leave the digit that is in the place of desired accuracy as it is and replace all digits to the right with zeros.

This rule is somewhat arbitrary. It is not used by grocers or other merchants in rounding off to pennies. If the price is any amount over a penny, they will charge the next higher cent.

EXAMPLES

1. Round off 7283.5 to the nearest hundred.

    Answer:
    7300.0, or 7300
    (The 2 is in the hundreds place. Just to the right of the 2 is 8. Eight

is more than 5 so 2 is changed to 3, and zeros replace the remaining digits to the right.)

2. Round off 5.749 to the nearest tenth.

Answer:
5.700, or 5.7
(The 7 is in the tenths place. Just to the right of the 7 is 4. 4 is less than 5 so the 7 remains unchanged, and zeros replace the remaining digits to the right.)

3. Round off 396.53 to the nearest unit.

Answer:
397.00, or 397

4. Round off 7.45796 to the nearest ten-thousandth.

Answer:
7.4580
(The 9 is in the ten-thousandths place. Just to the right of the 9 is 6. Six is more than 5 so the 9 is changed to one larger. Note that this affects two digits.)

If no decimal point is written, the number is a whole number and the decimal point is understood to be to the right of the right-hand digit. For example, 596 and 596. mean the same thing.

SELF QUIZ	Round off as indicated.	ANSWERS
	1. 572.3 (nearest ten)	1. 570
	2. 6.749 (nearest tenth)	2. 6.7
	3. 7558 (nearest thousand)	3. 8000
	4. 0.07921 (nearest thousandth)	4. .079

EXERCISES 4.2

Round off each of the following decimal numbers as indicated.

To the nearest tenth:

1. 4.763	2. 5.031	3. 76.349	4. 76.352	5. 89.015
6. 7.555	7. 18.009	8. 37.666	9. 14.3338	10. 0.036

Assign:
Odd Problems
1-67

To the nearest hundredth:

11. 0.385   **12.** 0.296   **13.** 5.722   **14.** 8.987   **15.** 6.996

16. 13.1346   **17.** 0.0782   **18.** 6.0035   **19.** 5.7092   **20.** 2.8347

To the nearest thousandth:

21. 0.0672   **22.** 0.05550   **23.** 0.6338   **24.** 7.6666

25. 32.4785   **26.** 9.4302   **27.** 17.36371   **28.** 4.44449

29. 0.00191   **30.** 20.76962

To the nearest whole number (or nearest unit):

31. 479.23   **32.** 6.872   **33.** 17.50   **34.** 19.999

35. 382.48   **36.** 649.66   **37.** 439.78   **38.** 701.413

39. 6333.11   **40.** 8122.825

To the nearest ten:

41. 5163.   **42.** 6475   **43.** 495   **44.** 572.5

45. 998.5   **46.** 378.92   **47.** 5476.2   **48.** 76,523.1

49. 92,540.9   **50.** 7007.7

To the nearest thousand:

51. 7398   **52.** 62,275   **53.** 47,823.4   **54.** 103,499

55. 217,480.2   **56.** 9872.5   **57.** 379,500   **58.** 4,500,762

59. 7,305,438   **60.** 573,333.3

61. .0005783   (nearest hundred-thousandth)

62. .5449   (nearest hundredth)

63. 473.8   (nearest ten)

64. 5.00632   (nearest thousandth)

65. 473.8   (nearest hundred)

66. 5750   (nearest thousand)

67. 3.2296   (nearest thousandth)

68. 15.548   (nearest tenth)

69. 78,419   (nearest ten thousand)

70. 78,419   (nearest ten)

4.3   ADDING AND SUBTRACTING DECIMAL NUMBERS

Consider the sum of decimal numbers $5.2 + 6.32 + 13.06$. In mixed number form,

$$5.2 + 6.32 + 13.06 = 5\frac{2}{10} + 6\frac{32}{100} + 13\frac{6}{100}$$

Now add the whole numbers and fractions in the following way.

$$5\frac{2}{10} + 6\frac{32}{100} + 13\frac{6}{100} = (5 + 6 + 13) + \left(\frac{2}{10} + \frac{30}{100} + \frac{2}{100} + \frac{6}{100}\right)$$

$$= (5 + 6 + 13) + \left(\frac{2}{10} + \frac{3}{10}\right) + \left(\frac{2}{100} + \frac{6}{100}\right)$$

$$= 24 + \frac{5}{10} + \frac{8}{100}$$

This is not the normal procedure, of course, but it does point out two things. First, decimal numbers are rational numbers and can be written in mixed number or fraction form. Second, to add rational numbers, common denominators are necessary. Notice how $\frac{32}{100}$ was broken up so that tenths and hundreds could be added separately.

$$\left(\frac{32}{100} = \frac{30}{100} + \frac{2}{100} = \frac{3}{10} + \frac{2}{100}\right)$$

The same result can be found by writing the decimal numbers one under the other. However, in order to be sure the correct fraction parts are being added, the decimal points must be in a vertical straight line. The decimal point in the sum must be in line with the other decimal points. The rules for adding decimal numbers are the same as for whole numbers once the decimal points are lined up.

None of the numbers in this section are approximate numbers; that is, none are the results of measurements. They are exact numbers, so zeros may be written to the right of the last digit in the fraction part to help keep digits in line.

EXAMPLES   1.   Find the sum $5.2 + 6.32 + 13.06$.

$$\begin{array}{r} 5.20 \\ 6.32 \\ \underline{13.06} \\ 24.58 \end{array}$$

[NOTE: Writing 5.2 as 5.20 helps keep the digits lined up.]

2.   Find the sum $8 + 3.76 + 47.689 + .2$.

$$
\begin{array}{r}
8.000 \\
3.760 \\
47.689 \\
0.200 \\
\hline
59.649
\end{array}
$$

[NOTE: The zeros are filled in to help keep the digits in line.]

Subtraction with decimal numbers also requires that the decimal points be in a vertical line when the numbers are written one under the other. The reason is so that fractions with the same denominator will be subtracted. Again, the rules are the same as with whole numbers after the decimal points have been lined up. The decimal point in the difference must be in line with the other decimal points.

EXAMPLES   1.   Find the difference $5.438 - 2.653$.

$$
\begin{array}{r}
5.438 \\
- 2.653 \\
\hline
2.785
\end{array}
$$

Since subtraction is the reverse of addition, the result may be checked by adding.

$$
\begin{array}{r}
2.785 \\
+ 2.653 \\
\hline
5.438
\end{array}
$$

2.   Find the difference $17.2 - 3.6954$.

$$
\begin{array}{r}
17.2000 \\
- 3.6954 \\
\hline
13.5046
\end{array}
$$

Check:

$$
\begin{array}{r}
13.5046 \\
+ 3.6954 \\
\hline
17.2000
\end{array}
$$

SELF QUIZ	Find each of the indicated sums or differences.	ANSWERS
	1. $46.2 + 3.07 + 2.6$	1. 51.87
	2. $9 + 5.6 + .58$	2. 15.18
	3. $6.4 - 3.7$	3. 2.7
	4. $18 - .4384$	4. 17.5616

EXERCISES 4.3

Find each of the indicated sums.

1. $.6 + .4 + 1.3$ 2. $5 + 6.1 + .4$ 3. $.59 + 6.91 + .05$

4. $3.488 + 16.593 + 25.002$ 5. $37.02 + 25 + 6.4 + 3.89$

6. $4.0086 + .034 + .6 + .05$ 7. $43.766 + 9.33 + 17 + 206$

8. $52.3 + 6 + 21.01 + 4.005$ 9. $2.051 + .2006 + 5.4 + 37$

10. $5 + 2.37 + 463 + 10.88$

11.	12.	13.	14.
47.3	1.007	4.128	5.0015
42.03	20.063	.02	2.443
29.003	.49	3.	.0469

15.	16.	17.	18.
75.2	107.39	34.967	4.156
3.682	5.061	50.6	3.7
14.995	23.54	8.562	25.682
	64.9801	9.3	13.405

19.	20.
74.	983.4
3.529	47.518
52.62	805.411
7.001	300.766

Find each of the indicated differences.

21. $5.2 - 3.76$ 22. $17.83 - 8.9$ 23. $29.5 - 13.61$

24. $1.0057 - .03$ 25. $78.015 - 13.068$

26.	27.	28.	29.	30.
22.418	4.8	31.009	4.	40.718
17.523	.0026	.534	1.0566	6.532

31. Find the sum of the three numbers: four hundred thirty-six and seventeen thousandths; five hundred and three tenths; one hundred seventy-seven and three hundred forty-eight thousandths.

32. Find the sum of the three numbers: seven hundred sixty-three ten-thousandths; fourteen thousandths; one thousand, four hundred twenty-nine ten-thousandths.

33. Find the difference between five hundred forty-eight and thirteen hundredths; and three hundred ninety-three and fifty-seven thousandths.

34. Find the difference between sixty-five and four hundred twenty-six ten-thousandths; and fifty-eight and two thousand, five hundred seventy-one ten-thousandths.

35. Find the difference between the two sums $8 + 13.59 + 6.05$ and $21.53 + 7.02$.

36. Mrs. Barclay charged the following items at a department store: hat—$35.12, dress—$47.50, shoes—$15.75, and purse—$6.88. What was the total amount of her charge at the store?

37. The inside radius of a pipe is 2.381 inches, and the outside radius is 2.634 inches. What is the thickness of the pipe?

38. An architect's scale drawing of a building gives the sides as 3.05 inches, 1.063 inches, 2.179 inches, 3.02 inches, and 2.738 inches. What is the sum of these measurements?

39. Find the difference between seven and thirty-nine thousandths; and four and one hundred six ten-thousandths. Also, find the difference between two hundred and seventeen hundredths; and one hundred five and nine hundredths. Find the sum of these two differences.

40. A discount of $2.10 was given on each item Mr. Sward bought at a sale. What was his cost for a man's shirt marked $5.00, a pair of shoes marked $7.50, and a robe marked $10.00?

4.4   MULTIPLYING DECIMAL NUMBERS

Multiplication with decimal numbers is similar to multiplication with whole numbers. The only difference is the placement of the decimal point in the product. To understand the procedure of placing the decimal point, consider the product (9.7) (.25) using fraction notation.

$$(9.7)\,(.25) = \left(9\frac{7}{10}\right)\left(\frac{25}{100}\right) = \frac{97}{10} \cdot \frac{25}{100} = \frac{97 \cdot 25}{1000} = \frac{2425}{1000}$$

$$= 2\frac{425}{1000} = 2.425$$

The decimal point is placed so that the product (2.425) has thousandths in it. This seems reasonable because one of the factors (9.7) has tenths in it and the other factor (.25) has hundredths in it. Multiplying tenths by hundredths $\left(\frac{1}{10} \cdot \frac{1}{100} = \frac{1}{1000}\right)$ gives thousandths.

Try another example: (3.62)(41.81). Each factor here has hundredths in it and since $\frac{1}{100} \cdot \frac{1}{100} = \frac{1}{10,000}$, the product should have ten-thousandths in it.

$$(3.62)(41.81) = \left(3\frac{62}{100}\right)\left(41\frac{81}{100}\right) = \frac{362}{100} \cdot \frac{4181}{100} = \frac{1,513,522}{10,000}$$

So $\qquad$ $(3.62)(41.81) = 151\dfrac{3522}{10{,}000} = 151.3522$

We were right. The product does have ten-thousandths in it.

RULE FOR MULTIPLYING TWO DECIMAL NUMBERS

1. Multiply the two numbers as if they were whole numbers.

2. Count the total number of places to the right of the decimal points in both factors.

3. This sum is the number of places to the right of the decimal point in the product.

If you were to continue in the manner of these two examples, you would soon develop the preceding rule (or shortcut) for placing the decimal point in products of decimal numbers.

EXAMPLES

1.
$$
\begin{array}{r}
2.432 \\
5.1 \\
\hline
2432 \\
12\ 160 \\
\hline
12.4032
\end{array}
$$

2.
$$
\begin{array}{r}
4.35 \\
12.6 \\
\hline
2\ 610 \\
8\ 70 \\
43\ 5 \\
\hline
54.810
\end{array}
$$

3.
$$
\begin{array}{r}
1.76 \\
25 \\
\hline
8\ 80 \\
35\ 2 \\
\hline
44.00
\end{array}
$$

Multiplication of decimal numbers by powers of ten can be shortened considerably from the usual procedure illustrated in Examples 1, 2, and 3. The powers of ten (10^0, 10^1, 10^2, 10^3, . . .) are whole numbers, so the placement of the decimal depends only on the decimal number being multiplied. Multiplication by powers of ten was discussed in Section 1.6 and involves writing zeros. For example,

$$10^3(3581) = 1000(3581) = 3{,}581{,}000$$

Now, using this procedure and the rule for placing decimal points in products, the product 10^3 (35.81) can be found in the following way.

$$10^3(35.81) = 35{,}810.00$$

EXAMPLES   1. $10(3.468) = 34.680 = 34.68$

2. $10^2(3.468) = 346.800 = 346.8$

In Examples 1 through 4, the decimal point is moved three places after the whole numbers have been multiplied because there are three places to the right of the decimal point in 3.468.

3. $10^3(3.468) = 3,468.000 = 3,468$

4. $10^4(3.468) = 34,680.000$
$= 34,680$

5. $10^6(.5813) = 581,300.0000$
$= 581,300$

The same results can be found simply by moving the decimal point to the right in the decimal number being multiplied. The number of places depends on the power of ten and is the same as the number of zeros in the power of ten or the exponent of the power of ten.

EXAMPLES   1. $10^2(5.734) = 100(5.734) = 573.4$

The decimal point is moved two places to the right, which corresponds to the two zeros in 100 and to the exponent 2.

2. $10^3(.9641) = 1000(.9641) = 964.1$

The decimal point is moved three places to the right, which corresponds to the three zeros in 1000 and to the exponent 3.

SELF QUIZ	Find each of the indicated products.	ANSWERS
	1. $(.8)(.2)$	1. .16
	2. $(5.6)(.04)$	2. .224
	3. $10^4(3.781)$	3. 37810.

EXERCISES 4.4

Find each of the indicated products.

1. $(.6)(.7)$	2. $3(2.5)$	3. $1.4(.2)$	4. $(3.5)(.6)$
5. $6(3.1)$	6. $.2(.02)$	7. $.5(.05)$	8. $.03(.03)$
9. $4.1(.06)$	10. $.7(.1)$	11. $.06(.01)$	12. $.23(.12)$
13. $4.7(.02)$	14. $.51(.13)$	15. $4.15(2.6)$	16. $5.9(.25)$
17. $4(.75)$	18. $16(.875)$	19. $8(.125)$	20. $.66(.33)$

21. $10^2(3.46)$ **22.** $10^2(20.57)$ **23.** $100(7.82)$ **24.** $100(16.1)$

25. $10^1(.435)$ **26.** $10^3(4.1782)$ **27.** $10^0(.38)$ **28.** $10^4(51.329)$

29. $10^4(7.12)$ **30.** $10^5(2.5148)$

31. .005   **32.** .137   **33.** 1.06   **34.** .01063   **35.** 71.222
   .009   .06   .14   .087   .111

36. To buy a boat, a man can pay $2036.50 cash, or he can put $400 down and make 18 monthly payments of $104.30. How much does he save by paying cash?

37. Mr. Stromm bought two suits for $65.30 each, one pair of shoes for $22.95, three pairs of socks for $1.25 a pair, four shirts for $5.97 each, and six ties for $2.65 each. What was his total bill?

38. If you drive south at 57.6 miles per hour for 3 hours, then west at 62.4 miles per hour for 4 hours, how far have you driven in the 7 hours? (Assume you started at least 300 miles east of the Pacific Ocean and 200 miles north of the Gulf of Mexico.)

39. If an architect makes a drawing to the scale that 1 inch represents 6.75 feet, what distance is represented by 5.5 inches?

40. If an automobile dealer makes $150.70 on each used car he sells and $425.30 on each new car he sells, how much did he make the month that he sold eleven used and six new cars?

41. Find the difference of the product of seven and eight tenths with thirty-two and five tenths, and the product of thirteen and fifty-six hundredths with three and one tenth.

42. Multiply the numbers 2.456 and 3.16, then round off the product to the nearest tenth. Next, round off each of the factors to the nearest tenth and then multiply and round off this product to the nearest tenth. Did you get the same answer?

43. Follow the same procedure as in Problem 42 with the numbers 12.81 and 9.32. Can you come to any conclusions about rounding off after these two exercises? Try the same procedure with the same two sets of numbers, rounding off to the nearest whole number. Now, can you come to any conclusions?

44. Suppose a tax assessor figures the tax at .07 of the assessed value of your home. If the assessed value is figured at the rate of .32 of the retail value, what taxes do you pay on your $36,500 home?

45. Write 3.0062 in words. Write .578 in words. Write the sum of these two numbers in words. Write their product in words. Write their difference in words.

4.5 DIVIDING DECIMAL NUMBERS

Division with decimal numbers is similar to division with whole numbers. The difference is in the placement of the decimal point. One of the main purposes of this section is to find a convenient way to assure correct placement of the decimal point in the quotient.

Suppose we want to divide 51.52 by 3.2. Using mixed numbers, the process can be written as follows:

$$\frac{51.52}{3.2} = \frac{51\dfrac{52}{100}}{3\dfrac{2}{10}} = \frac{\dfrac{5152}{100}}{\dfrac{32}{10}} = \frac{5152}{100} \cdot \frac{10}{32}$$

$$= \frac{5152}{32} \cdot \frac{10}{100}$$

$$= \frac{5152}{32} \cdot \frac{1}{10}$$

$$= \frac{5152}{320}$$

Thus, the division of decimal numbers can be related to division with whole numbers. The same results can be found by multiplying the numerator and denominator (dividend and divisor) by 100 so that they will both be whole numbers.

$$\frac{51.52}{3.2} = \frac{100}{100} \cdot \frac{51.52}{3.2} = \frac{5152}{320}$$

Dividing gives the following result.

$$3.2 \overline{)51.52} \quad \text{or} \quad 320 \overline{)5152}$$

$$\begin{array}{r} 16 \\ 320 \overline{)5152} \\ \underline{320} \\ 1952 \\ \underline{1920} \\ 32 \text{ remainder} \end{array}$$

But the division need not stop with whole numbers because the properties of decimal numbers are now available. The quotient may be a decimal number. So, put the decimal point in the quotient directly above the decimal point in the dividend, write one or more zeros to the right of the decimal point in the dividend, and continue dividing.

```
              16.1      Check:
      320)5152.0          16.1     or        16.1
          320             320                 3.2
         1952            322 0               3 22
         1920             483                48 3
          320           5152.0              51.52
          320
            0
```

Actually, placement of the decimal point can be accomplished easily if only the divisor is a whole number. That is, multiplication of divisor and dividend by 10 will be sufficient in this example. The multiplication is indicated by the arrows showing the *new* placements of the decimal points.

$$3.2\,\overline{)51.5\,2}$$

Dividing gives the same quotient, 16.1.

```
              16.1
      3.2.)51.5.2
          32
          19 5
          19 2
            3 2
            3 2
              0
```

This last technique is the one most commonly used.

EXAMPLES   1.   12.87 ÷ 4.5   2.   5.1 ÷ 1.36

```
                2.86                              3.75
      4.5.)12.8.70                    1.36.)5.10.00
          9 0                               4 08
          3 8 7                             1 02 0
          3 6 0                               95 2
            2 70                               6 80
            2 70                               6 80
               0                                  0

   Check:  2.86                       Check:  3.75
             4.5                                1.36
          1 430                                2250
         11 44                                1 125
         12.870                               3 75
                                             5.1000
```

The remainders will not always be 0. In such cases, agreement is made before the division is performed as to how accurate the quotient should be. The quotient is found to *one more place* than the desired accuracy, then rounded off. The answer is then only approximate and cannot be checked by the usual method.

EXAMPLE $2 \div 3.1$ (Round the quotient off to the nearest thousandth if the remainder is not 0.)

$$
\begin{array}{r}
.6451 \approx .645 \\
3.1.\overline{)2.0.0000} \\
186 \\
\hline
1\,40 \\
1\,24 \\
\hline
160 \\
155 \\
\hline
50 \\
31 \\
\hline
19
\end{array}
$$

[NOTE: The quotient was found to the ten-thousandths place before rounding off to the nearest thousandth. The sign $\approx$ is read "is approximately equal to."]

Dividing by powers of ten can be shortened by remembering that division is a reverse multiplication. For $\dfrac{473.6}{100}$, the question is what number times 100 gives 473.6? Or, $100(?) = 473.6$. (?) must have the same digits as 473.6 with the decimal point moved two places to the left, because

$$100(4.736) = 473.6$$

Therefore,

$$\frac{473.6}{100} = 4.736$$

Division by 10 moves the decimal point to the left one place.

Division by 100 moves the decimal point to the left two places.

Division by 1000 moves the decimal point to the left three places, and so on.

EXAMPLES   1. $\dfrac{583.9}{1000} = .5839$ 2. $\dfrac{583.9}{10,000} = .05839$ 3. $\dfrac{782}{10^2} = 7.82$

SELF QUIZ	Find each of the indicated quotients.	ANSWERS
	1. 4)$\overline{1.83}$ (nearest hundredth)	1. .46
	2. .06)$\overline{43.721}$ (nearest thousandth)	2. 728.683
	3. $\dfrac{42.31}{10^3}$	3. .04231

EXERCISES 4.5

ODD Prob 1-53

Find the quotients to the nearest thousandth if the remainders are not zero (0).

1. 2)$\overline{4.68}$ 2. 3)$\overline{1.71}$ 3. .5)$\overline{4.95}$ 4. .9)$\overline{1.62}$

5. .8)$\overline{.064}$ 6. .7)$\overline{.63}$ 7. .04)$\overline{82.24}$ 8. .03)$\overline{16.02}$

9. 2.4)$\overline{48}$ 10. 5.6)$\overline{28}$ 11. 1.8)$\overline{.0036}$ 12. .14)$\overline{.042}$

13. 7)$\overline{6.6}$ 14. 3.2)$\overline{.416}$ 15. 9)$\overline{7.6}$ 16. 1.6)$\overline{9.76}$

17. 6)$\overline{1}$ 18. 1.2)$\overline{1.56}$ 19. 3.02)$\overline{9.1506}$ 20. 4.6)$\overline{5}$

21. 7.05)$\overline{.49773}$ 22. .37)$\overline{4.683}$ 23. .21)$\overline{65.226}$ 24. 100)$\overline{5.682}$

25. 1.62)$\overline{34}$ 26. 4.6 ÷ .009   27. 5.2 ÷ .71

28. .03 ÷ .008   29. .71 ÷ .025   30. 29.3 ÷ 6.9

Find the exact quotients without rounding off.

31. 100)$\overline{78.4}$ 32. 1000)$\overline{16.4963}$ 33. .5036 ÷ 10^3

34. 45.621 ÷ 10^2 35. 7.682 ÷ 10^0 36. $\dfrac{.00167}{10^3}$

37. $\dfrac{.01826}{10^4}$ 38. $\dfrac{91.112}{10^0}$ 39. $\dfrac{.6122}{10^3}$

40. $\dfrac{10.413}{10^5}$

41. Find the average of the numbers 44.62, 57.8, 39.49, 76.2, 61.523 correct to the nearest hundredth.

42. If a store advertises three cans of beans for $1, what will be the charge for one can? [NOTE: If the cost of any item involves a fraction of a cent, a store will always charge the full penny.]

43. If an automobile averages 17.2 miles per gallon, how far will it go on 18 gallons of gasoline?

44. A road grader burns 8.4 gallons of fuel oil each hour that it operates. How long will the grader operate on 100.4 gallons of fuel oil?

45. If a motorcycle can travel 450.6 miles in 7.81 hours, what will be its average speed?

46. If the postal rates on educational material are 10¢ for the first pound and 5¢ for each additional pound, what will be the cost per pound to mail 8 pounds of books?

47. A frozen quarter of beef can be bought much cheaper than the same amount of meat purchased a few pounds at a time. If the charge for 150 pounds of meat is $90.32, what is the cost per pound?

48. Two men measured the length of a strip of metal three different times with three different instruments. The first man's measurements were 1.002 inches, 1.0019 inches, and 1.0022 inches. The measurements taken by the second man were 1.0018 inches, 1.0017 inches, and 1.0021 inches. What is the difference between the averages of their measurements?

49. In testing airplanes at high altitudes, a pilot flew one plane to heights of 40,000 feet; 42,000 feet; and 41,600 feet. He flew a second plane to heights of 41,000 feet; 39,400 feet; 45,200 feet; and 42,300 feet. What is the difference between the average altitudes?

50. To purchase a home for $75,000, a down payment of $15,000 is made. If a 30-year mortgage on a loan of $60,000 is going to cost a total of $189,558, what will the monthly payments be on the mortgage?

51. The Do-It-Yourself Lumberyard maintains an average merchandise inventory of $25,000 (valued at selling price). If the lumberyard wishes to have an annual merchandise turnover of $5\frac{1}{2}$ times during the year, what must the average monthly sales be?

52. Find the product of the sum of eighty-two and three hundredths and seventy-six and nine hundredths with the difference between sixteen and thirty-two hundredths and twelve and ninety-six thousandths.

53. A company bought twenty new cars and sold twenty of their used cars. Thirteen of the new cars were station wagons, and they cost $5280 each. The other seven new cars were two-door cars that cost $3560 each. The average selling price of the used cars was $1250. What did the company pay for the new cars? What did the company receive for the used cars? What was the net expense to the company for buying new cars?

54. A woman drove for 18 hours and 15 minutes (not counting stops to eat) at an average speed of 62.5 miles per hour. She was 36 years old. How far did she drive?

4.6   THE REAL NUMBERS

As in many exercises in the last section, division with decimal numbers does not always give a 0 remainder. In these situations, you were instructed to round off the quotient to some specified place of accuracy. What would happen if you continued to divide? Would the remainder ever be 0?

Consider the rational number $\frac{1}{7}$, which can mean $1 \div 7$. Divide and see what happens!

```
        .142857142857 . . .
    7)1.000000000000 . . .
       7
      ──
       30
       28
      ──
        20
        14
       ──
        60
        56
       ──
         40
         35
        ──
         50
         49
        ──
          10
           7
         ──
          30
          28
         ──
           20
           14
          ──
           60
           56
          ──
            40
            35
           ──
            50
            49
           ──
             1
```

Zeros can be written forever after the 1, and the pattern of digits 1, 4, 2, 8, 5, 7 will continue to repeat itself forever.

Thus, $\frac{1}{7} = .142857142857142857 \ldots$

Or $\frac{1}{7} = .\overline{142857}$

The three dots (. . .) and the bar both indicate that the pattern of digits is to repeat without end.

You may already know

$$\frac{1}{3} = .\bar{3} = .33333 \ldots \quad \text{and} \quad \frac{2}{3} = .\bar{6} = .66666 \ldots$$

You may not know

$$\frac{3}{11} = .\overline{27} = .2727272727 \ldots$$

Divide 3 by 11 to assure yourself that this statement is true.

In fact, *every rational number can be represented as an infinite repeating decimal.* The repeating part of such a decimal may not start at the decimal point, but once a pattern begins repeating, it will continue from then on.

EXAMPLES

1. $2\frac{1}{3} = 2.\bar{3} = 2.33333 \ldots$

2. $38\frac{1}{6} = 38.166666 \ldots = 38.1\bar{6}$

3. $1\frac{3}{4} = 1.7500000 \ldots = 1.75\bar{0}$ (Here, zeros are considered to be repeating digits.)

Those decimal numbers that repeat zeros are called **terminating decimals.** Thus, terminating decimals are just the decimal numbers. For example,

$$1.75 = 1.75\bar{0} = 1\frac{75}{100} = \frac{175}{100}$$

Are there infinite decimals that are nonrepeating? The answer is *yes.* Only the rational numbers can be written as **infinite repeating decimals.** Examples 4 and 5 illustrate some **infinite nonrepeating decimals.**

4. $.763198273486175 \ldots$ No pattern is indicated here. The next digit after 5 is not known from this information.

5. $4.01001000100001000001 \ldots$ Here there is a pattern, but it is not a repeating pattern.

The *infinite nonrepeating decimals* are called **irrational numbers.** The irrational numbers are more numerous than the rational numbers and are very important in further studies in mathematics. They hold a central place in such studies as trigonometry and calculus.

Two especially important irrational numbers for mathematicians and scientists are π (pi) and e.

$$\pi = 3.14159265358979 \ldots$$

and $\qquad e = 2.718281828459045 \ldots$

Pi (π) relates to circles and angular measurements, and e relates to the growth and decay of bacteria, minerals, and other elements.

Other irrational numbers are $\sqrt{2}$ (square root of 2), $\sqrt{3}$ (square root of 3), $\sqrt[3]{5}$ (cube root of 5), and so on. Generally, irrational numbers are rounded off, or **truncated,** just as rational numbers are for practical or daily uses. That is,

$$\pi \approx 3.14 \text{ to the nearest hundredth}$$
$$e \approx 2.718 \text{ to the nearest thousandth}$$
$$\frac{1}{3} \approx .33 \text{ to the nearest hundredth}$$

The rational and irrational numbers are considered to be one set of numbers called the **real numbers.**

DEFINITION   A **real number** is any number that can be written as an infinite decimal *(repeating or nonrepeating).*

The statement was made earlier that any rational number can be written as an infinite repeating decimal. This is so because for $\frac{a}{b}$, or $a \div b$, the remainders will always be less than b. Therefore, if we keep dividing, eventually one of the remainders will repeat itself, and a pattern will develop giving a repeating decimal in the quotient.

Suppose an infinite repeating decimal is given. How does one find the fraction form of the rational number? For example, what rational number in fraction form is equal to .4545454545 . . .?

If the repeating part starts at the decimal point, the formula

$$\frac{P}{10^N - 1}$$

will give the corresponding fraction. P is the whole number given by just the digits that repeat, and N is the number of digits in the repeating pattern. The technique for finding the fraction when the repeating part begins later will not be discussed here. If there are digits to the left of the decimal point, they represent a whole number, and this number may be added to the fraction found using the formula on the repeating decimal part.

EXAMPLES

1. Write .123123123123 . . . in fraction form.

$P = 123$ and $N = 3$

$$\frac{P}{10^N - 1} = \frac{123}{10^3 - 1} = \frac{123}{1000 - 1} = \frac{123}{999} = \frac{\cancel{3} \cdot 41}{\cancel{3} \cdot 333} = \frac{41}{333}$$

2. Write $.\overline{78}$ in fraction form.

$P = 78$ and $N = 2$

$$\frac{P}{10^N - 1} = \frac{78}{10^2 - 1} = \frac{78}{100 - 1} = \frac{78}{99} = \frac{\cancel{3} \cdot 26}{\cancel{3} \cdot 33} = \frac{26}{33}$$

3. Write $32.\overline{45}$ as a mixed number with $.\overline{45}$ in fraction form.

$P = 45$ and $N = 2$

$$\frac{P}{10^N - 1} = \frac{45}{10^2 - 1} = \frac{45}{100 - 1} = \frac{45}{99} = \frac{\cancel{9} \cdot 5}{\cancel{9} \cdot 11} = \frac{5}{11}$$

So,

$$32.\overline{45} = 32\frac{5}{11}$$

SELF QUIZ   ANSWERS

1. Find an infinite decimal representation for $\frac{6}{11}$.

     1. $.\overline{54}$

2. Represent $.\overline{639}$ in fraction form.

     2. $\frac{71}{111}$

3. Truncate 8.3471621 . . . . at the hundredths place.

     3. 8.35

EXERCISES 4.6

1. Define *real number*.

2. Every rational number can be written as a/an _____ _____.

3. In a book of mathematical tables (try the library), find decimal representations for $\sqrt{2}$, $\sqrt{3}$, and $\sqrt{5}$ accurate to four decimal places.

Find the infinite repeating decimal representation for each of the following rational numbers.

4. $\dfrac{3}{4}$ 5. $\dfrac{1}{5}$ 6. $\dfrac{2}{3}$ 7. $\dfrac{7}{8}$ 8. $\dfrac{1}{6}$ 9. $\dfrac{1}{3}$ 10. $\dfrac{1}{9}$

11. $\dfrac{4}{9}$ 12. $\dfrac{2}{11}$ 13. $\dfrac{1}{12}$ 14. $\dfrac{3}{32}$ 15. $\dfrac{3}{16}$ 16. $\dfrac{5}{8}$ 17. $\dfrac{7}{9}$

18. $\dfrac{6}{11}$ 19. $\dfrac{7}{30}$ 20. $\dfrac{1}{13}$ 21. $\dfrac{29}{7}$ 22. $\dfrac{31}{10}$ 23. $\dfrac{17}{14}$ 24. $\dfrac{43}{8}$

25. $\dfrac{17}{6}$

Represent each infinite repeating decimal in fraction form.

26. $.\overline{3}$ 27. $.\overline{21}$ 28. $.\overline{414}$ 29. $.\overline{5}$ 30. $.\overline{6}$

31. $.\overline{7}$ 32. $.\overline{567}$ 33. $.\overline{9}$ 34. $.\overline{123}$ 35. $.\overline{51}$

36. $.1\overline{8}$ 37. $.1\overline{29}$ 38. $.2\overline{88}$ 39. $.78\overline{33}$ 40. $.357\overline{9}$

41. Using the formula $\dfrac{P}{10^N - 1}$,

$$.99999 \ldots = \dfrac{9}{10^1 - 1} = \dfrac{9}{9} = 1$$

Does this make sense to you? Explain your reasoning and discuss this in class.

Truncate each of the following real numbers at the thousandths place.

42. $5.716894 \ldots$ 43. $72.983726 \ldots$ 44. $6.55549321 \ldots$

45. $.77352\overline{1}$ 46. $7.23\overline{7}$ 47. $97.186\overline{32}$

48. $.027635 \ldots$ 49. $.000\overline{4}$ 50. $.5\overline{6}$

51. The number π is used to find the circumference (distance around) of circles and the area of circles. Some people use $\pi = \dfrac{22}{7}$, others use $\pi = 3.14$, and still others use $\pi = 3.1416$. How close is each of these numbers to the true value of π? Which number $\left(\dfrac{22}{7}, 3.14, \text{or } 3.1416\right)$ is the closest value to π?

52. The circumference (distance around) of a circle is given by the formula $C = \pi \cdot d$ where d is the length of the diameter of the circle. A diameter is a segment joining two points on the circle and passing through the center of the circle. If the diameter of a circle is 10 centi-

meters, which will give a better approximation of the circumference, using 3.14 for π or $\frac{22}{7}$ for π? Why? What is this approximation?

SUMMARY: CHAPTER 4

DEFINITION   A **decimal number** is a rational number that has a power of ten as its denominator.

SPECIAL NOTES

A.   The *th* at the end of a word indicates a fraction part (a part to the right of the decimal point).

B.   The hyphen (-) indicates one word.

To **round off** a number means to find another number close to the original number.

RULE FOR ROUNDING OFF DECIMAL NUMBERS

1.   Look at the single digit just to the right of the digit that is in the place of desired accuracy.

2.   If this digit is 5 or greater, make the digit in the desired place of accuracy one larger and replace all digits to the right with zeros.

3.   If this digit is less than 5, leave the digit that is in the place of desired accuracy as it is and replace all digits to the right with zeros.

RULE FOR MULTIPLYING TWO DECIMAL NUMBERS

1.   Multiply the two numbers as if they were whole numbers.

2.   Count the total number of places to the right of the decimal points in both factors.

3.   This sum is the number of places to the right of the decimal point in the product.

Every **rational number** can be represented as an infinite repeating decimal.

Every **irrational number** can be represented as an infinite nonrepeating decimal.

DEFINITION   A **real number** is any number that can be written as an infinite decimal (repeating or nonrepeating).

If the repeating part of a decimal number starts at the decimal point, the formula

$$\frac{P}{10^N - 1}$$

will give the corresponding fraction.

REVIEW QUESTIONS · CHAPTER 4

1. A decimal number is a _____ that has a power of ten as its denominator.

2. (True or False?) π is a real number. Explain your answer briefly.

3. (True or False?) π is a rational number. Explain your answer briefly.

4. (True or False?) π is an irrational number. Explain your answer briefly.

5. Write 56.49 in expanded notation.

Write the following in words.

6. 4.008   7. 900.5   8. 6.5781

Write the following in decimal notation.

9. two hundred and seventeen hundredths

10. eighty-four and seventy-five thousandths

11. three thousand three and three thousandths

Round off as indicated.

12. 5863   (nearest hundred)   13. 7.629   (nearest tenth)

14. .0385   (nearest thousandth)   15. 72.997   (nearest hundredth)

Add or subtract as indicated.

16. $5.4 + 7.34 + 14.08$ 17. $3 + 7.86 + 52.891 + .4$

18. $34.967 + 40.8 + 9.451 + 8.2$ 19. $32.5 - 14.71$

20. $16.92 - 7.9$ 21. $5 - 1.0377$

Multiply.

22. 6.31(2.5) **23.** (19.2)(6.6) **24.** $10^3(4.5681)$

Divide (round off to the nearest hundredth).

25. $.52\overline{)7.8}$ **26.** $.02 \div .008$ **27.** $1000\overline{).461}$

28. Multiply $10^4(36.528)$.

29. Divide $\dfrac{76.41}{10^4}$.

30. Find the average (to the nearest tenth) of 62.5, 28.4, and 31.29.

Write each of the following rational numbers as infinite repeating decimals.

31. $\dfrac{1}{6}$ **32.** $\dfrac{2}{7}$ **33.** $\dfrac{3}{5}$ **34.** $\dfrac{5}{11}$ **35.** $\dfrac{17}{13}$

Write each infinite repeating decimal in fraction form.

36. $.\overline{1}$ **37.** $.\overline{32}$ **38.** $.5\overline{05}$ **39.** $.\overline{27}$ **40.** $.\overline{321}$

41. Truncate 7.662983516 . . . at the hundred-thousandths place.

42. A motorcycle averages 42.8 miles per gallon of gas. How many miles can the motorcycle travel on 3.5 gallons of gas?

43. Find the difference between sixty-four and five hundred thirty-six ten-thousandths; and fifty-nine and three thousand, six hundred eighty-one ten-thousandths.

44. On a certain map, 1 inch represents 35.5 miles. What distance is represented by 4.7 inches?

45. If a board 16.5 feet long costs $3.57, what is the cost of 1 foot of the board?

46. A part-time clerk earned $220 the first month on a new job. This was .4 of what he had anticipated. What had he thought he would make in a month? If he continued at the current rate, what would he make in a year? How much does this differ from what he had hoped to make in a year?

RATIO AND PROPORTION

5.1   RATIO

(1) FRACTION
(2) DIVISION
(3) RATIO

A rational number such as $\frac{3}{4}$ has been shown to have two meanings: one, to represent so many equal parts of a whole; and two, to mean division. A third meaning is to indicate a comparison of two quantities.

DEFINITION   A **ratio** is a comparison of two quantities.

WHAT DOES

RATIO $\frac{5}{2}$ MEAN

If the odds on a horse race are $\frac{5}{2}$ (also written 5 to 2 and 5:2), then, for every $2 bet, $5 may be won. Thus, $\frac{5}{2}$ represents a comparison of money won to money bet. In such a situation, $\frac{5}{2}$ does *not* mean 5 of 2 equal parts, and it does *not* mean division.

The order of the numbers is, of course, very important. If the odds in the race are $\frac{2}{5}$, then only $2 can be won on a $5 bet — a completely different situation from odds of $\frac{5}{2}$.

Batting averages of major league players are in the newspapers every day during baseball season. These are ratios of hits to times at bat. A player with an average of .300 gets 3 hits every 10 times at bat; the ratio of hits to times at bat is $\frac{3}{10}$.

$\frac{HITS}{TIMES \ AT \ BAT}$

In a ratio, the units of the numerator and denominator should be labeled or otherwise explained to avoid confusion. A ratio of 1 foot to 3 yards, or

$$\frac{1 \text{ foot}}{3 \text{ yards}}$$

$\frac{1 \ INCH}{50 \ MILES}$ what about

is not wrong, but it can be misleading. Using the same units, say feet, gives a more meaningful comparison. Thus, since 9 feet is the same as 3 yards, the ratio 1 foot to 3 yards is more meaningful as 1 foot to 9 feet, or

WORK SEVERAL
EVEN IN CLASS

$$\frac{1 \text{ foot}}{9 \text{ feet}} \quad \text{or} \quad \frac{1}{9}$$

NOTE   *Common units should be used in the numerator and denominator of a ratio whenever possible.*

ASSIGN ODD

EXERCISES 5.1

1-29

Write the following comparisons as ratios reduced to lowest terms. Be sure to use common units for the numerator and denominator whenever possible.

WORK EVEN IN CLASS.

1. 14 red balls to 13 white balls

2. 6 oranges to 5 apples

3. one dime to two nickels

4. one quarter to five cents

5. two one-dollar bills to one five-dollar bill

6. 30 chairs to 25 students

7. 25 students to 30 chairs

8. 7 girls to 15 boys

9. 15 boys to 7 girls

10. 16 children to 6 families

11. 20 children to 5 families

12. 1000 television sets to 900 homes

13. 3500 homes to 3500 refrigerators

14. 75 gold medals to 75 Olympic events

15. 2 yards to 5 feet

16. 18 inches to 2 feet

17. five basketball games won to seven basketball games played

18. four football games won to three football games lost

19. 75 hits to 200 times at bat

20. 16 home runs to 300 times at bat

21. earnings of $56 to investments totaling $400

22. expenses of $5240 to sales of $7336

23. 20 shots made to 40 shots attempted in a basketball game

24. three book shelves to 18 feet of lumber

25. 360° for every complete rotation of the hour hand on a clock

26. 18 miles to 1 gallon of gas

27. 10 centimeters to 1 decimeter

28. 10 decimeters to 1 meter

29. 100 centimeters to 1 meter

30. $500 to four men's suits

5.2   PROPORTION

The statement $\frac{3}{5} = \frac{1}{4}$ is an equation because it says that two numbers are equal. However, it is a *false* equation. Naturally, we are more interested in true equations. How can we tell if an equation such as $\frac{a}{b} = \frac{c}{d}$ is true or false?

DEFINITION

Go Over

A **proportion** is a statement that two ratios are equal. In symbols, $\frac{a}{b} = \frac{c}{d}$ is a proportion. [NOTE: A proportion is an equation, and $\frac{a}{b}$ and $\frac{c}{d}$ are ratios.

In a proportion, the numbers are called **terms** and are named as follows:

Go Over

$$\begin{array}{l} \text{first term} \rightarrow \\ \text{second term} \rightarrow \end{array} \frac{a}{b} = \frac{c}{d} \begin{array}{l} \leftarrow \text{ third term} \\ \leftarrow \text{ fourth term} \end{array}$$

a and d are called the **extremes.** b and c are called the **means.**
  A proportion is true if the *product of the extremes equals the product of the means;* that is,

a : b = c : d

From First Definition

$$\frac{a}{b} = \frac{c}{d} \quad \text{if and only if} \quad a \cdot d = b \cdot c \quad \text{where } b \neq 0 \text{ and } d \neq 0$$

Cannot have denominators equal to zero

Now we can tell $\frac{3}{5} = \frac{1}{4}$ is false because $3 \cdot 4 \neq 5 \cdot 1$.

  Consider the true proportion $\frac{3}{5} = \frac{6}{10}$. Three other true proportions may be written as follows using the same four numbers:

Do not stress extremes or means

$$\frac{10}{5} = \frac{6}{3} \qquad \text{The extremes 3 and 10 are exchanged.}$$

$$\frac{3}{6} = \frac{5}{10} \qquad \text{The means 5 and 6 are exchanged.}$$

$$\frac{10}{6} = \frac{5}{3} \qquad \text{The means and extremes are exchanged.}$$

Show cross products are the same

Work several problems

All four proportions are true because $3 \cdot 10 = 5 \cdot 6$. In all four proportions, the extremes are 3 and 10, and the means are 5 and 6.

EXAMPLES

1. $\dfrac{9}{13} = \dfrac{4.5}{6.5}$ is true because $9(6.5) = 58.5$, and $13(4.5) = 58.5$.

2. $\dfrac{3}{7} = \dfrac{6}{15}$ is false because $3 \cdot 15 = 45$, $7 \cdot 6 = 42$, and $45 \neq 42$.

3. $\dfrac{12}{54} = \dfrac{14}{63}$ Reducing each ratio helps here.

$\dfrac{12}{54} = \dfrac{\cancel{6} \cdot 2}{\cancel{6} \cdot 9} = \dfrac{2}{9}$ and $\dfrac{14}{63} = \dfrac{\cancel{7} \cdot 2}{\cancel{7} \cdot 9} = \dfrac{2}{9}$ so $\dfrac{12}{54} = \dfrac{14}{63}$ is true.

SELF QUIZ   Determine whether the following proportions are true or false.

ANSWERS

1. $\dfrac{6}{21} = \dfrac{10}{35}$

1. True: $6 \cdot 35 = 21 \cdot 10$

2. $\dfrac{5\frac{1}{2}}{2\frac{3}{4}} = \dfrac{1\frac{1}{10}}{\frac{11}{20}}$

2. True: $\dfrac{11}{2} \cdot \dfrac{11}{20} = \dfrac{11}{4} \cdot \dfrac{11}{10}$

ASSIGN: ODD 1-29

EXERCISES 5.2

Determine whether the following proportions are true or false.

WORK IN
EVEN IN
CLASS
2-20

1. $\dfrac{5}{6} = \dfrac{10}{12}$ 2. $\dfrac{2}{7} = \dfrac{5}{17}$ 3. $\dfrac{7}{21} = \dfrac{4}{12}$ 4. $\dfrac{6}{15} = \dfrac{2}{5}$

5. $\dfrac{5}{8} = \dfrac{12}{17}$ 6. $\dfrac{12}{15} = \dfrac{20}{25}$ 7. $\dfrac{5}{3} = \dfrac{15}{9}$ 8. $\dfrac{6}{8} = \dfrac{15}{20}$

9. $\dfrac{2}{5} = \dfrac{4}{10}$ 10. $\dfrac{3}{5} = \dfrac{60}{100}$ 11. $\dfrac{125}{1000} = \dfrac{1}{8}$ 12. $\dfrac{3}{8} = \dfrac{375}{1000}$

13. $\dfrac{1}{4} = \dfrac{25}{100}$ 14. $\dfrac{7}{8} = \dfrac{875}{1000}$ 15. $\dfrac{3}{16} = \dfrac{9}{48}$ 16. $\dfrac{2}{3} = \dfrac{66}{100}$

17. $\dfrac{1}{3} = \dfrac{33}{100}$ 18. $\dfrac{14}{6} = \dfrac{21}{8}$ 19. $\dfrac{4}{9} = \dfrac{7}{12}$ 20. $\dfrac{19}{16} = \dfrac{20}{17}$

For each proportion below, write three other proportions using the same four numbers. Show that each of these is true if the original proportion is true and false if the original proportion is false.

21. $\dfrac{3}{6} = \dfrac{4}{8}$ 22. $\dfrac{12}{18} = \dfrac{14}{21}$ 23. $\dfrac{5}{6} = \dfrac{7}{8}$ 24. $\dfrac{7.5}{10} = \dfrac{3}{4}$

25. $\dfrac{6.2}{3.1} = \dfrac{10.2}{5.1}$ **26.** $\dfrac{8\frac{1}{2}}{2\frac{1}{3}} = \dfrac{4\frac{1}{4}}{1\frac{1}{6}}$ **27.** $\dfrac{6\frac{1}{5}}{1\frac{1}{7}} = \dfrac{3\frac{1}{10}}{\frac{8}{14}}$ **28.** $\dfrac{6}{24} = \dfrac{10}{48}$

29. $\dfrac{7}{16} = \dfrac{3\frac{1}{2}}{8}$ **30.** $\dfrac{10}{17} = \dfrac{5}{8\frac{1}{2}}$

5.3   FINDING THE UNKNOWN TERM IN A PROPORTION

Many problems can be solved with the use of proportions. In such problems, three terms of a proportion will be known, and the problem will be solved when the fourth term is found. The purpose of this section is to develop the skills necessary to find unknown terms in a proportion.

EXAMPLES

1. Let x represent the unknown term and suppose $\dfrac{2}{4} = \dfrac{5}{x}$.

If $\dfrac{2}{4} = \dfrac{5}{x}$ is true, then

$2 \cdot x = 4 \cdot 5$ is true.   The product of the extremes must equal the product of the means.

$\dfrac{2 \cdot x}{2} = \dfrac{20}{2}$ Divide both sides of the equation by 2, the number next to the variable x. (This number is called the **coefficient** of x.)

$\dfrac{\overset{1}{\cancel{2}} \cdot x}{\cancel{2}} = \dfrac{\overset{10}{\cancel{20}}}{\cancel{2}}$ Simplify by reducing both sides.

$x = 10$

Check:
Substitute 10 for x in the proportion.

$\dfrac{2}{4} = \dfrac{5}{10}$ is true because $2 \cdot 10 = 5 \cdot 4$.

2. Find z if $\dfrac{4}{z} = \dfrac{3}{8}$.

$\dfrac{4}{z} = \dfrac{3}{8}$

$4 \cdot 8 = 3 \cdot z$ The product of the extremes must equal the product of the means.

$\dfrac{32}{3} = \dfrac{\cancel{3} \cdot z}{\cancel{3}}$ Divide both sides of the equation by 3, the coefficient of z.

$$\frac{32}{3} = 1 \cdot z \qquad \text{Reduce.}$$

$$\frac{32}{3} = z \quad \text{or} \quad 10\frac{2}{3} = z \qquad \text{Check:}$$

Substitute $\frac{32}{3}$ for z in the proportion.

$$\frac{4}{\frac{32}{3}} = \frac{3}{8} \text{ since } 4 \cdot 8 = 32, \text{ and } \frac{32}{3} \cdot 3 = 32.$$

3. Find y if $\frac{6}{24} = \frac{y}{72}$.

$$\frac{6}{24} = \frac{y}{72}$$

$$\frac{1}{4} = \frac{y}{72} \qquad \text{Reduce } \frac{6}{24} \text{ to simplify the procedure.}$$

$$1 \cdot 72 = 4 \cdot y$$

$$\frac{72}{4} = \frac{\cancel{4} \cdot y}{\cancel{4}}$$

$$18 = y \qquad \text{Check:}$$

$$\frac{6}{24} = \frac{18}{72} \text{ since } 6 \cdot 72 = 432, \text{ and } 18 \cdot 24 = 432.$$

4. Find w if $\frac{5}{3\frac{1}{2}} = \frac{w}{21}$.

$$\frac{5}{3\frac{1}{2}} = \frac{w}{21}$$

$$5 \cdot 21 = 3\frac{1}{2} \cdot w \qquad \text{The product of the extremes must equal the product of the means.}$$

$$105 = \frac{7}{2} \cdot w$$

$$\frac{2}{7} \cdot 105 = \frac{2}{7} \cdot \frac{7}{2} \cdot w \qquad \left\{ \begin{array}{l} \text{Multiply both sides by the} \\ \text{reciprocal of the coefficient of } w. \\ \text{Multiplying by } \frac{2}{7} \text{ is the same as} \\ \text{dividing by } \frac{7}{2}. \text{ That is, } \frac{105}{\frac{7}{2}} = 105 \cdot \frac{2}{7}. \end{array} \right.$$

$$\frac{2}{7} \cdot 7 \cdot 15 = 1 \cdot w$$

$$30 = w$$

Check:

$$\frac{5}{3\frac{1}{2}} = \frac{30}{21} \text{ since } 5 \cdot 21 = 105, \text{ and } 30 \cdot \frac{7}{2} = 105.$$

SELF QUIZ   Find the unknown term in each proportion that will   ANSWERS
make the proportion true.

1. $\dfrac{5}{7} = \dfrac{x}{49}$ 1. $x = 35$

2. $\dfrac{3}{5} = \dfrac{R}{100}$ 2. $R = 60$

3. $\dfrac{2\frac{1}{2}}{6} = \dfrac{1\frac{1}{2}}{y}$ 3. $y = \dfrac{18}{5}$

EXERCISES 5.3

Odd 1-49

Find the unknown term in each proportion that will make the proportion true.

1. $\dfrac{4}{10} = \dfrac{5}{x}$ 2. $\dfrac{3}{6} = \dfrac{6}{x}$ 3. $\dfrac{7}{21} = \dfrac{y}{6}$ 4. $\dfrac{5}{7} = \dfrac{z}{28}$

5. $\dfrac{8}{z} = \dfrac{6}{20}$ 6. $\dfrac{7}{x} = \dfrac{5}{15}$ 7. $\dfrac{10}{y} = \dfrac{15}{48}$ 8. $\dfrac{6}{9} = \dfrac{14}{x}$

9. $\dfrac{x}{4} = \dfrac{12}{16}$ 10. $\dfrac{y}{3} = \dfrac{7}{2}$ 11. $\dfrac{x}{5} = \dfrac{2}{3}$ 12. $\dfrac{7}{10} = \dfrac{35}{z}$

13. $\dfrac{1}{8} = \dfrac{1\frac{1}{4}}{y}$ 14. $\dfrac{3}{8} = \dfrac{y}{100}$ 15. $\dfrac{1}{6} = \dfrac{x}{100}$ 16. $\dfrac{1}{6} = \dfrac{16\frac{2}{3}}{y}$

17. $\dfrac{x}{4} = \dfrac{1\frac{1}{4}}{5}$ 18. $\dfrac{1}{3} = \dfrac{x}{100}$ 19. $\dfrac{2}{5} = \dfrac{y}{100}$ 20. $\dfrac{z}{100} = \dfrac{1}{20}$

21. $\dfrac{1}{2} = \dfrac{x}{100}$ 22. $\dfrac{3}{5} = \dfrac{60}{z}$ 23. $\dfrac{1}{200} = \dfrac{y}{100}$ 24. $\dfrac{x}{15} = \dfrac{1}{15}$

25. $\dfrac{3}{16} = \dfrac{9}{x}$ 26. $\dfrac{4}{25} = \dfrac{x}{125}$ 27. $\dfrac{9}{x} = \dfrac{4\frac{1}{2}}{11}$ 28. $\dfrac{y}{6} = \dfrac{2\frac{1}{2}}{12}$

29. $\dfrac{36}{z} = \dfrac{40}{15}$ 30. $\dfrac{45}{63} = \dfrac{x}{7}$ 31. $\dfrac{A}{4} = \dfrac{75}{100}$ 32. $\dfrac{15}{B} = \dfrac{3}{100}$

33. $\dfrac{1}{8} = \dfrac{R}{100}$ 34. $\dfrac{A}{30} = \dfrac{33\frac{1}{3}}{100}$ 35. $\dfrac{20}{B} = \dfrac{400}{100}$ 36. $\dfrac{3}{8} = \dfrac{R}{100}$

37. $\dfrac{R}{100} = \dfrac{5}{4}$ 38. $\dfrac{91}{B} = \dfrac{7}{100}$ 39. $\dfrac{A}{300} = \dfrac{15}{100}$ 40. $\dfrac{A}{50} = \dfrac{50}{100}$

41. $\dfrac{6.2}{5} = \dfrac{x}{15}$ **42.** $\dfrac{750}{2000} = \dfrac{R}{100}$ **43.** $\dfrac{860}{P} = \dfrac{2}{100}$ **44.** $\dfrac{I}{2000} = \dfrac{2\frac{1}{2}}{100}$

45. $\dfrac{I}{2000} = \dfrac{9}{100}$ **46.** $\dfrac{25}{\pi} = \dfrac{75}{x}$ **47.** $\dfrac{16}{1} = \dfrac{C}{\pi}$ **48.** $\dfrac{\pi}{C} = \dfrac{1}{3}$

49. $\dfrac{72}{2.4} = \dfrac{10}{y}$ **50.** $\dfrac{w}{360} = \dfrac{\pi}{2 \cdot \pi}$

5.4 SOLVING WORD PROBLEMS USING PROPORTIONS

New tires advertised in the newspaper were two for $52.50. A man who preferred this particular type of tire needed five for his car, including a new spare tire. What would he pay for new tires?

This type of problem is quite common and is easily solved using proportions. In setting up a proportion to solve the problem, one term is unknown, in this case the price of five new tires. Any letter may be used to represent this unknown term. The ratios in the proportion should follow one of two patterns.

1.   The terms of each individual ratio may be in the same unit, but the numerators must correspond, and the denominators must correspond.

$$\dfrac{2 \text{ tires}}{5 \text{ tires}} = \dfrac{\$52.50}{\$x}$$ (corresponding price for 2 tires)
(corresponding price for 5 tires)

One ratio is tires to tires, and the other ratio is dollars to dollars.

2.   The terms of each individual ratio may be in different units, but the terms must be corresponding, and the other ratio must have the units in the same order.

$$\dfrac{2 \text{ tires}}{\$52.50} = \dfrac{5 \text{ tires}}{\$x}$$ The price in the denominator corresponds to the number of tires in the numerator.

Each ratio is tires to dollars.

Solving either proportion gives the price of five tires.

1. $\dfrac{2}{5} = \dfrac{52.50}{x}$ 2. $\dfrac{2}{52.50} = \dfrac{5}{x}$

$2 \cdot x = 262.50$ $2 \cdot x = 262.50$

$\dfrac{2 \cdot x}{2} = \dfrac{262.50}{2}$ $\dfrac{2 \cdot x}{2} = \dfrac{262.50}{2}$

$x = \$131.25$ $x = \$131.25$

Thus, using either proportion, we find that five tires cost $131.25. Try another problem.

An architect draws house plans using a scale of $\frac{1}{2}$ inch to represent 10 feet. How many inches would represent 36 feet?

Solution:

Either of the following two proportions will give the solution.

$$\frac{\frac{1}{2}\text{ inch}}{10\text{ feet}} = \frac{y\text{ inches}}{36\text{ feet}} \qquad \text{or} \qquad \frac{\frac{1}{2}\text{ inch}}{y\text{ inches}} = \frac{10\text{ feet}}{36\text{ feet}}$$

$$\frac{1}{2}\cdot 36 = 10\cdot y \qquad\qquad\qquad \frac{1}{2}\cdot 36 = 10\cdot y$$

$$\frac{18}{10} = \frac{10\cdot y}{10} \qquad\qquad\qquad \frac{18}{10} = \frac{10\cdot y}{10}$$

$$\frac{9}{2} = y \qquad\qquad\qquad\qquad \frac{9}{2} = y$$

So, $\frac{9}{2}$ inches (or $4\frac{1}{2}$ inches) represents 36 feet.

EXERCISES 5.4

1. If 6 white shirts sell for $25, what would be the cost of 10 white shirts?

2. If a dozen oranges cost 70¢, what would $1\frac{1}{2}$ dozen cost?

3. Suppose gasoline sells for 60¢ per gallon. How many gallons could be bought for $558?

4. If the price of a certain type of cloth is $1.75 per yard, how many yards could a dressmaker buy with $35?

5. A company buys cars by the fleet every six months. At the first of the year, it bought 25 cars at a total price of $88,500. If the price per car was the same in July, what did it pay for 30 cars then? What was the price per car?

6. The Slo-Motion Construction Co. needs 35 men to build 10 houses in a certain amount of time. How many men would it need to build 18 homes in the same amount of time?

7. On a scale drawing on an 8 inch by 11 inch piece of paper, five centimeters represents $1\frac{1}{2}$ miles. How many miles does seven centimeters on the drawing represent?

8. A building 14 stories high casts a shadow of 30 feet at a certain time of day. What would be the length of the shadow of a 20-story building at the same time of day in the same city?

9. If $\frac{1}{2}$ inch represents 10 miles on a map, how many inches would represent 25 miles?

10. An architect drew plans for a city park using a scale of $\frac{1}{4}$ inch to represent 25 feet. How many feet would 2 inches represent?

11. Two numbers are in the ratio of 5 to 2. The number eight is in that same ratio with a fourth number. What is the fourth number?

12. If a machine processes 300 sheets of paper in 90 seconds, how many sheets will it process in 2 minutes?

13. If 2 units of a gas weigh 175 grams, what is the weight of 5 units of the same gas?

14. An old car burns $1\frac{1}{2}$ quarts of oil when it is driven 1500 miles. How many quarts of oil will it burn in 4000 miles?

15. A salesman drove 900 miles in two weeks. How far could he expect to drive in fifty-two weeks?

16. A supermarket sells three cans of beans for $1.00. What would be the charge for two cans of beans?

17. A fifteen-year-old student typist took a timed test and typed 105 words in three minutes. How many words were typed in twelve seconds? What was the rate per minute?

18. An artist, working hard, figures he can do three paintings every two weeks. How long would he take to do 18 paintings?

19. The Plenty-Good Candy Co. opened five stores in one year and showed profits of $175,000 that year. How many stores would be needed to produce a yearly profit of one million dollars for the company?

20. A mixture of antifreeze and water is made so that $\frac{1}{2}$ gallon of anti-freeze is mixed with 3 gallons of water. How many gallons of water would be mixed with 2 gallons of antifreeze? What would be the minimum capacity of the container for this mixture?

21. An investor thinks she should make $6 for every $100 she invests. How much would she expect to make on an investment of $7000?

22. A Little League baseball team bought 15 bats for $37.50. What would it have paid for 10 bats?

23. An electric fan makes 180 revolutions per minute. How many revolutions will the fan make if it runs for 24 hours?

24. Playing the Hilly Golf Club course three times, Mr. Michael used his driver 42 times, his putter 102 times, and hit 12 shots out of sand traps. How many drives, putts, and sand shots would he hit if he played the course 10 times?

25. Driving steadily, a woman made a trip of 200 miles in $4\frac{1}{2}$ hours. How long would she take to drive 500 miles at the same rate of speed?

26. Property valued at $70,000 is insured with Company A for $30,000 and with Company B for $40,000. A fire caused damage of $28,000. What amount will be paid by each company if each pays a proportional amount?

27. A man earned $72 in one year on a $1000 investment. What would he have earned if his investment had been $4525?

28. The Sloppy Paint Store had expenses totaling $20,000 for six months and sales of $28,000 for the same period. If the store is to show the same ratio of expenses to sales, what would the sales have to be if expenses are $24,000 for the next six-month period?

29. Find two numbers that are in the same ratio as $3:2$.

30. There are one hundred centimeters in one meter and ten millimeters in one centimeter. How many millimeters are in seven meters?

31. There are one thousand grams in one kilogram. How many grams are in five and three tenths kilograms?

32. Suppose you bet $2 on a horse race and you won. If the teller paid you $10 at the pay window, what were the odds? If the teller paid you $3, what were the odds?

33. If a 21-year-old typist can type eight pages of a manuscript in 56 minutes, how many pages can be typed in 2 hours? How long would it take for a complete manuscript of 300 pages to be typed?

34. If gasoline costs $62\frac{9}{10}$¢ per gallon, what does it cost to fill a 21-gallon gas tank?

35. Three astronauts have traveled 80,000 miles on their mission. They figure this is three fifths of the total distance they will travel. What is the total distance of the mission?

SUMMARY: CHAPTER 5

DEFINITION   A **ratio** is a comparison of two quantities.

DEFINITION   A **proportion** is a statement that two ratios are equal. In symbols, $\frac{a}{b} = \frac{c}{d}$ is a proportion. [NOTE: A proportion is an equation, and $\frac{a}{b}$ and $\frac{c}{d}$ are ratios.]

In a proportion, the numbers are called **terms** and are named as follows:

$$\text{first term} \rightarrow \frac{a}{b} = \frac{c}{d} \leftarrow \text{third term}$$
$$\text{second term} \rightarrow \quad\quad \leftarrow \text{fourth term}$$

a and d are called the **extremes.** b and c are called the **means.**

$$\frac{a}{b} = \frac{c}{d} \quad \text{if and only if} \quad a \cdot d = b \cdot c \quad \text{where } b \neq 0 \text{ and } d \neq 0$$

REVIEW QUESTIONS · CHAPTER 5

1. State three different meanings for the rational number $\frac{5}{8}$.

2. A proportion is a statement that two _____ are _____.

3. A proportion has _____ terms.

Write the following comparisons as ratios reduced to lowest terms.

4. 75 students to 80 chairs in a classroom

5. 20 inches to three feet

6. earnings of $500 to investments of $5500

Determine whether each of the following proportions is true or false. If a proportion is true, write three other true proportions using the same four numbers.

7. $\dfrac{4}{12} = \dfrac{7}{21}$ 8. $\dfrac{2}{3} = \dfrac{66}{100}$ 9. $\dfrac{4}{5} = \dfrac{80}{100}$ 10. $\dfrac{9}{32} = \dfrac{4\frac{1}{2}}{16}$

Find the unknown term in each of the following proportions.

11. $\dfrac{45}{9} = \dfrac{x}{5}$ 12. $\dfrac{y}{72} = \dfrac{10}{36}$ 13. $\dfrac{7\frac{1}{8}}{1} = \dfrac{x}{8}$ 14. $\dfrac{52}{26} = \dfrac{39}{z}$

15. If a machine produces 5000 hairpins in 2 hours, how many will it produce in two 8-hour days?

16. An automobile was slowing down at the rate of 5 miles per hour (mph) for every 3 seconds. If the automobile was going 65 mph when it began to slow down, how fast was it going at the end of 12 seconds?

17. If apples sell for 39¢ a pound, what would be the cost of 5 pounds of apples?

18. If five pizzas cost $9.25, what would three pizzas cost?

19. To buy a house for $56,500, a down payment of $11,300 is needed. What down payment is needed to buy another house for $45,600?

20. On a map, $\frac{3}{4}$ inch represents 30 miles. What is the actual distance between two cities that are marked 5 inches apart on the map?

PERCENT

6

6.1   UNDERSTANDING PERCENT

Suppose two men make investments in the stock market, and we would like to know which man made the better investment. We need some way to compare the two investments. A common way is to compare the ratios of amounts of money made to amounts invested. The amount of money made is called the **interest** on the investment.

If the men invest the same amount of money, say $1000, then the better investment is obviously the one that makes more interest. If Man 1 makes $200 and Man 2 makes $300, then the ratios of interest to amount invested are $\dfrac{200}{1000}$ and $\dfrac{300}{1000}$, respectively. The denominators are both 1000. Therefore, since 300 is greater than 200, Man 2 made the better investment.

The problem is more difficult if the amounts invested are not the same. If Man 1 invests $500 and makes $200, and Man 2 invests $300 and makes $150, we see the ratios of interest to amount invested are $\dfrac{200}{500}$ and $\dfrac{150}{300}$. Man 1 made more money since $200 is greater than $150, but this does not tell us which investment is better since the amounts invested are different. Reducing the two ratios gives us

$$\frac{200}{500} = \frac{2 \cdot 100}{5 \cdot 100} = \frac{2}{5} \quad \text{and} \quad \frac{150}{300} = \frac{1 \cdot 150}{2 \cdot 150} = \frac{1}{2}$$

Now, a common denominator for $\dfrac{2}{5}$ and $\dfrac{1}{2}$ is 10; however, for our purposes in this chapter, we will always make the common denominator 100. This may not be the simplest procedure, but it gives a sort of standard comparison for all fractions. So,

$$\frac{2}{5} = \frac{2 \cdot 20}{5 \cdot 20} = \frac{40}{100} \quad \text{and} \quad \frac{1}{2} = \frac{1 \cdot 50}{2 \cdot 50} = \frac{50}{100}$$

Therefore, Man 2 made the better investment. He made $50 for every $100 he invested, and Man 1 made $40 for every $100 he invested. We say Man 2 made 50 percent on his investment, and Man 1 made 40 percent on his investment.

The word **percent** comes from the Latin *per centum* meaning "per hundred." So, **percent** is simply a ratio of some number to 100. The following analogy might be helpful. There are 100 pennies, or cents, in a dollar; therefore, one cent is $\dfrac{1}{100}$ of a dollar. Likewise, one percent is the ratio $\dfrac{1}{100}$. We can also think of the percent sign, %, as being a rearrangement of the digits of 100 by putting the 1 between the two 0's and raising the left 0 a little and lowering the right 0 a little, giving $^01_0$, which readily leads us to % as the sign for percent. In other words, percent is just another way of discussing fractions with denominators of 100; or, **percent means hundredths.**

EXAMPLES

1. $\dfrac{50}{100} = 50\%$

2. $\dfrac{60}{100} = 60\%$

3. $\dfrac{16\frac{2}{3}}{100} = 16\frac{2}{3}\%$

4. $\dfrac{25}{100} = 25\%$

5. $\dfrac{17.51}{100} = 17.51\%$

6. $\dfrac{.32}{100} = .32\%$

EXERCISES 6.1

Change the following fractions to percents.

1. $\dfrac{30}{100}$

2. $\dfrac{16}{100}$

3. $\dfrac{96}{100}$

4. $\dfrac{98}{100}$

5. $\dfrac{2\frac{1}{2}}{100}$

6. $\dfrac{53}{100}$

7. $\dfrac{1.8}{100}$

8. $\dfrac{2.6}{100}$

9. $\dfrac{73}{100}$

10. $\dfrac{12.5}{100}$

11. $\dfrac{37.5}{100}$

12. $\dfrac{42}{100}$

13. $\dfrac{5\frac{1}{3}}{100}$

14. $\dfrac{.6}{100}$

15. $\dfrac{.5}{100}$

16. $\dfrac{39}{100}$

17. $\dfrac{\frac{1}{2}}{100}$

18. $\dfrac{\frac{1}{4}}{100}$

19. $\dfrac{\frac{2}{3}}{100}$

20. $\dfrac{50}{100}$

21. $\dfrac{132}{100}$

22. $\dfrac{500}{100}$

23. $\dfrac{625}{100}$

24. $\dfrac{69}{100}$

25. $\dfrac{95}{100}$

26. $\dfrac{148}{100}$

27. $\dfrac{950}{100}$

28. $\dfrac{120}{100}$

29. $\dfrac{1000}{100}$

30. $\dfrac{1500}{100}$

6.2   DECIMALS AND PERCENTS

CHANGE DECIMALS TO PERCENT

In Section 6.1 we said *percent means hundredths,* and all the fractions had denominators of 100. Thus,

$$\frac{50}{100} = 50\% \quad \text{and} \quad \frac{183}{100} = 183\%$$

But decimals are simply fractions with powers of ten as denominators. So, the relationship between decimals and percent should be easy. How would you change .42 to percent? or 1.25 to percent? Did you think

$$.42 = \frac{42}{100} = 42\% \qquad \text{(.42 is read ``42 hundredths'' so } .42 = \frac{42}{100}$$
$$\text{is natural.)}$$

and $\qquad 1.25 = 1\frac{25}{100} = \frac{125}{100} = 125\%$

What about changing .832 to percent? The following technique is useful.

$$.832 = \frac{100}{100}\left(\frac{.832}{1}\right) = \frac{83.2}{100} = 83.2\%$$

Moving the decimal over two places corresponds to multiplication by 100. Writing the % sign corresponds to dividing by 100. We have the following rule.

To change a decimal to a percent, move the decimal point two places to the right and write the % sign.

EXAMPLES

1. $.2 = 20\%$

2. $1.75 = 175\%$

3. $.005 = .5\%$

4. $.01 = 1\%$

5. $.02\frac{1}{3} = 2\frac{1}{3}\%$

The relationship between pennies and dollars is a good illustration of percent.

$$200 \text{ pennies} = \$2.00 = 200\% \text{ of a dollar}$$
$$100 \text{ pennies} = \$1.00 = 100\% \text{ of a dollar}$$
$$50 \text{ pennies} = \$\ .50 = 50\% \text{ of a dollar}$$
$$25 \text{ pennies} = \$\ .25 = 25\% \text{ of a dollar}$$
$$1 \text{ penny} \ \ = \$\ .01 = 1\% \text{ of a dollar}$$
$$\frac{1}{2} \text{ penny} \ \ = \$\ .005 = .5\% \text{ of a dollar}$$

Now consider changing a percent to a decimal. How would you

change 36% to a decimal? Simply reverse the procedure for changing decimals to percents.

$$36\% = .36$$

Move the decimal point two places to the left and drop the % sign.

> **To change a percent to a decimal,** move the decimal point two places to the left and drop the % sign.

EXAMPLES

1. $42\% = .42$

2. $52.3\% = .523$

3. $23\frac{1}{4}\% = .23\frac{1}{4}$ or $.2325$

4. $425\% = 4.25$

5. $.25\% = .0025$

SELF QUIZ	Change from decimals to percents.	ANSWERS
	1. .57	1. 57%
	2. 2.35	2. 235%
	Change from percents to decimals.	
	3. 25%	3. .25
	4. 1.4%	4. .014

EXERCISES 6.2

Change the following decimals to percents.

1. .02	**2.** .09	**3.** .1	**4.** .52	**5.** .36	**6.** .7
7. .5	**8.** .98	**9.** .05	**10.** .016	**11.** .005	**12.** .25
13. .025	**14.** .0025	**15.** .175	**16.** .235	**17.** .476	**18.** .045
19. 1.25	**20.** 1.08	**21.** 1.002	**22.** 2.1	**23.** 5.6	**24.** 7.4
25. 4.50	**26.** 7	**27.** 23	**28.** 14	**29.** 1	**30.** 10

Do Min Even / Class Assign odd 1-69

Change the following percents to decimals.

31. 10%	**32.** 3%	**33.** 6%	**34.** 27%	**35.** 32%
36. 57%	**37.** 60%	**38.** 70%	**39.** 12%	**40.** 9%
41. .6%	**42.** 1.3%	**43.** 59.1%	**44.** 72.8%	**45.** 16.9%
46. 175%	**47.** 200%	**48.** 134%	**49.** 125.2%	**50.** 100%

51. $3\frac{1}{2}\%$ **52.** $5\frac{1}{4}\%$ **53.** $7\frac{1}{4}\%$ **54.** $1\frac{1}{2}\%$ **55.** $2\frac{1}{8}\%$

56. $17\frac{1}{3}\%$ **57.** $25\frac{2}{3}\%$ **58.** $50\frac{1}{6}\%$ **59.** $35\frac{3}{8}\%$ **60.** $70\frac{1}{5}\%$

61. $16\frac{3}{5}\%$ **62.** $10\frac{1}{20}\%$ **63.** $15\frac{1}{10}\%$ **64.** $14\frac{1}{7}\%$ **65.** $63\frac{1}{9}\%$

66. 4900%   **67.** 5132%   **68.** $6333\frac{1}{3}\%$ **69.** $72\frac{2}{5}\%$

70. $6514\frac{3}{100}\%$

6.3   CHANGING FRACTIONS TO PERCENTS

Since *percent means hundredths,* changing fractions with denominator 100 to percents is easy. We did this in Section 6.1. But, what if the denominator is not 100? How would you change $\frac{3}{4}$ or $\frac{5}{8}$ to a percent? One way would be to change the number to a decimal by dividing, then change to a percent. For example,

$$\frac{3}{4} = .75 = 75\%$$

$$\begin{array}{r} .75 \\ 4\overline{)3.00} \\ 2\,8 \\ \hline 20 \\ 20 \\ \hline 0 \end{array}$$

$$\frac{5}{8} = .625 = 62.5\%$$

$$\begin{array}{r} .625 \\ 8\overline{)5.000} \\ 4\,8 \\ \hline 20 \\ 16 \\ \hline 40 \\ 40 \\ \hline 0 \end{array}$$

Another way is to use proportions because percent means hundredths.

$$\frac{3}{4} = \frac{x}{100} \qquad\qquad \frac{5}{8} = \frac{x}{100}$$

$$300 = 4x \qquad\qquad 500 = 8x$$

$$\frac{300}{4} = \frac{4x}{4} \qquad\qquad \frac{500}{8} = \frac{8x}{8}$$

$$75 = x \qquad\qquad 62\frac{1}{2} = x$$

So, $\dfrac{3}{4} = \dfrac{75}{100} = 75\%$ So, $\dfrac{5}{8} = \dfrac{62\frac{1}{2}}{100} = 62\frac{1}{2}\%$ or 62.5%

EXAMPLES   Change the following fractions to percents. Use either the decimal method or the proportion method.

Decimal Method   *Proportion Method*

1. $\dfrac{3}{10}$

$$\frac{3}{10} = .3 = 30\% \qquad\qquad \frac{3}{10} = \frac{x}{100}$$

$$\frac{300}{10} = \frac{10x}{10} \qquad 30 = x$$

So, $\dfrac{3}{10} = \dfrac{30}{100} = 30\%$

2. $2\dfrac{1}{3}$

$$2\frac{1}{3} = \frac{7}{3} \qquad\qquad\qquad \frac{7}{3} = \frac{x}{100}$$

$$\begin{array}{r} 2.33\frac{1}{3} \\ 3\overline{)7.00} \\ \underline{6} \\ 10 \\ \underline{9} \\ 10 \\ \underline{9} \\ 1 \end{array} \qquad\qquad \frac{700}{3} = \frac{3x}{3} \qquad 233\frac{1}{3} = x$$

So, $2\dfrac{1}{3} = \dfrac{7}{3} = \dfrac{233\frac{1}{3}}{100} = 233\frac{1}{3}\%$

$$2\frac{1}{3} = \frac{7}{3} = 2.33\frac{1}{3} = 233\frac{1}{3}\%$$

SELF QUIZ	Change from fractions to percents.	ANSWERS
	1. $\frac{1}{4}$	1. 25%
	2. $\frac{1}{8}$	2. $12\frac{1}{2}$% or 12.5%
	3. $1\frac{1}{4}$	3. 125%

EXERCISES 6.3

Change the following numbers to percents.

1. $\frac{3}{100}$ 2. $\frac{16}{100}$ 3. $\frac{7}{100}$ 4. $\frac{29}{100}$ 5. $\frac{70}{100}$ 6. $\frac{87}{100}$

7. $\frac{83}{100}$ 8. $\frac{75}{100}$ 9. $\frac{63}{100}$ 10. $\frac{97}{100}$ 11. $\frac{102}{100}$ 12. $\frac{136}{100}$

13. $\frac{257}{100}$ 14. $\frac{200}{100}$ 15. $\frac{300}{100}$ 16. $\frac{1000}{100}$ 17. $\frac{125}{100}$ 18. $\frac{250}{100}$

19. $\frac{1}{10}$ 20. $\frac{7}{10}$ 21. $\frac{1}{2}$ 22. $\frac{1}{4}$ 23. $\frac{3}{4}$ 24. $\frac{1}{3}$

25. $\frac{2}{3}$ 26. $\frac{1}{8}$ 27. $\frac{5}{8}$ 28. $\frac{7}{8}$ 29. $\frac{1}{9}$ 30. $\frac{5}{9}$

31. $\frac{1}{5}$ 32. $\frac{3}{5}$ 33. $\frac{1}{25}$ 34. $\frac{1}{50}$ 35. $\frac{51}{50}$ 36. $\frac{3}{25}$

37. $\frac{4}{5}$ 38. $1\frac{1}{2}$ 39. $2\frac{1}{5}$ 40. $\frac{2}{5}$ 41. $\frac{3}{20}$ 42. $\frac{33}{20}$

43. $4\frac{1}{20}$ 44. $\frac{1}{6}$ 45. $\frac{5}{6}$ 46. $\frac{7}{9}$ 47. $1\frac{1}{10}$ 48. $2\frac{3}{10}$

49. $1\frac{7}{100}$ 50. $\frac{1}{15}$ 51. $\frac{1}{12}$ 52. $\frac{1}{16}$ 53. $\frac{15}{12}$ 54. $\frac{2}{7}$

55. $\frac{17}{15}$

6.4   CHANGING PERCENTS TO FRACTIONS

How would you change 60% to a fraction? Since *percent means hundredths,* we know $60\% = \frac{60}{100}$. That is, every percent can be written as a

WRITE AS A

fraction with denominator 100. The fraction should then be reduced if possible.

EXAMPLES

1. $60\% = \dfrac{60}{100} = \dfrac{3 \cdot 20}{5 \cdot 20} = \dfrac{3}{5}$

2. $70\% = \dfrac{70}{100} = \dfrac{7 \cdot 10}{10 \cdot 10} = \dfrac{7}{10}$

3. $16\dfrac{2}{3}\% = \dfrac{16\dfrac{2}{3}}{100} = \dfrac{\dfrac{50}{3}}{100} = \dfrac{\overset{1}{\cancel{50}}}{3} \cdot \dfrac{1}{\underset{2}{\cancel{100}}} = \dfrac{1}{6}$

4. $7\dfrac{1}{4}\% = \dfrac{7\dfrac{1}{4}}{100} = \dfrac{\dfrac{29}{4}}{100} = \dfrac{29}{4} \cdot \dfrac{1}{100} = \dfrac{29}{400}$

SELF QUIZ   Change from percents to fractions.   ANSWERS

1. 12%   1. $\dfrac{3}{25}$

2.   8%   2. $\dfrac{2}{25}$

3. $3\dfrac{1}{3}\%$ 3. $\dfrac{1}{30}$

MAKE INTO A FRACTION
PLACE OVER 100 AND REDUCE

EXERCISES 6.4

WORK FROM IN CLASS
EVEN IN CLASS
ASSIGN: ODD 1- 49

Change the following percents to fractions or mixed numbers.

1. 10%	**2.** 5%	**3.** 15%	**4.** 17%	**5.** 25%
6. 50%	**7.** 30%	**8.** 32%	**9.** 80%	**10.** 75%
11. $66\dfrac{2}{3}\%$	**12.** $33\dfrac{1}{3}\%$	**13.** 33%	**14.** 20%	**15.** $12\dfrac{1}{2}\%$
16. $37\dfrac{1}{2}\%$	**17.** 60%	**18.** 90%	**19.** 95%	**20.** 1%
21. 2%	**22.** $62\dfrac{1}{2}\%$	**23.** $2\dfrac{1}{4}\%$	**24.** $16\dfrac{2}{3}\%$	**25.** $\dfrac{1}{4}\%$
26. $\dfrac{1}{2}\%$	**27.** $11\dfrac{1}{9}\%$	**28.** 44%	**29.** $44\dfrac{4}{9}\%$	**30.** 100%
31. 200%	**32.** 300%	**33.** 120%	**34.** 175%	**35.** 55%

36. $212\frac{1}{2}\%$ **37.** 1000%   **38.** 5200%   **39.** 4820%   **40.** 4825%

41. $5\frac{1}{2}\%$ **42.** $\frac{3}{4}\%$ **43.** 216%   **44.** 304%   **45.** 72%

46. 36%   **47.** $5\frac{1}{7}\%$ **48.** $87\frac{1}{2}\%$ **49.** 24%   **50.** $58\frac{1}{3}\%$

6.5   PROBLEMS INVOLVING PERCENT

Have you tried problems like

Find 45% of 60?

or,   13 is 25% of what number?

or,   What percent of 65 is 19.5%?

What do you do for these problems? Are they all alike? Do you add, subtract, multiply, or divide?

Percent problems come in *only three* basic forms. We will use *one* basic proportion to solve all three types of problems.

Using A = amount, B = base, and R = percent, the following proportion is the key to the solution of any of the three basic types of percent problems.

$$\frac{A}{B} = \frac{R}{100}$$

Base is the number we are finding the percent of. **Amount** is the result of multiplying the base by the percent. Amount is also called **percentage**. For example, we know that

50% of 30 is 15
↑   ↑   ↑
R   B   A

R is the percent, and R is 50. $\left(\text{Thus, } \dfrac{R}{100} = \dfrac{50}{100}.\right)$

B is the base, and this is the number we are finding the percent of. B is 30.

A is the amount, or the result of multiplying the percent times the base. A is 15.

Note that

$$\frac{A}{B} = \frac{R}{100} \quad \text{or} \quad \frac{15}{30} = \frac{50}{100}$$

The three letters A, B, and R correspond to the three basic types of percent problems. In each problem, we know two of these quantities and need to find the third.

Problem Type 1

45% of 60 is _____ $R = 45, B = 60, A$ is unknown
$\uparrow$ $\uparrow$ $\uparrow$
R B A

$$\frac{A}{60} = \frac{45}{100}$$

$$100 \cdot A = 60 \cdot 45$$

$$\frac{100 \cdot A}{100} = \frac{60 \cdot 45}{100}$$

$$A = \frac{60 \cdot 45}{100} = \frac{2 \cdot 3 \cdot 2 \cdot 5 \cdot 3 \cdot 3 \cdot 5}{2 \cdot 2 \cdot 5 \cdot 5} = 27$$

So, 45% of 60 is 27.

Check:

Does $\frac{27}{60} = \frac{45}{100}$?

Yes, since $27 \cdot 100 = 2700$, and $60 \cdot 45 = 2700$.

Problem Type 2

25% of _____ is 13 $R = 25, A = 13, B$ is unknown
$\uparrow$ $\uparrow$ $\uparrow$
R B A

$$\frac{13}{B} = \frac{25}{100}$$

$$\frac{13}{B} = \frac{1}{4}$$

$$13 \cdot 4 = B \cdot 1$$

$$52 = B$$

So, 25% of 52 is 13.

Check:

Does $\dfrac{13}{52} = \dfrac{25}{100}$?

Yes, since $13 \cdot 100 = 1300$, and $25 \cdot 52 = 1300$.

Problem Type 3

_____ % of 65 is 19.5 $A = 19.5$, $B = 65$, R is unknown
  ↑   ↑   ↑
  R B A

$$\frac{19.5}{65} = \frac{R}{100}$$

$$19.5 \cdot 100 = 65 \cdot R$$

$$\frac{1950}{65} = \frac{65 \cdot R}{65}$$

$$30 = R$$

So, 30% of 65 is 19.5.

Check:

Does $\dfrac{19.5}{65} = \dfrac{30}{100}$?

Yes, since $19.5 \cdot 100 = 1950$, and $65 \cdot 30 = 1950$.

In Problem Type 1 where R and B are known, the following procedure is sometimes easier than using the proportion.

$$\frac{A}{B} = \frac{R}{100}$$

$$B \cdot \frac{A}{B} = \frac{R}{100} \cdot B \qquad \text{Multiply both sides by } B.$$

$$A = \frac{R}{100} \cdot B$$

This means that A (the amount or percentage) can be found just by multiplying the base by the percent (as a decimal). For example, to find 45% of 60 is ___A___,

$$60 = B$$
$$\underline{.45} = R \text{ (as a decimal)}$$
$$3\ 00$$
$$\underline{24\ 0}$$
$$27.00 = A$$

This decimal method is not easier for Types 2 and 3. The key features of the proportion method are that (a) you don't have to decide whether to multiply or divide, and (b) the one method works for all three types.

SELF QUIZ	Solve each problem for the unknown quantity.	ANSWERS
	1. 5% of 70 is _____.	1. 3.5
	2. _____% of 80 is 9.6.	2. 12%
	3. 15% of _____ is 10.	3. $66\frac{2}{3}$

EXERCISES 6.5

ASSIGN ODD 1-19

Solve each problem for the unknown quantity. (Use either the (pro-portion) method or the decimal method for Type 1 problems.)

1. 10% of 70 is _____.

2. 5% of 62 is _____.

3. 15% of 60 is _____.

4. 25% of 72 is _____.

5. 2% of _____ is 3.

6. 20% of _____ is 17.

7. 3% of _____ is 14.

8. 30% of _____ is 14.

9. _____% of 75 is 15.

10. _____% of 100 is 50.

11. _____% of 34 is 17.

12. _____% of 12 is 4.

13. _____% of 48 is 16.

14. _____% of 30 is 6.

15. _____% of 150 is 60.

16. 35% of _____ is 24.

17. 42% of _____ is 10.

18. 60% of 95 is _____.

19. 58% of 120 is _____.

20. 75% of 12 is _____.

In the following problems, remember that B is the number you are finding the percent of, regardless of the other wording.

LESSON 11 ↓ *ODD PROBS 21-49*

21. _12.5_ is 50% of 25.

22. 14 is $5\frac{1}{2}$% of _____.

23. 16 is $25\frac{1}{2}$% of _62 $\frac{38}{51}$_

24. _____ is 31.4% of 76.

25. _2.109_ is 11.1% of 19.

26. $14\frac{2}{7}$% of 28 is _____.

27. 13.5 is _84$\frac{3}{8}$_% of 16.

28. $20\frac{1}{4}$% of _____ is 30.

29. 150% of _25⅓_ is 38.

30. 200% of 75 is _____.

31. 48 is 300% of _16_.

32. $33\frac{1}{3}$% of _____ is 24.

33. _900_ % of 20 is 180.

34. $66\frac{2}{3}$% of 27 is _____.

35. _150_ % of 70 is 105.

36. 32 is _____ % of 24.

37. _6_ is $16\frac{2}{3}$% of 36.

38. $37\frac{1}{2}$% of 56 is _____.

39. 40 is $12\frac{1}{2}$% of _320_.

40. 15% of _____ is 92.1.

41. _76_ is 100% of 76.

42. _____ is 150% of 20.

43. 100% of _82_ is 82.

44. 90% of 90 is _____.

45. 150% of 50 is _75_.

46. 10% of 100 is _____.

47. 200% of 6 is _12_.

48. _____ is 40% of 5.

49. 13 is _130_ % of 10.

50. 35 is _____ % of $17\frac{1}{2}$.

6.6   APPLICATIONS (DISCOUNT, COMMISSION, AND OTHERS)

Percent is part of our daily lives. We need percent to understand sales tax, discounts, salaries based on commission, profit and loss, and scores on math tests.

Whether or not you can solve a particular problem depends on your own past experiences, your attitudes, your abilities to reason, your math skills, and many other things. Do not become discouraged. The following attack plan should help.

ATTACK PLAN FOR PERCENT PROBLEMS

1. Read the problem carefully.

2. Decide what is unknown (R, A, or B).

3. Write down the values you do know for R, A, or B. (You must know two of these.)

4. Set up and solve the proportion $\dfrac{A}{B} = \dfrac{R}{100}$.

5. Check to see that the answer seems reasonable to you.

EXAMPLES

1. A refrigerator sells for $450 and is on sale at a 20% discount. What is the amount of the discount? What is the sale price?

    Solution:
    There are two questions here.

    A. Find the discount.
        20% of $450 is _____ (Type 1 problem)

$$\begin{array}{r} \$\ 450 \\ .20 \\ \hline \$90.00 \end{array}$$

    The discount is $90.

    B. Find the sale price by subtracting the discount from the original price. (No percent is involved here.)

$$\begin{array}{r} \$450 \\ -\ 90 \\ \hline \$360 \end{array} \text{ is the sale price.}$$

2. If the sales tax rate is 6%, what would be the final cost of the refrigerator in Example 1?

    Solution:
    We want the final cost including tax. So, first find the amount of tax.

    6% of $360 is _____ (Type 1 problem)

$$\begin{array}{r} \$\ \ 360 \\ .06 \\ \hline \$21.60 \end{array} \text{ is the amount of tax.}$$

    Now add the tax to the cost of the refrigerator. (This involves your knowledge of how tax is computed. It does not involve percent.)

$$\begin{array}{r} \$360.00 \\ +\ 21.60 \\ \hline \$381.60 \end{array} \text{ is the final cost.}$$

3. A salesman earns a salary of $600 a month plus a commission of 8% on his sales over $5000. What did he earn the month that he sold $9400 worth of merchandise?

    Solution:
    Again, the problem involves more than percent. Reasoning is important!

A. What do we take 8% of?
We are to take 8% of his sales *over* $5000.

$$\begin{array}{r} \$\ 9400 \\ -\ 5000 \\ \hline \$\ 4400 \end{array} \text{ in sales over } \$5000$$

8% of $4400 is _____   (Type 1 problem)

$$\begin{array}{r} \$\ \ 4400 \\ .08 \\ \hline \$352.00 \end{array} \text{ is the commission earned.}$$

B. What is his monthly pay?
He earns $600 plus the commission.

$$\begin{array}{r} \$\ 600 \\ +\ 352 \\ \hline \$\ 952 \end{array} \text{ is the total earned for the month.}$$

4. A woman paid $4410 for a used car. She was allowed a 2% discount off the original asking price because she paid cash. What was the original asking price?

Solution:
Do *not* take 2% of $4410. The discount was not based on $4410. It was based on the asking price. In fact, we do *not* take 2% of anything.

We must *reason* that the $4410 represents 98% of the asking price. (100% − 2% = 98%) So, 98% of _____ is $4410   (Type 2 problem)

$$\frac{4410}{B} = \frac{98}{100}$$

$$\frac{441000}{98} = \frac{98 \cdot B}{98}$$

$$\$4500 = B$$

$$\begin{array}{r} 4500 \\ 98{\overline{)441000}} \\ \underline{392} \\ 490 \\ \underline{490} \\ 00 \end{array}$$

The asking price was $4500.

5. A company manufactures certain fittings. These fittings cost the company $5.00 each to produce, and they retail for $10.00 each. However, the company charges only $7.00 for each fitting to a particularly good customer. How much profit is made on each item sold to the good customer? What percent of profit is made?

Solution:

The profit is obvious. (Price − cost = profit)

$$\begin{array}{rl} \$\ 7.00 & \text{price} \\ -\ 5.00 & \text{cost} \\ \hline \$\ 2.00 & \text{profit} \end{array}$$

The question of percent of profit is not obvious. In fact, *it is purposely worded ambiguously in this problem.*

Using ratio,

$$\frac{\$\ 2\ \text{profit}}{\$\ 5\ \text{cost}} = .40 = 40\%\ \text{profit based on cost}$$

$$\frac{\$2\ \text{profit}}{\$7\ \text{selling price}} = .28\frac{4}{7} = 28\frac{4}{7}\%\ \text{profit based on selling price}$$

Which profit percent is correct, 40% or $28\frac{4}{7}$%? They are *both* correct.

The question is not complete unless you know what the profit is based on. (The retail price of $10.00 does not affect the company's profit.)

EXERCISES 6.6

1. A real estate salesman works on a 6% commission. What is his commission on a house he sold for $58,000?

2. A manufacturer gave a store owner a discount of 3% because she bought $3500 worth of dresses. What was the amount of the discount? What did the store owner pay for the dresses?

3. A computer programmer was told he would be given a bonus of 5% of any money his programs could save the company. How much would he have to save the company to earn a bonus of $500?

4. The property taxes on a house were $350. What was the tax rate if the house was valued at $12,500? What was the estimated retail value of the property if the base value for calculating taxes is $\frac{1}{4}$ the estimated retail value?

5. A department store sold sheets and towels at a discount of 30% during a white sale. If sheets were originally $8.50 each, and towels were originally $4.50 each, what were the discount prices? One type of towel was marked at $3.01 for the sale. What was the original price of this type of towel?

6. A carpet salesman was told she would make a bonus of 15% on her sales over $3,000 for one week. How much did she sell to make a bonus of $75?

7. At South Junior High School, the seventh graders had 340 cavities. This was 85% of the cavities of the eighth graders. How many cavities did the eighth gráders have if the average age of the eighth graders was 13.4 years?

8. A shoes saleswoman worked on a straight 9% commission. Her friend worked on a salary of $300 a month plus a 5% commission. How much did each woman make the month each sold $4500 worth of shoes?

9. A student missed 3 problems on a math test and was given a grade of 85%. If all the problems were of equal value, how many problems were on the test? A friend received a grade of 80%.

10. A basketball player made 120 of 300 shots he attempted. What percent of his shots did he make? What percent did he miss? He did not date the coach's daughter.

11. A fur coat was discounted $150. This was a 20% discount. What was the original selling price of the coat?

12. Golf clubs were marked for sale by a golf pro at $160, which was a 20% discount based on the original selling price. They cost the pro $120. What was the original selling price? What was the amount of profit? What was the percent of profit based on cost? What was the percent of profit based on the original selling price?

13. The discount on men's suits was $50, and they were on sale for $200. What was the original selling price? What was the rate of discount? If the suits cost the store owner $150 each, what was his percent of profit based on cost? What was his percent of profit based on the selling price? Vests were included in all the prices and were valued at $35 each.

14. A secretary decided to spend 20 minutes of her lunch hour shopping. What percent of her lunch hour was left for dining? She worked an 8-hour day.

15. If the sales tax is 6¢ on a dollar, what was the tax on a purchase of $30.20? What was paid for the purchase?

16. If the taxes on a gallon of gasoline are 15¢ per gallon, and the price is 60.0¢ per gallon, what percent of the price is taxes? If the gas cost the station owner 40¢ a gallon, what percent of profit does he make based on his cost?

17. In one year, Mr. Jacobs, who is 27 years old and married, earned $12,000. He spent $2400 on rent, $4800 on food, and $1800 on taxes. What percent of his income went for rent? for food? for taxes? His wife's income was $8,000. They used 50% of her income for entertainment.

18. The author of a book was told she would have to cut the number of pages by 12% in order for the book to sell at a popular price and still show a profit. What percent of the pages were in the final form? If the book contained 220 pages in final form, how many pages were in it originally? How many pages were cut?

19. In order to get more subscribers, a book club offered three books whose total selling price was originally $17.55 for $7.02. What was the amount of the discount? Based on the original selling price, what was the rate of discount on these three books?

20. The cost of a television set was $250 for the store owner, and he sold this set for $345. What was his percent of markup based on cost? (Markup is the difference between selling price and cost. That is, markup is another name for the amount of profit.) What was his percent of profit based on selling price? Why are these two percents not the same?

21. By buying a complete set of pots and pans, a young houswife received a discount of 10% of the original selling price. She paid $216. What was the original selling price? What was the amount of discount? The set cost the dealer $192. What was his percent of profit based on cost? What was his percent of profit based on the selling price?

22. A car dealer bought a used car for $600. He wanted to make a profit of 25% based on his cost. What was the selling price of the car? What was the dealer's markup? What was his percent of profit based on selling price?

23. A man weighed 200 pounds. He lost 20 pounds in three months. What percent did he lose? Then he gained back 20 pounds one month later. What percent did he gain? Are these two percents the same or different? Why? What did he weigh at the end of the four-month period?

24. An office supply store buys pens from the manufacturer at $18.00 a dozen and receives a discount of 20% on any amount over 8 dozen. What was the cost to the store for 10 dozen pens? What will be the retail asking price for each pen if the store marks up the price so that it makes 40% profit based on cost?

25. An auto supply store received a shipment of auto parts together with the bill for $845.30. Some of the parts were not as ordered, however, and were returned at once. The value of the parts returned was $175.50. The terms of the billing provide the store with a 2% discount if it pays cash within two weeks. What did the store finally pay for the parts it kept if it paid cash within two weeks?

6.7   APPLICATIONS (SIMPLE INTEREST)

Interest is money paid for the use of money. The money invested or borrowed is called the **principal.** The **rate** is the **percent of interest** and is generally stated as an annual (yearly) rate. That is, if you borrow $40,000 at 9% interest to buy your home, the 9% interest is an annual interest rate. The principal is $40,000.

Interest is either paid or earned depending on whether you are the borrower or the lender. In either case, the concept of money paid for the use of money is the same. The purpose of this section is to explain how interest is calculated.

There are two kinds of interest, **simple interest** and **compound interest.** If you put $500 in a savings account at 6% interest and *leave it there for one year,* then the simple interest is .06 × $500 = $30.00.

If you leave the $30.00 in the savings account, then you will be earning interest on the original $500 and the $30.00 interest. For the next year, *interest will be paid on the interest.* This is *compound interest,* or *interest compounded annually.* The interest you earn the second year will be .06 × $530 = $31.80.

In many cases, interest is not computed on a yearly basis because people may leave their money in a savings account or borrow money for only a few days. In these cases, the *simple interest* is calculated with the following formula.

FORMULA FOR CALCULATING SIMPLE INTEREST

$I = P \times R \times T$ where I = interest
P = principal
R = rate
T = time (in years)

We will agree to use 360 days in one year. (This is a common practice in business and banking.) *T is a fractional part of a year because the rate R is a yearly rate.*

EXAMPLES

1. A woman borrows $500 for 90 days at 10% interest. How much interest did she pay? [NOTE: She makes only one payment—at the end of 90 days. If she made monthly payments, then she would be paying compound interest, and the rate would not be 10%.]

Solution:

$$I = P \times R \times T \quad P = \$500, R = 10\% = \frac{10}{100}, T = 90 \text{ days} = \frac{90}{360} \text{ year}$$

$$I = 500 \times \frac{10}{100} \times \frac{90}{360}$$

$$= 500 \times \frac{\overset{5}{\cancel{10}}}{\underset{1}{\cancel{100}}} \times \frac{\overset{1}{\cancel{90}}}{\underset{2}{\cancel{360}}} = \frac{25}{2} = \$12.50$$

She paid $12.50 interest.

2. You loan $1000 to a friend for 6 months at an interest rate of 7%. How much will you be paid at the end of 6 months?

Solution:

$$I = P \times R \times T \quad P = \$1000, R = 7\% = \frac{7}{100}, T = 6 \text{ months} = \frac{6}{12} \text{ year}$$

$$I = 1000 \times \frac{7}{100} \times \frac{6}{12}$$

$$= \overset{\overset{5}{\cancel{10}}}{\cancel{1000}} \times \frac{7}{\underset{1}{\cancel{100}}} \times \frac{\overset{1}{\cancel{6}}}{\underset{\underset{1}{2}}{\cancel{12}}} = \$35.00$$

The interest is $35.00. You will be paid the principal plus the interest.

$$\$1000 + \$35 = \$1035$$

Suppose you would like to invest some money for 30 days, and you would like to make $50 interest. You know you can get 10% interest. How much do you need to invest?

Here we know the interest I, the rate R, and the time T. We do not know the principal P. The formula $I = P \times R \times T$ can be rewritten in the following forms.

$$P = \frac{I}{R \cdot T} \qquad R = \frac{I}{P \cdot T} \qquad T = \frac{I}{P \cdot R}$$

In this problem,

$$P = \frac{I}{R \cdot T} \quad I = \$50, R = 10\% = \frac{10}{100}, T = 30 \text{ days} = \frac{30}{360} \text{ year}$$

$$P = \frac{50}{\dfrac{10}{100} \cdot \dfrac{30}{360}}$$

$$= \frac{50}{\dfrac{1}{10} \cdot \dfrac{1}{12}} = \frac{50}{\dfrac{1}{120}} = \frac{50}{1} \cdot \frac{120}{1} = \$6000$$

You would have to invest \$6000 for 30 days at 10% to earn \$50.

EXERCISES 6.7

1. What would be the amount of interest in one year on a savings account of \$800 if the bank pays 5% interest?

2. If interest is paid at 7% for one year, what would a principal of \$600 earn?

3. If a principal of \$300 is invested at a rate of 7% for 90 days, what would be the interest earned?

4. A loan of \$5000 is made at 8% for a period of 6 months. How much interest is paid?

5. If you borrow \$750 for 30 days at 18%, how much interest will you pay? [NOTE: 18% is the annual interest rate paid when a rate of $1\frac{1}{2}\%$ is charged per month.]

6. How much interest is paid on a 60-day loan of \$500 at 10%?

7. Find the interest paid on a savings account of \$1800 left for 120 days at 6%.

8. A savings account of \$2300 is left for 90 days drawing interest at a rate of 6%. How much interest is earned? What is the amount in the account?

9. A bank pays interest at 5% on savings accounts. How much will be in an account of \$14,600 left for 6 months?

10. One thousand dollars worth of merchandise is charged at a local department store for 60 days at 18% interest. How much is owed at the end of 60 days?

11. You buy an oven on sale from \$500 to \$450, but you don't pay the bill for 60 days and are charged interest at a rate of 18%. How much do you pay for the oven by waiting 60 days to pay? How much did you save by buying the oven on sale?

12. A friend borrows $500 from you for a period of 8 months and pays you interest at 6%. How much interest are you paid? Suppose you ask 8% instead. Then how much interest are you paid?

13. How much would you have to invest at 8% for 60 days to earn interest of $500?

14. How many days must you leave $1000 in a savings account at $5\frac{1}{2}\%$ to earn $11.00?

15. What is the rate of interest charged if a loan of $2500 for 90 days is paid off with $2562.50?

16. Determine the missing item in each row.

PRINCIPAL	RATE	TIME	INTEREST
$ 400	6%	90 days	?
?	5%	120 days	$ 5.00
$ 560	$4\frac{1}{2}\%$	?	$ 3.15
$2700	?	40 days	$25.50

17. Determine the missing item in each row.

PRINCIPAL	RATE	TIME	INTEREST
$500	18%	30 days	?
$500	18%	?	$15.00
$500	?	90 days	$22.50
?	18%	30 days	$ 1.50

18. If you have a savings account of $25,000 drawing interest at 8%, how much interest will you earn in 6 months? How long must you leave the money in the account to earn $1500?

19. You have accumulated $50,000, and you want to live off the interest each year. If you need $500 a month to live on, what interest rate must you be earning with your $50,000? At this interest rate, what would be the annual interest on $75,000? Could you live on this interest?

20. Your $2500 savings account draws interest at $5\frac{1}{2}\%$. How many days will it take for you to earn $68.75? If the interest rate is then raised to 6%, what will your money earn in the next six months?

6.8   APPLICATIONS (COMPOUND INTEREST)

Interest paid on interest is called **compound interest.** Suppose you deposit $1000 in a savings account, and interest is paid at 6% compounded semiannually. This means interest is paid twice a year, or every six months. The 6% is an annual rate. So, to figure the yearly interest earned, we calculate the simple interest twice (every six months).

$$I = P \times R \times T \quad P = \$1000, R = 6\%, T = 6 \text{ months} = \frac{1}{2} \text{ year}$$

$$= 1000 \cdot \frac{6}{100} \cdot \frac{1}{2}$$

$$= \$30$$

Now, at the end of six months, your new principal is $1000 + $30 = $1030.

$$I = P \times R \times T$$

$$= 1030 \cdot \frac{6}{100} \cdot \frac{1}{2}$$

$$= \overset{10.3}{\underset{1}{\cancel{1030}}} \cdot \frac{\overset{3}{\cancel{6}}}{100} \cdot \frac{1}{\underset{1}{\cancel{2}}} = \$30.90$$

The interest for the second six months is $30.90. You made $.90 more because you had a larger principal. The total interest paid is

$$\$30.00 + \$30.90 = \$60.90$$

Then, if you were to leave your principal and interest in the account, the principal for the first six months of the second year would be

$$\$1030.00 + \$30.90 = \$1060.90$$

Many savings and loan associations and banks now pay interest compounded daily. That is, your interest is figured daily, and you are paid interest on your interest every day. To calculate interest compounded daily, we would need either a computer or a large set of tables. The purpose here is just to teach the concept of compound interest so we will work only with interest compounded annually (once a year), semiannually (twice a year), and quarterly (four times a year).

EXAMPLE

1. $2000 is deposited in a savings account, and interest is computed quarterly at 8%. How much interest will be earned in one year?

Solution:

This problem involves four calculations because interest is to be accumulated every 3 months $\left(\text{or } \frac{1}{4} \text{ year}\right)$.

A. $I = 2000 \times \dfrac{8}{100} \times \dfrac{1}{4}$ $\qquad\qquad$ $(P = 2000)$

$\qquad = \overset{20}{\cancel{2000}} \times \dfrac{\overset{2}{\cancel{8}}}{\underset{1}{\cancel{100}}} \times \dfrac{1}{\underset{1}{\cancel{4}}} = \$40$

B. $I = 2040 \times \dfrac{8}{100} \times \dfrac{1}{4}$ $\qquad\qquad$ $(P = 2000 + 40 = 2040)$

$\qquad = \overset{20.4}{\cancel{2040}} \times \dfrac{\overset{2}{\cancel{8}}}{\underset{1}{\cancel{100}}} \times \dfrac{1}{\underset{1}{\cancel{4}}} = \$40.80$

C. $I = 2080.80 \times \dfrac{8}{100} \times \dfrac{1}{4}$ $\qquad\qquad$ $(P = 2040 + 40.80$
$\qquad\qquad\qquad\qquad\qquad\qquad\qquad\qquad\qquad\quad = 2080.80)$

$\qquad = \overset{20.808}{\cancel{2080.80}} \times \dfrac{\overset{2}{\cancel{8}}}{\underset{1}{\cancel{100}}} \times \dfrac{1}{\underset{1}{\cancel{4}}} = \$41.62$ $\quad$ (rounded off)

D. $I = 2122.42 \times \dfrac{\overset{2}{\cancel{8}}}{\cancel{100}} \times \dfrac{1}{\underset{1}{\cancel{4}}}$ $\qquad\qquad$ $(P = 2080.80 + 41.62$
$\qquad\qquad\qquad\qquad\qquad\qquad\qquad\qquad\qquad\quad = 2122.42)$

$\qquad \begin{array}{r} \$2122.42 \\ \underline{.02} \\ \$42.4484 \end{array}$ $\quad$ (using $\dfrac{2}{100} = .02$)

$\qquad I = 42.45$ $\qquad$ (rounded off)

The total interest earned is

$$\begin{array}{r} \$\ 40.00 \\ 40.80 \\ 41.62 \\ \underline{42.45} \\ \$164.87 \end{array}$$

Inflation (the fact that it will take more money next year to buy what you did this year) is a form of annual compound interest. Tables A and B show inflation at 6% and at 8% on a \$10,000 principal.

TABLE A   INFLATION AT 6%

YEAR	PRINCIPAL	INTEREST	TOTAL
1	$10,000.00	$ 600.00	$10,600.00
2	10,600.00	636.00	11.236.00
3	11,236.00	674.16	11,910.16
4	11,910.16	714.61	12,624.77
5	12,624.77	754.49	13,382.26
6	13,382.26	802.93	14,185.19
7	14,185.19	851.11	15,036.30
8	15,036.30	902.20	15,938.50
9	15,938.50	956.29	16,894.79
10	16.894.79	1,003.68	17,908.47
11	17,908.47	1,074.51	18,982.98
12	18,982.98	1,138.99	20,121.96

TABLE B   INFLATION AT 8%

YEAR	PRINCIPAL	INTEREST	TOTAL
1	$10,000.00	$ 800.00	$10,800.00
2	10,800.00	864.00	11,664.00
3	11,664.00	933.12	12,597.12
4	12,597.12	1,007.77	13,604.89
5	13,604.89	1,088.39	14,693.28
6	14,693.28	1,175.46	15,868.74
7	15,868.74	1,269.50	17,138.24
8	17,138.24	1,371.06	18,509.30
9	18,509.30	1,480.74	19,990.04

The tables can be interpreted as follows. If inflation continues at 6%, and you have $10,000 in a savings account drawing interest at 6% for twelve years, your $10,000 will double to $20,000, but you will not be able to buy any more than you could twelve years ago.

If your interest rate is 8%, and inflation is at 8%, your savings will double in nine years, but you will be just keeping up with inflation, and your buying power will not be any greater than it was nine years ago.

These tables are in no way a prediction of what will happen to our economy. They are simply factual data to help you understand compound interest.

EXERCISES 6.8

1. If a bank compounds interest at 4% semiannually, what will your $13,000 deposit be worth in 6 months? in 2 years?

2. $9,000 is deposited in a savings and loan association and is compounded annually at 5%. What will be the amount in the account at the end of three years?

3. If an account is compounded quarterly at a rate of 6%, what will be the interest earned on $5,000 in one year? What will be the total amount in the account? How much more interest is earned in the first year by compounding quarterly than if the account were compounded annually?

4. If $5,000 is deposited in a savings account at $5\frac{1}{2}$% compounded annually, what will be the amount of interest earned in two years? What will be the amount of the account?

5. How much interest will be earned on a savings account of $3,000 in two years if interest is compounded annually at 6%? if interest is compounded semiannually? if interest is compounded quarterly?

6. If interest is calculated at 7% compounded annually, what will be the value of $4,000 at the end of 5 years?

7. You borrowed $4,000 and agreed to make equal payments of $1,000 each plus interest over the next four years. Interest is at a rate of 8% based only on what you owe. How much interest did you pay? How much interest would you have paid if you had not made the annual payments and paid only the $4,000 plus interest compounded annually at the end of four years?

8. Continue Table A for two more years (i.e., years 13 and 14). How many years altogether do you think it would take the original $10,000 to triple?

9. What will be the value of a $15,000 savings account at the end of three years if interest is calculated at 10% compounded annually? Is this the same as if the interest were calculated at 5% compounded semiannually? What is the difference, if there is any? Suppose the semiannual compounding is at 5% for six years. Now will the value be the same?

10. If a bank pays interest at 4% compounded semiannually, how much interest will be earned on $3,200 in one year? How much will be earned if the interest is compounded quarterly? What is the difference in interest earned in one year?

11. If a savings account of $10,000 draws interest at a rate of 10% compounded annually, how many years will it take for the $10,000 to double in value? Make a table similar to Tables A and B. Call it *Table C: Inflation at 10%*.

SUMMARY: CHAPTER 6

Percent means hundredths.

> **To change a decimal to a percent,** move the decimal point two places to the right and write the % sign.

> **To change a percent to a decimal,** move the decimal point two places to the left and drop the % sign.

> Using A = amount, B = base, and R = percent, the following proportion is the key to the solution of any of the three basic types of percent problems.
>
> $$\frac{A}{B} = \frac{R}{100}$$

ATTACK PLAN FOR PERCENT PROBLEMS

1. Read the problem carefully.

2. Decide what is unknown (R, A, or B).

3. Write down the values you do know for $R, A,$ or B. (You must know two of these.)

4. Set up and solve the proportion $\dfrac{A}{B} = \dfrac{R}{100}$.

5. Check to see that the answer seems reasonable to you.

Interest is money paid for the use of money. The money invested or borrowed is called the **principal.**

FORMULA FOR CALCULATING SIMPLE INTEREST

$I = P \times R \times T$ I = interest
P = principal
R = rate
T = time (in years)

$$P = \frac{I}{R \cdot T} \qquad R = \frac{I}{P \cdot T} \qquad T = \frac{I}{P \cdot R}$$

Compound interest is interest paid on interest.

REVIEW QUESTIONS · CHAPTER 6

Change the following rational numbers to percents.

1. $\dfrac{85}{100}$ 2. $\dfrac{37}{100}$ 3. $\dfrac{6}{10}$ 4. $\dfrac{13}{10}$ 5. $\dfrac{6}{5}$ 6. $3\dfrac{1}{4}$

7. $5\dfrac{3}{4}$ 8. $\dfrac{5}{12}$

Change the following percents to fractions or mixed numbers.

9. 14%   10. 40%   11. 66%   12. 400%   13. 2500%

14. 27%   15. $16\dfrac{2}{3}\%$ 16. $\dfrac{1}{4}\%$

Change the following decimals to percents.

17. .06   18. .3   19. .67   20. .027   21. 4.59   22. 5

Change the following percents to decimals.

23. 35%   24. 4%   25. .25%   26. $\dfrac{1}{4}\%$ 27. $13\dfrac{3}{4}\%$ 28. 7.1%

Solve each of the following problems using either the proportion method or the decimal method.

29. 30% of 52 is _____.   30. 15% of 17 is _____.

31. 3% of _____ is 7.   32. 42% of _____ is 18.

33. _____% of 36 is 7.2.   34. _____% of 48 is 16.

35. 75 is _____% of 300.   36. _____ is $\dfrac{1}{4}\%$ of 8.

37. 5 is 10% of _____.   38. 14 is $5\dfrac{1}{2}\%$ of _____.

39. _____ is $6\dfrac{2}{3}\%$ of 15.   40. 62 is _____% of 31.

41. A real estate salesman works on a 6% commission. What would be his commission on the sale of an apartment building for $67,400?

42. If a principal of $1000 is invested at a rate of 9% for 30 days, what would be the interest earned?

43. The salesman in Problem 41 earned a commission of $1524 on the sale of a home. What was the selling price of the home?

44. What is the percent of markup (to the nearest 1%) based on cost of an automobile that cost $1500 and sold for $2000? What is the percent of markup based on the selling price?

45. By purchasing four tires, a customer is given a discount of 10% on each tire. If he paid $108 for the tires, what was the original price of each tire?

46. George Blackwood made a short-term investment of $2000 at a rate of 15% and earned $50 interest. For what period of time did he have his money invested?

47. Determine the missing item in each row.

PRINCIPAL	RATE	TIME	INTEREST
$ 200	12%	180 days	?
$ 300	18%	?	$ 81
$1000	?	1 year	$ 85
?	9%	18 months	$270

48. How much interest will be earned on a savings account of $6500 in two years if interest is compounded annually at 6%? if interest is compounded semiannually? if interest is compounded quarterly?

MEASUREMENT:
THE METRIC SYSTEM

The metric system of measurement is *much easier* to understand and work with than the U.S. customary system (formerly called the English system). Consider the following comparison in difficulty of computation in figuring the area of a rectangle by each system.

U.S. Customary System

Find the area of a rectangle (in square feet) whose length is 4 feet 5 inches (4′5″) and whose width is 6 feet 8 inches (6′8″). (See Figure 7.1.)

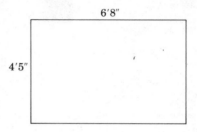

Figure 7.1

Area = length · width

$$\text{Area} = 4\frac{5}{12} \cdot 6\frac{8}{12} = \frac{53}{12} \cdot \frac{80}{12} = \frac{4240}{144} = 29\frac{64}{144} = 29\frac{4}{9} \text{ square feet}$$

[NOTE: To figure square inches, either $29\frac{4}{9}$ would have to be multiplied by 144, or 4′5″ and 6′8″ would have to be changed to inches before multiplying.]

Metric System

Find the area of a rectangle (in square meters, abbreviated m²) that has length 4 meters 5 centimeters (4 m 5 cm) and width 6 meters 8 centimeters (6 m 8 cm). (See Figure 7.2.)

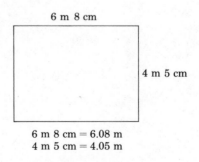

6 m 8 cm = 6.08 m
4 m 5 cm = 4.05 m

Figure 7.2

$$\text{Area} = (4.05)(6.08) = 24.624 \ \text{m}^2$$

[NOTE: To figure square centimeters, simply move the decimal point four places to the right: $24.624 \ \text{m}^2 = 246{,}240 \ \text{cm}^2$.]

At a distinct disadvantage in international trading (about 90% of the people in the world use the metric system), the United States is the only major industrialized country in the world still committed to the U.S. customary system (even England is in the process of changing to the metric system). However, even here, the metric system has been used for years in such areas as medicine, science, and military activities. The adoption of the metric system by the United States now has strong backing by many organizations and is generally considered to be a distinct probability.

The purpose of this chapter is to help the student become familiar with the metric units of measure and become skillful in manipulating within the metric system. Although tables of equivalent measures in the metric and U.S. customary systems and related graphs are included at the end of the chapter, they are for reference only. There will be little discussion and no exercises in converting from one system to the other. The student is encouraged to think in the metric system with as little thought of relating it to the U.S. customary system as possible.

7.1   LENGTH

The basic unit of length in the metric system is the **meter,** slightly longer than a yard. To relate to other units of length, the meter is simply multiplied by 10, 100, 1000, and so on, or by 0.1, 0.01, 0.001, and so on. Full advantage is taken of our knowledge of multiplying and dividing by powers of 10. (Remember that dividing by 10 is the same as multiplying by $\frac{1}{10}$ or 0.1, and dividing by 100 is the same as multiplying by $\frac{1}{100}$ or 0.01, and so on.) The prefixes* milli, centi, deci, deka, hecto, and kilo are simply attached to the basic unit to indicate which power of 10 or $\frac{1}{10}$ of the basic unit is involved. These prefixes are the same throughout the metric system, regardless of whether the units are of length (meter), mass (gram), or volume (liter).† (See Tables 7.1 and 7.2.) Although there are other units in the metric system, only these three and area will be discussed in this text. They are the most commonly used, and the methods and prefixes are the same for the other units.

---

*Other prefixes that indicate extremely small units are micro, nano, pico, femto, and atto. Prefixes that indicate extremely large units are mega, giga, and tera. These prefixes will not be used in this text.

†Optional spellings for meter and liter are metre and litre, respectively.

TABLE 7.1   METRIC PREFIXES AND
THEIR VALUE

PREFIX	VALUE	
milli	0.001	— thousandths
centi	0.01	— hundredths
deci	0.1	— tenths
basic unit	1	— ones
deka	10	— tens
hecto	100	— hundreds
kilo	1000	— thousands

TABLE 7.2   MEASURES OF LENGTH

1 *milli*meter   (mm) = 0.001 meter
1 *centi*meter   (cm) = 0.01 meter
1 *deci*meter   (dm) = 0.1 meter
1 meter   (m) = 1.0 meter
1 *deka*meter   (dam) = 10 meters
1 *hecto*meter   (hm) = 100 meters
1 *kilo*meter   (km) = 1000 meters

In order to have a universal physical standard of length, which is required for accurate scientific computations, the meter is defined as 1,650,763.73 wavelengths in vacuum of the orange-red line of the spectrum of krypton 86 (see Figure 7.3).

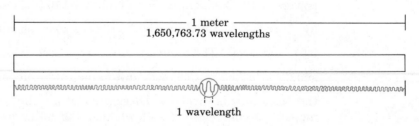

1 meter
1,650,763.73 wavelengths

1 wavelength

Figure 7.3

Copy the ruler marked in centimeters in Figure 7.4 on a piece of paper and measure the length of something with it (a book, desk top, pencil, shoe size). Estimate the length to the nearest tenth of a centimeter. You have just used the metric system.

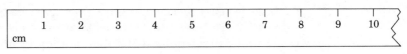

Centimeter ruler

Figure 7.4

Your measurement, however, might not have been as accurate as you would have liked. In Figure 7.5, the same ruler is now marked in millimeters. Copy the millimeter ruler and measure the same object you measured before. Its length has not changed so whatever measure you had in centimeters (considering an accurate estimate in tenths of centimeters) will be equal to the measure you had in millimeters. For example, if the book you measured was 23.4 cm with the centimeter ruler and 234 mm with the millimeter ruler, then 23.4 cm = 234 mm.

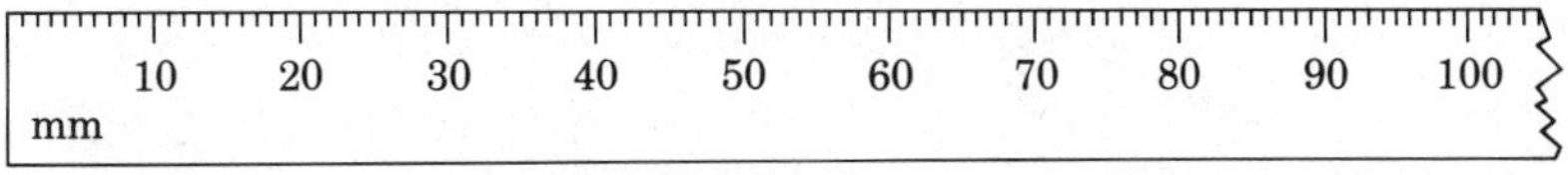

Millimeter ruler

Figure 7.5

The relationship between centimeters and millimeters is quite clear both from the rulers and from Table 7.2; that is, 10 mm = 1 cm. In fact, 10 of any unit are needed to make 1 of the next larger unit. Or, thinking in reverse, 0.1 of any unit is needed to make 1 of the next smaller unit (for example, 0.1 cm = 1 mm). (See Table 7.3.)

TABLE 7.3   RELATIONSHIPS BETWEEN
SMALLER AND LARGER MEASURES
OF LENGTH

10 mm = 1 cm	or	0.1 cm = 1 mm
10 cm = 1 dm		0.1 dm = 1 cm
10 dm = 1 m		0.1 m = 1 dm
10 m = 1 dam		0.1 dam = 1 m
10 dam = 1 hm		0.1 hm = 1 dam
10 hm = 1 km		0.1 km = 1 hm

EXAMPLES

1. 50 mm = 5 cm
2. 30 dm = 3 m
3. 42 hm = 4.2 km
4. 0.3 cm = 3 mm
5. 0.6 m = 6 dm
6. 0.2 hm = 2 dam

It is also easy to compare units that are more than one "jump" apart on Table 7.2. To "jump up," that is, to express a certain value given in a smaller unit as the equivalent value in a larger unit, multiply the original value as follows:

To "jump up"
one jump — multiply by 0.1.
two jumps — multiply by 0.01.
three jumps — multiply by 0.001.
four jumps — multiply by 0.0001, and so on.

EXAMPLE

Change 500 mm to meters.

Three jumps are involved: (1) millimeters to centimeters, (2) centimeters to decimeters, (3) decimeters to meters. Therefore,

$$500(0.001) = 0.5$$

and
$$500 \text{ mm} = 0.5 \text{ m}$$

To "jump down"
one jump — multiply by 10.
two jumps — multiply by 100.
three jumps — multiply by 1000.
four jumps — multiply by 10,000, and so on.

EXAMPLE

Change 42 m to centimeters.

Two jumps are involved: (1) meters to decimeters, (2) decimeters to centimeters. Therefore,

$$42(100) = 4200$$

and
$$42 \text{ m} = 4200 \text{ cm}$$

Close observation of the techniques described shows that **changing from one unit of length to another unit of length in the metric system can be accomplished by proper adjustment of the decimal point.** Indeed, this is one of the important features of the metric system. For larger units, the decimal point is moved left to make the number smaller. *(Think of the smaller number compensating for the larger unit.)* The decimal point should be moved to the right for a change to smaller units. *(Think of the larger number compensating for the smaller unit.)*

Although the units dekameter and hectometer are part of the metric system, they have very little practical use and will not appear in the exercises. In addition, since some countries use a comma in the same way that a decimal point is used in the United States, groups of three digits (either to the left or the right of the decimal position) are now separated by a space rather than a comma. Four digits need not be separated unless they are in tabular form.

EXAMPLES

1. Change 15 m to millimeters (three jumps down).

    15 m = 15 000 mm

2. Change 15 m to hectometers (two jumps up).

    15 m = 0.15 hm

3. 16.3 mm = _____ cm (one jump up).

    16.3 mm = 1.63 cm

4. 3050 km = _____ m (three jumps down).

    3050 km = 3 050 000 m

The chart in Figure 7.6 might be helpful for changing units. Each unit is listed across the top, and each vertical line separates the digits and is a place for a decimal point.

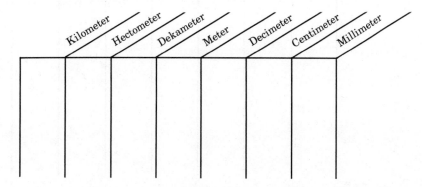

Figure 7.6

> **TO CHANGE FROM ONE UNIT TO ANOTHER USING THE CHART METHOD**
>
> 1. Enter the number so that each digit is in one column and the decimal point is on the given unit line.
> 2. Move the decimal point to the desired unit line.
> 3. Fill in all spaces with 0's.

For example, 15 m = 15 000 mm. The chart technique gives the same result. Figure 7.7 shows that 15 m = 15 000 mm.

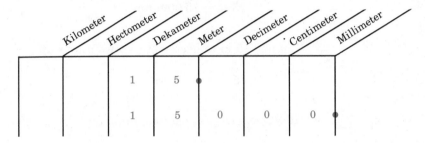

Figure 7.7

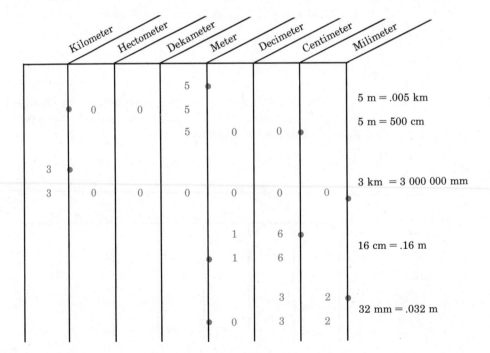

Figure 7.8

The chart in Figure 7.8 shows several unit changes. The technique illustrates the simplicity of the metric system.

SELF QUIZ	Change the following units as indicated.	ANSWERS
	1. 3m = ____ mm	1. 3000 mm
	2. 12 km = ____ m	2. 12 000 m
	3. 5 cm = ____ m	3. .05 m
	4. 25 m = ____ km	4. .025 km

EXERCISES 7.1

1. Measure your height in centimeters (to the nearest centimeter), in millimeters. (Meter sticks are usually available at hardware stores and lumberyards.)

2. Measure the lengths of five objects in your home with a meter stick to the nearest millimeter.

3. Measure the width of the classroom to the nearest centimeter.

4. Measure the length of a driveway to the nearest meter.

First estimate the length of each of the following line segments to the nearest centimeter. Then, measure each segment to see how close your estimates were.

5. ————————

6. ————————————————————————————————

7. ————————————————————

First estimate the length of each of the following line segments to the nearest millimeter. Then, measure each segment to see how close your estimates were.

8. ————————————————

9. ————————

10. ————————————————————————

First estimate the length of the perimeter (distance around) of each of the following figures to the nearest millimeter. Then, measure the perimeters to see how close your estimates were.

11.   **12.**   **13.**

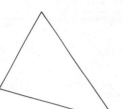
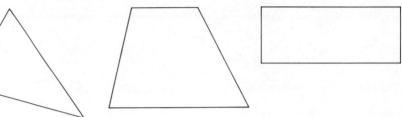

Long distances, such as distances between towns, are usually measured in kilometers. On the following "map," the scale is such that 1 cm represents 5 km. The letters represent towns, and the lines are roads. Using this scale, answer the following questions.

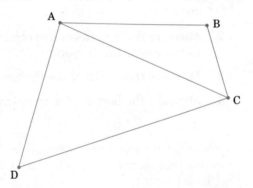

14. What is the distance in kilometers from *A* to *B*?

15. What is the distance in kilometers from *A* to *D*?

16. Which route is shorter, from *A* to *B* to *C* or from *A* to *C* to *B*? How much shorter?

17. Which route is shorter, from *D* to *A* to *B* or from *D* to *C* to *B*? How much shorter?

Change the following units as indicated.

18. 1 m = ___ cm   **19.** 12 m = ___ cm

20. 3 m = ____ cm

21. 4 m = ____ mm

22. 0.7 m = ____ mm

23. 0.16 m = ____ mm

24. 13 cm = ____ m

25. 75 mm = ____ cm

26. 36 mm = ____ m

27. 25 mm = ____ m

28. 37 km = ____ m

29. 15 km = ____ m

30. 17 dm = ____ mm

31. 25 cm = ____ m

32. 0.3 m = ____ km

33. 1 km = ____ mm

34. 3000 m = ____ km

35. 0.4 cm = ____ mm

36. 3 dm = ____ cm

37. 1.5 dm = ____ m

38. 0.561 mm = ____ cm

39. 750 mm = ____ dm

40. 3.51 m = ____ dm

41. 170 m = ____ km

42. 1.3 km = ____ m

43. 75 km = ____ m

44. 3.8 km = ____ dm

45. 786 cm = ____ mm

46. 342 km = ____ m

47. 59 m = ____ dm

48. 362 dm = ____ km

49. 57 cm = ____ dm

50. 896 mm = ____ m

7.2   MASS (WEIGHT)

Two closely related concepts are **mass** and **weight. Mass** is the amount of material in an object. If two objects balance on an equal arm balance, they have the same mass (Figure 7.9). The objects have the same mass regardless of where each one is in space. The amount of material in each object does not change because of its position in space.

If, however, one of the objects is on earth, and the other is in a satellite 100 kilometers above the earth, the objects have different weights.

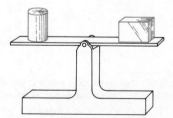

The two objects have the same *mass*

Figure 7.9

Weight is the force of the earth's gravitational pull on an object. The further an object is from earth, the less the gravitational pull of the earth. Thus, astronauts experience weightlessness in space. If an object was on a planet with stronger gravitational pull than the earth's, it would weigh more there than on earth, but its mass would be unchanged.

For practical purposes, because most of us do not stray far from the earth's surface, weight and mass are used interchangeably. In this text, they will be treated interchangeably. A mass of 20 kilograms will be said to weigh 20 kilograms.

In the metric system, the standard base unit of mass is the **kilogram,** * about 2.2 pounds. It is the mass of a certain cylinder of platinum-iridium alloy kept by the International Bureau of Weights and Measures in Paris. The National Bureau of Standards has a duplicate that serves as the mass standard for the United States.

In some fields, such as medicine and science, the **gram,** about the mass of a paper clip, is more convenient as a basic unit than the kilogram. The **metric ton** is useful for measuring heavy objects such as loaded trucks and railroad cars.

TABLE 7.4   MEASURES OF MASS

---

1 *milli*gram (mg) = 0.001 gram
1 gram   (g) = 1.0 gram
1 *kilo*gram   (kg) = 1000 grams
1 metric ton   (t) = 1000 kilograms

Many units of mass in the metric system do not have practical use. Therefore, the centigram, decigram, dekagram, and hectogram have been omitted from Table 7.4. As with length and volume, the units of mass are each 10 times the previous unit and 0.1 of the following unit (see Table 7.5).

TABLE 7.5   RELATIONSHIPS BETWEEN SMALLER AND LARGER MEASURES OF MASS

---

1 g = 1000 mg
1 g = 0.001 kg
1 kg = 0.001 t

---

*Originally, the basic unit was a gram, defined to be the mass of 1 cm³ of distilled water at 4° Centigrade. This mass is still considered accurate for many purposes, so that

1 cm³ of water has a mass of 1 g.
1 dm³ of water has a mass of 1 kg.
1 m³ of water has a mass of 1000 kg, or 1 metric ton.

Cubic units indicated by cm³, dm³, and m³ will be discussed in Section 7.4.

EXAMPLES

1. 60 mg = 0.06 g
2. 78 g = 78 000 mg
3. 135 mg = 0.135 g
4. 100 g = 0.1 kg
5. 5700 kg = 5.7 t
6. 3.4 t = 3400 kg

As in changing units of length, a chart is useful in changing units of mass (see Figure 7.10). The only difference in the chart is that the headings are parts of a gram instead of a meter. The use of the chart is the same. Simply enter the digits and decimal point in the correct places, then move the decimal point to the new unit line. Fill in any spaces with 0's.

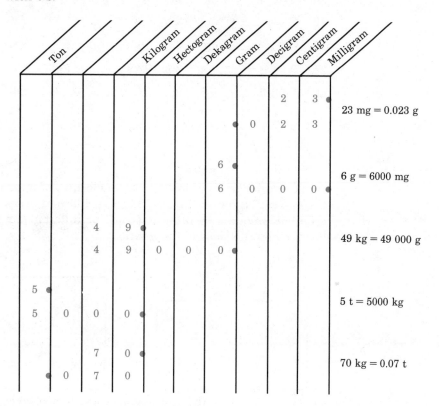

Figure 7.10

EXERCISES 7.2

Change the following units as indicated.

1. 7 g = ___ mg
2. 2 kg = ___ g
3. 34.5 mg = ___ g

4. 3700 kg = ___ t
5. 4000 kg = ___ t
6. 5600 g = ___ kg

7. 73 kg = ___ mg
8. 91 kg = ___ t
9. 0.54 g = ___ mg

10. 0.7 g = ___ mg
11. 5 t = ___ kg
12. 17 t = ___ kg

13. 2 t = ___ kg
14. 896 mg = ___ g
15. 896 g = ___ mg

16. 342 kg = ___ g
17. 75 000 g = ___ kg
18. 3000 mg = ___ g

19. 7 t = ___ g
20. 0.4 t = ___ g
21. .34 g = ___ kg

22. .78 g = ___ mg
23. 16 mg = ___ g
24. 2.5 g = ___ mg

25. 3.94 g = ___ mg
26. 92.3 g = ___ kg
27. 5.6 t = ___ kg

28. 7.58 t = ___ kg
29. 3547 kg = ___ t
30. 2963 kg = ___ t

7.3   AREA

Area is a measure of interior, or enclosure, of a surface. For example, the two rectangles in Figure 7.11 have different areas because they have different amounts of interior space, or different amounts of space are enclosed by the sides of the figures.

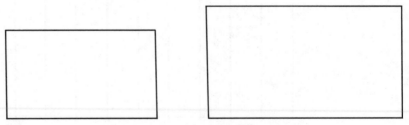

These two rectangles have different *areas*

Figure 7.11

Area is measured in square units. A square that is 1 centimeter long on each side is said to have an area of 1 square centimeter, or the area is 1 cm². A rectangle that is 7 cm on one side and 4 cm on the other side encloses 28 squares that have area 1 cm². So the rectangle is said to have an area of 28 square centimeters or 28 cm². (See Figure 7.12.)

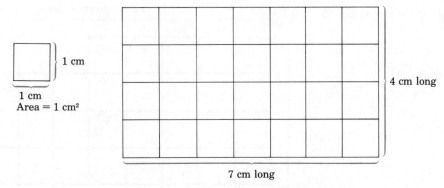

Area = 7 cm × 4 cm = 28 cm²

There are 28 squares that are each 1 cm² in the large rectangle

Figure 7.12

In the metric system, a square that is 1 centimeter on a side encloses 100 square millimeters, as shown in Figure 7.13.

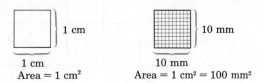

Figure 7.13

If a square is 10 cm on a side, then it contains 10 cm × 10 cm = 100 cm². But this same square is 1 dm on a side. So, it contains 1 dm². This means that

$$1 \text{ dm}^2 = 100 \text{ cm}^2$$

We already know that each square centimeter contains 100 square millimeters. So,

$$1 \text{ dm}^2 = 100 \text{ cm}^2 = 10\ 000 \text{ mm}^2$$

This relationship is illustrated in Figure 7.14 at the top of the next page.

Square millimeters, square centimeters, square decimeters, and square meters are useful measures for relatively small areas. For exam-

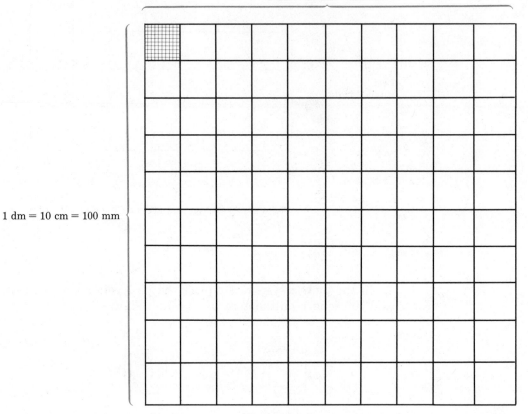

Figure 7.14

TABLE 7.6   MEASURES OF AREA

$$1 \text{ cm}^2 = 100 \text{ mm}^2$$
$$1 \text{ dm}^2 = \quad 100 \text{ cm}^2 = 10\ 000 \text{ mm}^2$$
$$1 \text{ m}^2 = 100 \text{ dm}^2 = 10\ 000 \text{ cm}^2 = 1\ 000\ 000 \text{ mm}^2$$

[NOTE: Each smaller unit of area is 100 times the previous unit of area — *not* just 10 times.]

ple, the area of the floor of your classroom could be measured in square meters, and the area of this page of paper could be measured in square centimeters or square millimeters.

In the U.S. customary system, large areas of land are measured in acres. In the metric system, large areas are measured in ares and hectares. (*Are* is pronounced "air" and abbreviated a.) A square with each

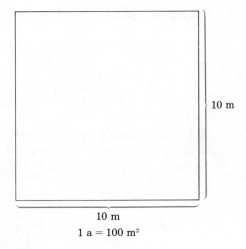

10 m

10 m

1 a = 100 m²

Figure 7.15

side 10 m long has an area of 1 are which is equal to 10 m × 10 m = 100 m². (See Figure 7.15.) A hectare (ha) is 100 ares. That is, 1 ha = 100 a.

TABLE 7.7 MEASURES OF LAND AREA

1 a = 100 m²
1 ha = 100 a = 10 000 m²

EXAMPLES

Change the following units as indicated.

1. 5 cm² = _____ mm²
    5 cm² = 500 mm²

2. 3 dm² = _____ cm² = _____ mm²
    3 dm² = 300 cm² = 30 000 mm²

3. 1.4 m² = _____ dm² = _____ cm² = _____ mm²
    1.4 m² = 140 dm² = 14 000 cm² = 1 400 000 mm²

4. 3.2 a = _____ m²
    3.2 a = 320 m²

5. 7.63 ha = _____ a = _____ m²
    7.63 ha = 763 a = 76 300 m²

6. How many ares are in 1 km²?
    [NOTE: One km is about .6 mile, so 1 km² is about .6 × .6 = .36 square mile.]

Remember that 1 km = 1000 m, so

$$1 \text{ km}^2 = (1000 \text{ m}) \times (1000 \text{ m})$$
$$= 1\ 000\ 000 \text{ m}^2$$
$$= 10\ 000 \text{ a} \quad \text{(Divide m}^2 \text{ by 100 to get}$$

(Divide m² by 100 to get ares because every 100 m² is equal to one are.)

7. A farmer plants corn and beans as shown in the figure. How many ares and how many hectares are planted in corn? in beans? (From Example 6 we know 1 km² = 10 000 a.)

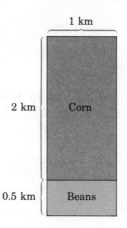

Corn: (2 km)(1 km) = 2 km² = 20 000 a
= 200 ha
Beans: (0.5 km)(1 km) = 0.5 km² = 5000 a
= 50 ha

SELF QUIZ	Change the following units as indicated.	ANSWERS
1.	22 cm² = ____ mm²	1. 2200 mm²
2.	500 mm² = ____ cm²	2. 5 cm²
3.	3.7 dm² = ____ cm² = ____ mm²	3. 370 cm² = 37 000 mm²
4.	3.6 a = ____ m²	4. 360 m²
5.	0.73 ha = ____ a = ____ m²	5. 73 a = 7300 m²

The formulas for the areas of common geometric figures are given here for reference and for use in the exercises.

Square
$A = s^2$

Rectangle
$A = lw$

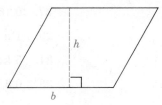

Parallelogram
$A = bh$

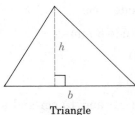

Triangle
$A = \frac{1}{2}bh$

Circle
$A = \pi r^2$

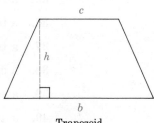

Trapezoid
$A = \frac{1}{2}h(b + c)$

EXERCISES 7.3

Change the following units as indicated.

1. $3 \text{ cm}^2 = $ ___ mm^2 **2.** $5.6 \text{ cm}^2 = $ ___ mm^2

3. $8.7 \text{ cm}^2 = $ ___ mm^2 **4.** $3.61 \text{ cm}^2 = $ ___ mm^2

5. $600 \text{ mm}^2 = $ ___ cm^2 **6.** $28 \text{ mm}^2 = $ ___ cm^2

7. $1400 \text{ mm}^2 = $ ___ cm^2 **8.** $20\ 000 \text{ mm}^2 = $ ___ cm^2

9. $4 \text{ dm}^2 = $ ___ $\text{cm}^2 = $ ___ mm^2 **10.** $7.3 \text{ dm}^2 = $ ___ $\text{cm}^2 = $ ___ mm^2

11. $57 \text{ dm}^2 = $ ___ $\text{cm}^2 = $ ___ mm^2

12. $0.6 \text{ dm}^2 = $ ___ $\text{cm}^2 = $ ___ mm^2

13. $17 \text{ m}^2 = $ ___ $\text{dm}^2 = $ ___ $\text{cm}^2 = $ ___ mm^2

14. $2.9 \text{ m}^2 = $ ___ $\text{dm}^2 = $ ___ $\text{cm}^2 = $ ___ mm^2

15. $0.03 \text{ m}^2 = $ ___ $\text{dm}^2 = $ ___ $\text{cm}^2 = $ ___ mm^2

16. $0.5 \text{ m}^2 = $ ___ $\text{dm}^2 = $ ___ $\text{cm}^2 = $ ___ mm^2

17. $142 \text{ mm}^2 = $ ___ cm^2 **18.** $5800 \text{ mm}^2 = $ ___ cm^2

19. $200 \text{ dm}^2 = $ ___ m^2 **20.** $35 \text{ dm}^2 = $ ___ m^2

21. $7.8 \text{ a} = $ ___ m^2 **22.** $300 \text{ a} = $ ___ m^2

23. $0.04 \text{ a} = $ ___ m^2 **24.** $0.53 \text{ a} = $ ___ m^2

25. $8.69 \text{ ha} = $ ___ $\text{a} = $ ___ m^2 **26.** $7.81 \text{ ha} = $ ___ $\text{a} = $ ___ m^2

27. 0.16 ha = ___ a = ___ m²   28. 0.02 ha = ___ a = ___ m²

29. 1 a = ___ ha   30. 15 a = ___ ha

31. 5 km² = ___ a = ___ ha   32. 4.76 km² = ___ a = ___ ha

33. 0.3 km² = ___ a = ___ ha   34. 0.532 km² = ___ a = ___ ha

Find the area of each of the following figures.

35. a rectangle 35 cm long and 25 cm wide

36. a triangle with base 2 cm and altitude 6 cm

37. a triangle with base 5 mm and altitude 8 mm

38. a circle of radius 5 m (use $\pi \approx 3.14$)

39. a circle of radius 1.5 cm (use $\pi \approx 3.14$)

40. a trapezoid with parallel sides of 8 cm and 10 cm and altitude of 35 cm

41. a trapezoid with parallel sides of 3.5 mm and 4.2 mm and altitude 1 cm

42. a parallelogram of altitude 10 cm to a base of 5 mm

Find the areas of each of the following figures with the indicated dimensions (use $\pi \approx 3.14$).

43.

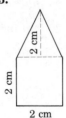

2 cm

2 cm

2 cm

44.

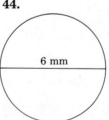

6 mm

45.

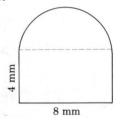

4 mm

8 mm

46.

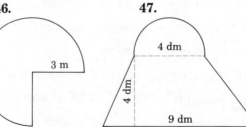

3 m

47.

4 dm

4 dm

9 dm

48.

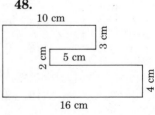

10 cm

3 cm

2 cm   5 cm

16 cm

4 cm

Find the areas of the shaded portions in ares (use $\pi \approx 3.14$).

49.

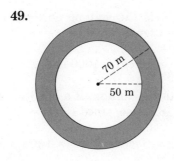

50.

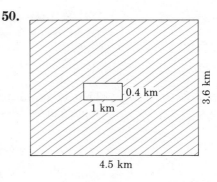

7.4 VOLUME

Volume is a measure of the space enclosed by a three-dimensional figure. The volume or space contained within a cube that is 1 cm on each edge is *one cubic centimeter,* or 1 cm³, as shown in Figure 7.16. A cubic centimeter is about the size of a sugar cube.

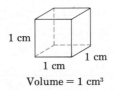

Volume = 1 cm³

Figure 7.16

A rectangular solid that has edges of 3 cm and 2 cm and 5 cm has a volume of 3 cm × 2 cm × 5 cm = 30 cm³. We can think of the rectangular solid as being three layers of ten cubic centimeters, as shown in Figure 7.17.

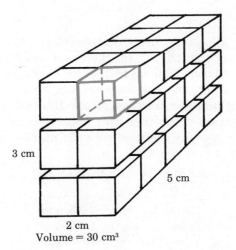
Volume = 30 cm³

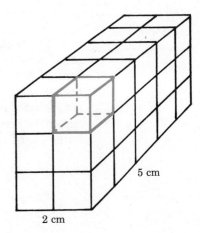

Figure 7.17

A **liter** (abbreviated ℓ) is the volume enclosed in a cube that is 10 cm on each edge. So, one liter is equal to

$$10 \text{ cm} \times 10 \text{ cm} \times 10 \text{ cm} = 1000 \text{ cm}^3$$

That is, one liter is equal to the volume of a box that holds about 1000 sugar cubes. (See Figure 7.18.) Also, since 10 cm = 1 dm,

$$10 \text{ cm} \times 10 \text{ cm} \times 10 \text{ cm} = 1 \text{ dm} \times 1 \text{ dm} \times 1 \text{ dm} = 1 \text{ dm}^3$$

Thus, $1 \ell = 1 \text{ dm}^3 = 1000 \text{ cm}^3$

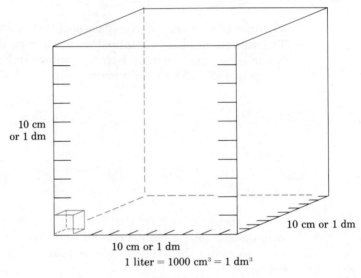

10 cm
or 1 dm

10 cm or 1 dm

10 cm or 1 dm

1 liter = 1000 cm³ = 1 dm³

Figure 7.18

The prefixes kilo, hecto, deka, deci, centi, and milli all indicate the same parts of a liter that they do for the meter and the gram. The same type of chart used before will be helpful for changing units. However, the centiliter, deciliter, and dekaliter are not commonly used and are not included in any tables or exercises. [NOTE: Liter will be abbreviated ℓ except in combination with other unit abbreviations such as ml. The letter l used alone is easily confused with the number 1.]

TABLE 7.8   MEASURES OF VOLUME

1 *milli*liter	(ml)	= 0.001 liter
1 liter	(ℓ)	= 1.0 liter
1 *hecto*liter	(hl)	= 100 liters
1 *kilo*liter	(kl)	= 1000 liters

TABLE 7.9 EQUIVALENTS
FOR METRIC MEASURES
OF VOLUME

$$1 \text{ ml} = 1 \text{ cm}^3$$
$$1000 \text{ ml} = 1 \ell = 1 \text{ dm}^3$$
$$10 \text{ hl} = 1 \text{ kl} = 1 \text{ m}^3$$

The chart in Figure 7.19 shows several unit changes.

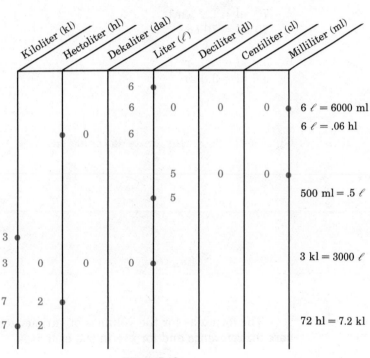

Figure 7.19

EXAMPLES   Change the following units as indicated.

1. 5000 ml = _____ ℓ
    5000 ml = 5 ℓ

2. 3.2 ℓ = _____ ml
    3.2 ℓ = 3200 ml

3. 60 hl = _____ kl
    60 hl = 6 kl

4. 637 ml = _____ ℓ
    637 ml = 0.637 ℓ

NOTE: Since 1 ℓ = 1000 ml   and   1 ℓ = 1000 cm³,

1 cm³ = 1 ml

Also, 1 kl = 1000 ℓ = 1 000 000 cm³   and   1 000 000 cm³ = 1 m³. So,

1 kl = 1 m³

5. 70 ml = ____ cm³
    70 ml = 70 cm³

6. 3.8 kl = ____ m³
    3.8 kl = 3.8 m³

SELF QUIZ	Change the following units as indicated.	ANSWERS
	1. 2 ml = ____ ℓ	1. 0.002 ℓ
	2. 3.6 kl = ____ ℓ	2. 3600 ℓ
	3. 500 ml = ____ ℓ	3. 0.5 ℓ
	4. 500 ml = ____ cm³	4. 500 cm³
	5. 42 hl = ____ kl	5. 4.2 kl

The formulas for the volumes of various geometric solids are given here for reference and for use in the exercises.

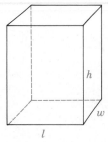

Rectangular solid
$V = lwh$

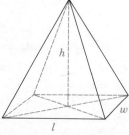

Rectangular pyramid
$V = \frac{1}{3}lwh$

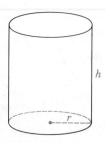

Right circular cylinder
$V = \pi r^2 h$

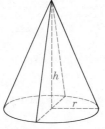

Right circular cone
$V = \frac{1}{3}\pi r^2 h$

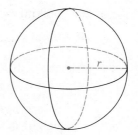

Sphere
$V = \frac{4}{3}\pi r^3$

EXERCISES 7.4

Copy and complete the following tables.

1.
$1 \text{ cm}^3 = \underline{\hspace{1cm}} \text{ mm}^3$
$1 \text{ dm}^3 = \underline{\hspace{1cm}} \text{ cm}^3$
$1 \text{ m}^3 = \underline{\hspace{1cm}} \text{ dm}^3$
$1 \text{ km}^3 = \underline{\hspace{1cm}} \text{ m}^3$

2.
$1 \text{ dm} = \underline{\hspace{1cm}} \text{ cm}$
$1 \text{ dm} = \underline{\hspace{1cm}} \text{ mm}$
$1 \text{ dm}^2 = \underline{\hspace{1cm}} \text{ cm}^2$
$1 \text{ dm}^2 = \underline{\hspace{1cm}} \text{ mm}^2$
$1 \text{ dm}^3 = \underline{\hspace{1cm}} \text{ cm}^3$
$1 \text{ dm}^3 = \underline{\hspace{1cm}} \text{ mm}^3$

3.
$1 \text{ m} = \underline{\hspace{1cm}} \text{ dm}$
$1 \text{ m} = \underline{\hspace{1cm}} \text{ cm}$
$1 \text{ m}^2 = \underline{\hspace{1cm}} \text{ dm}^2$
$1 \text{ m}^2 = \underline{\hspace{1cm}} \text{ cm}^2$
$1 \text{ m}^3 = \underline{\hspace{1cm}} \text{ dm}^3$
$1 \text{ m}^3 = \underline{\hspace{1cm}} \text{ cm}^3$

4.
$1 \text{ km} = \underline{\hspace{1cm}} \text{ m}$
$1 \text{ km}^2 = \underline{\hspace{1cm}} \text{ m}^2$
$1 \text{ km}^3 = \underline{\hspace{1cm}} \text{ m}^3$
$1 \text{ km} = \underline{\hspace{1cm}} \text{ dm}$
$1 \text{ km}^2 = \underline{\hspace{1cm}} \text{ ha}$
$1 \text{ km}^3 = \underline{\hspace{1cm}} \text{ kl}$

Change the following units as indicated.

5. $73 \text{ kl} = \underline{\hspace{1cm}} \ell$ **6.** $0.9 \text{ kl} = \underline{\hspace{1cm}} \ell$ **7.** $400 \text{ ml} = \underline{\hspace{1cm}} \ell$

8. $525 \text{ ml} = \underline{\hspace{1cm}} \ell$ **9.** $63 \ell = \underline{\hspace{1cm}} \text{ ml}$ **10.** $8.7 \ell = \underline{\hspace{1cm}} \text{ ml}$

11. $5 \text{ hl} = \underline{\hspace{1cm}} \text{ kl}$ **12.** $69 \text{ hl} = \underline{\hspace{1cm}} \text{ kl}$ **13.** $19 \text{ ml} = \underline{\hspace{1cm}} \text{ cm}^3$

Change the following units as indicated.

14. $5 \text{ cm}^3 = \underline{\hspace{1cm}} \text{ mm}^3$ **15.** $2 \text{ dm}^3 = \underline{\hspace{1cm}} \text{ cm}^3$ **16.** $76.4 \text{ ml} = \underline{\hspace{1cm}} \ell$

17. $5.3 \ell = \underline{\hspace{1cm}} \text{ ml}$ **18.** $30 \text{ cm}^3 = \underline{\hspace{1cm}} \text{ ml}$ **19.** $30 \text{ cm}^3 = \underline{\hspace{1cm}} \ell$

20. $5.3 \text{ ml} = \underline{\hspace{1cm}} \ell$ **21.** $48 \text{ kl} = \underline{\hspace{1cm}} \ell$ **22.** $72\,000 \ell = \underline{\hspace{1cm}} \text{ kl}$

23. $32 \ell = \underline{\hspace{1cm}} \text{ hl}$ **24.** $80 \ell = \underline{\hspace{1cm}} \text{ ml}$ **25.** $290 \ell = \underline{\hspace{1cm}} \text{ kl}$

26. 569 ml = ___ ℓ　　**27.** 72 hl = ___ ml　　**28.** 7 ℓ = ___ ml

29. 95 hl = ___ ℓ　　**30.** 72 ℓ = ___ hl

Find the value of each of the following solids in a convenient unit from Table 7.8 (use $\pi \approx 3.14$).

31. a rectangular solid with a length of 5 dm, a width of 2 dm, and a height of 7 dm

32. a right circular cylinder 15 cm high and 1 dm in diameter (A diameter of a circle is a segment through the center with end points on the circle.)

33. a sphere with radius 4.5 cm

34. a sphere with diameter 12 dm

35. a right circular cone 3 dm high with a 2-dm radius

36. a rectangular pyramid with a length of 8 cm, a width of 10 mm, and a height of 3 dm

Find the volume of each of the solids with the dimensions indicated.

37.　　　　　　　　**38.**　　　　　　　　**39.**

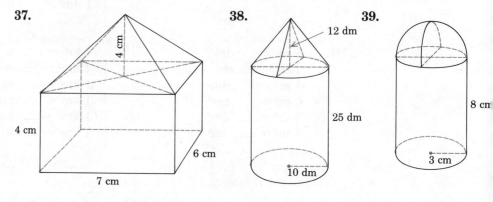

7.5   U.S. CUSTOMARY AND METRIC EQUIVALENTS

In Tables 7.10 – 7.13, the equivalents are rounded off.

TABLE 7.10   LENGTH EQUIVALENTS

U.S. TO METRIC	METRIC TO U.S.
1 in. = 2.54 cm (exact)	1 cm = .394 in.
1 ft = 0.305 m	1 m = 3.28 ft
1 yd = 0.915 m	1 m = 1.09 yd
1 mi = 1.61 km	1 km = .62 mi

TABLE 7.11   AREA EQUIVALENTS

U.S. TO METRIC	METRIC TO U.S.
1 in.^{2} = 6.45 cm^2	1 cm^2 = .155 in.^{2}
1 ft^2 = 0.093 m^2	1 m^2 = 10.764 ft^2
1 yd^2 = 0.836 m^2	1 m^2 = 1.196 yd^2
1 acre = 0.405 ha	1 ha = 2.47 acres

TABLE 7.12   VOLUME EQUIVALENTS

U.S. TO METRIC	METRIC TO U.S.
1 in.^{3} = 16.387 cm^3	1 cm^3 = .06 in.^{3}
1 ft^3 = 0.028 m^3	1 m^3 = 35.315 ft^3
1 qt = 0.946 ℓ	1 ℓ = 1.06 qt
1 gal = 3.785 ℓ	1 ℓ = .264 gal

TABLE 7.13   MASS EQUIVALENTS

U.S. TO METRIC	METRIC TO U.S.
1 oz = 28.35 g	1 g = .035 oz
1 lb = 0.454 kg	1 kg = 2.205 lb

Various equivalents can be read conveniently from the graphs in Figures 7.20 – 7.27.

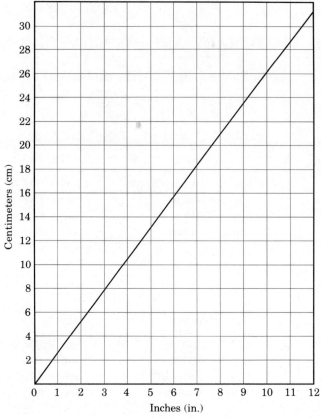

Figure 7.20   Centimeters/Inches

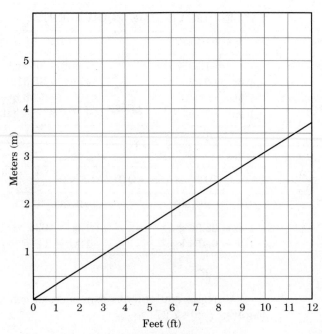

Figure 7.21   Meters/Feet

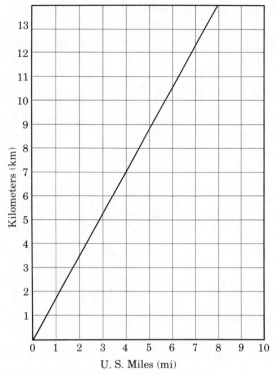

Figure 7.22   Kilometers/U.S. Miles

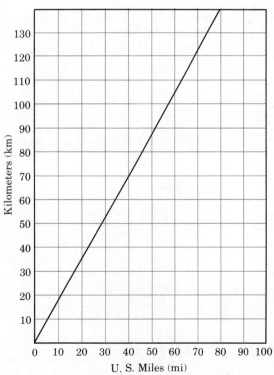

Figure 7.23   Kilometers/U.S. Miles

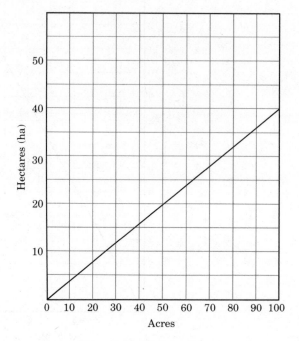

Figure 7.24   Hectares/Acres

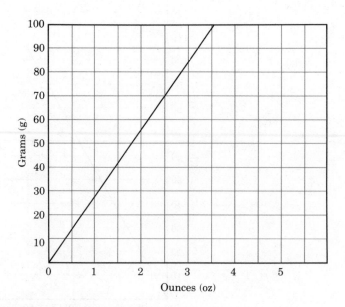

Figure 7.25   Grams/Ounces

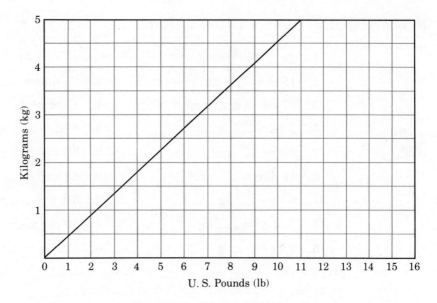

Figure 7.26   Kilograms/U.S. Pounds

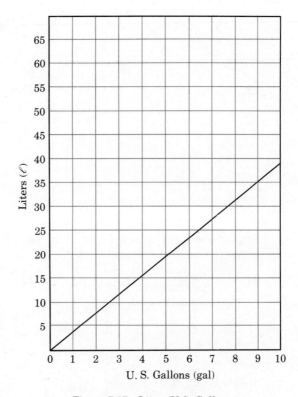

Figure 7.27   Liters/U.S. Gallons

REVIEW QUESTIONS · CHAPTER 7

Change the following units as indicated.

1. 15 m = ＿＿ cm

2. 35 mm = ＿＿ dm

3. 37 cm² = ＿＿ mm²

4. 17 mm² = ＿＿ cm²

5. 3 ha = ＿＿ a

6. 3 ha = ＿＿ m²

7. 5 ℓ = ＿＿ cm³

8. 36 ℓ = ＿＿ ml

9. 13 dm³ = ＿＿ cm³

10. 68 cm³ = ＿＿ mm³

11. 5 kg = ＿＿ g

12. 3.4 g = ＿＿ mg

13. 6.71 t '= ＿＿ kg

14. 19 mg = ＿＿ g

15. 8 kg = ＿＿ g

16. 4290 g = ＿＿ kg

Find the area of each of the following figures with dimensions indicated in convenient metric units (use $\pi \approx 3.14$).

17.

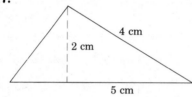

18.

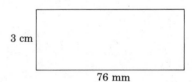

19.

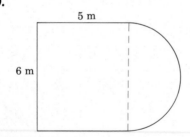

20.

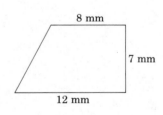

Find the volume in liters of each of the following solids with dimensions indicated (use $\pi \approx 3.14$).

21.

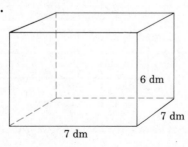

22.

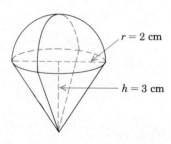

SQUARE ROOT

8

A number is **squared** when it is multiplied by itself. If a natural number is squared, the result is called a **perfect square number.** For example, squaring 3 gives $3^2 = 9$, and 9 is a perfect square number. Table 8.1 gives the perfect square numbers found by squaring the natural numbers from 1 through 20. The table should be memorized for convenience.

TABLE 8.1 PERFECT SQUARE NUMBERS FOR NATURAL NUMBERS FROM 1 THROUGH 20

$1^2 = 1$	$6^2 = 36$	$11^2 = 121$	$16^2 = 256$
$2^2 = 4$	$7^2 = 49$	$12^2 = 144$	$17^2 = 289$
$3^2 = 9$	$8^2 = 64$	$13^2 = 169$	$18^2 = 324$
$4^2 = 16$	$9^2 = 81$	$14^2 = 196$	$19^2 = 361$
$5^2 = 25$	$10^2 = 100$	$15^2 = 225$	$20^2 = 400$

The set of perfect squares is $\{1, 4, 9, 16, 25, 36, 49, \ldots\}$.

8.1 SQUARE ROOTS

What number multiplied by itself gives 25? What number is 49 the square of? The answers are 5 and 7, respectively. Both questions refer to the reverse of squaring, finding the **square root.** In other words, *5 is the square root of 25, and 7 is the square root of 49*.

DEFINITION   The **square root** of a given number is the number whose square is the given number.

The symbol for square root is $\sqrt{}$, called a **square root sign,** or a **radical sign.** Thus, because $5^2 = 25$ and $7^2 = 49$, $\sqrt{25} = 5$ and $\sqrt{49} = 7$.

Table 8.2 contains the square roots of the perfect square numbers from 1 through 400. Close inspection of Tables 8.1 and 8.2 will show the relationship between squares and square roots.

TABLE 8.2 SQUARE ROOTS OF PERFECT SQUARE NUMBERS FROM 1 THROUGH 400

$\sqrt{1} = 1$	$\sqrt{36} = 6$	$\sqrt{121} = 11$	$\sqrt{256} = 16$
$\sqrt{4} = 2$	$\sqrt{49} = 7$	$\sqrt{144} = 12$	$\sqrt{289} = 17$
$\sqrt{9} = 3$	$\sqrt{64} = 8$	$\sqrt{169} = 13$	$\sqrt{324} = 18$
$\sqrt{16} = 4$	$\sqrt{81} = 9$	$\sqrt{196} = 14$	$\sqrt{361} = 19$
$\sqrt{25} = 5$	$\sqrt{100} = 10$	$\sqrt{225} = 15$	$\sqrt{400} = 20$

Any real number can be squared, and each real number greater than or equal to 0 has a square root. The following examples indicate the relationship between squares and square roots for rational numbers related to the numbers in Tables 8.1 and 8.2.

EXAMPLES

1. $\left(\dfrac{1}{2}\right)^2 = \dfrac{1}{4}$ so $\sqrt{\dfrac{1}{4}} = \dfrac{1}{2}$

2. $\left(\dfrac{2}{3}\right)^2 = \dfrac{4}{9}$ so $\sqrt{\dfrac{4}{9}} = \dfrac{2}{3}$

3. $\left(\dfrac{4}{5}\right)^2 = \dfrac{16}{25}$ so $\sqrt{\dfrac{16}{25}} = \dfrac{4}{5}$

In general, if a and b are natural numbers,

$$\sqrt{\dfrac{a}{b}} = \dfrac{\sqrt{a}}{\sqrt{b}}$$

---

4. $(1.5)^2 = 2.25$ so $\sqrt{2.25} = 1.5$

5. $(1.3)^2 = 1.69$ so $\sqrt{1.69} = 1.3$

6. $(.8)^2 = .64$ so $\sqrt{.64} = .8$

The square roots of some numbers are not as easily found as those in the examples or in the tables. In fact, *most* square roots are irrational numbers (nonrepeating infinite decimals); that is, *most* square roots can only be approximated with decimals. Two methods for approximating square roots will be discussed in the next two sections. To illustrate the need for these methods, consider the following decimal approximations to $\sqrt{2}$.

1.4	1.41	1.414	1.4142	1.41421
1.4	1.41	1.414	1.4142	1.41421
56	141	5656	28284	141421
1 4	564	1414	56568	282842
1.96	1 41	5656	14142	565684
	1.9881	1 414	56568	141421
		1.999396	1 4142	565684
			1.99996164	1 41421
				1.9999899241

1.5	1.42	1.415	1.4143	1.41422
1.5	1.42	1.415	1.4143	1.41422
65	284	7075	42429	282844
1 5	568	1415	56572	282844
2.15	1 42	5660	14143	565688
	2.0164	1 415	56572	141422
		2.002225	1 4143	565688
			2.00024449	1 41422
				2.0000182084

So, $\sqrt{2}$ is between 1.41421 and 1.41422.

EXERCISES 8.1

Square the numbers as indicated.

ASSIGN ODD
1 - 69

1. 3^2 2. 7^2 3. 8^2 4. 12^2 5. 16^2

6. 17^2 7. 18^2 8. 19^2 9. 20^2 10. 25^2

11. $\left(\frac{1}{2}\right)^2$ 12. $\left(\frac{1}{4}\right)^2$ 13. $\left(\frac{2}{3}\right)^2$ 14. $\left(\frac{5}{6}\right)^2$ 15. $\left(\frac{10}{11}\right)^2$

16. $\left(\frac{14}{17}\right)^2$ 17. $\left(\frac{25}{31}\right)^2$ 18. $(.1)^2$ 19. $(.4)^2$ 20. $(.6)^2$

21. $(.9)^2$ 22. $(1.3)^2$ 23. $(1.5)^2$ 24. $(1.8)^2$ 25. $(2.4)^2$

26. $(3.6)^2$ 27. $(6.1)^2$ 28. $(.01)^2$ 29. $(.05)^2$ 30. $(.08)^2$

31. $(.12)^2$ 32. $(.14)^2$ 33. $(.19)^2$ 34. $(2.01)^2$ 35. $(3.52)^2$

Using the table of square roots and a basic knowledge of fractions and decimals, find the following square roots.

36. $\sqrt{16}$ 37. $\sqrt{25}$ 38. $\sqrt{81}$ 39. $\sqrt{169}$ 40. $\sqrt{121}$

41. $\sqrt{324}$ 42. $\sqrt{400}$ 43. $\sqrt{1.69}$ 44. $\sqrt{3.61}$ 45. $\sqrt{2.89}$

46. $\sqrt{.04}$ 47. $\sqrt{.0004}$ 48. $\sqrt{.09}$ 49. $\sqrt{.0009}$ 50. $\sqrt{.01}$

51. $\sqrt{.0324}$ 52. $\sqrt{.0225}$ 53. $\sqrt{.0144}$ 54. $\sqrt{.0064}$ 55. $\sqrt{.0036}$

56. $\sqrt{\frac{9}{25}}$ 57. $\sqrt{\frac{16}{25}}$ 58. $\sqrt{\frac{49}{36}}$ 59. $\sqrt{\frac{9}{49}}$ 60. $\sqrt{\frac{4}{81}}$

61. $\sqrt{\frac{100}{121}}$ 62. $\sqrt{\frac{324}{169}}$ 63. $\sqrt{\frac{400}{49}}$ 64. $\sqrt{\frac{121}{144}}$ 65. $\sqrt{\frac{256}{361}}$

66. Approximate $\sqrt{3}$ to the nearest hundredth by "trial and error."

67. Approximate $\sqrt{5}$ to the nearest hundredth by "trial and error."

68. What is the number whose square is .81?

69. 3.24 is the square of what number?

70. If the square of 25 is added to the square root of 49, what is the sum?

8.2   THE SQUARE ROOT ALGORITHM

The following examples illustrate a method for finding (or approximating) square roots called the **square root algorithm.*** The process can be learned by studying the examples and comments carefully. Algebraic techniques are involved in the proof of the algorithm, so the proof will not be given here.

EXAMPLES   1.   Find $\sqrt{120409}$.

$\sqrt{12|04|09.}$ STEP 1. Beginning at the decimal point and moving left, mark off the digits in pairs. Each pair of digits will determine a single digit in the square root.

$\sqrt{12|04|09.}$
$\quad\ 9$ STEP 2. Find the largest perfect square number less than or equal to the leftmost pair of digits. (In this case, 9 is the largest perfect square less than 12.)

$\quad\quad 3\quad\quad .$
$\sqrt{12|04|09.}$
$\quad\ 9$
$\quad\ 3$ STEP 3. Write the square root of this number above the leftmost pair of digits and subtract the perfect square from the pair of digits. (In this case, $\sqrt{9} = 3$, and $12 - 9 = 3$.)

$\quad\quad 3\quad\quad .$
$\sqrt{12|04|09.}$
$\quad\ 9$
$6\quad 3\ 04$ STEP 4. Bring down the next two digits and double the root found in Step 3. (In this case, we bring down 04 and double 3.)

$\quad\quad 3\quad\quad .$
$\sqrt{12|04|09.}$
$\quad\ 9$
$6\_\quad 3\ 04$ STEP 5. Place a blank (_) next to the doubled digit, indicating a missing digit. This missing digit must also be in the square root.

$\quad\quad 3\ \ 5\quad .$
$\sqrt{12|04|09.}$
$\quad\ 9$
$6\underline{5}\quad 3\ 04$ STEP 6. Try to find the missing digit by trial division. (In this case, trial divide 60 into 304 or 6 into 30.) Now, $30 \div 6 = 5$ so put 5 in both the blank and the square root.

---

*An algorithm is a process or pattern of steps to be followed in working with numbers.

$$\begin{array}{r} 3\ \ 5\ \ . \\ \sqrt{12\mathbf{|}04\mathbf{|}09.} \\ 9 \\ \hline 65 \quad 3\ 04 \\ \underline{5} \quad 3\ 25 \end{array}$$

STEP 7. Multiplying 5 by 65 gives a number larger than 304, so try 4. Multiply $4 \cdot 64 = 256$ and subtract. [NOTE: To aid in multiplying, write 5 directly below 65 and 4 directly below 64.]

$$\begin{array}{r} 3\ \ 4\ \ . \\ \sqrt{12\mathbf{|}04\mathbf{|}09.} \\ 9 \\ \hline 64 \quad 3\ 04 \\ \underline{4} \quad 2\ 56 \\ \hline 48 \end{array}$$

$$\begin{array}{r} 3\ \ 4\ \ 7. \\ \sqrt{12\mathbf{|}04\mathbf{|}09.} \\ 9 \\ \hline 64 \quad 3\ 04 \\ \underline{4} \quad 2\ 56 \\ \hline 687 \quad 48\ 09 \\ \underline{7} \quad 48\ 09 \end{array}$$

STEP 8. Again double the digits in the square root and bring down the next two digits and repeat the process of trial division to find the missing digit. (In this case, doubling 34 gives 68 and trial division gives 7 as the missing digit.)

$$\sqrt{120409} = 347$$

2. Find $\sqrt{63001}$.

(Some of the steps in Example 1 will be combined.)

$$\sqrt{6\mathbf{|}30\mathbf{|}01.}$$

STEP 1. In marking off pairs of digits, the left-most "pair" may be a single digit, as in this case.

$$\begin{array}{r} 2 \\ \sqrt{6\mathbf{|}30\mathbf{|}01.} \\ 4 \\ \hline 2\ 30 \end{array}$$

STEP 2. The largest perfect square less than 6 is 4. $\sqrt{4} = 2$, and $6 - 4 = 2$. Bring down 30.

$$\begin{array}{r} 2\ \ 5 \\ \sqrt{6\mathbf{|}30\mathbf{|}01.} \\ 4 \\ \hline 45 \quad 2\ 30 \\ \underline{5} \quad 2\ 25 \end{array}$$

STEP 3. Double the square root, 2, and trial divide to find the missing digit. In this case, 5 works since $5 \cdot 45 = 225$, which is less than 230.

$$\begin{array}{r} 2\ \ 5\ \ 1. \\ \sqrt{6\mathbf{|}30\mathbf{|}01.} \\ 4 \\ \hline 45 \quad 2\ 30 \\ \underline{5} \quad 2\ 25 \\ \hline 501 \quad 5\ 01 \\ \underline{1} \quad 5\ 01 \end{array}$$

STEP 4. Subtracting, bringing down the last pair of digits, and doubling the square root, the missing digit is 1.

$$\sqrt{63001} = 251$$

The square root of any decimal can be found, or at least approximated by this process. Simply write zeros to the right of the decimal point as far as necessary and *mark the digits off in pairs*. To find the square root of a number correct to a certain decimal place, carry the process one step beyond that place and round off.

3.   Find $\sqrt{14.3}$ correct to the nearest hundredth.

$$
\begin{array}{r}
3.\ 7\ 8\ 1 \approx 3.78 \\
\sqrt{14.30\,00\,00}
\end{array}
$$

$$
\begin{array}{rr}
 & 9 \\
67 & 5\,30 \\
7 & 4\,69 \\
748 & 61\,00 \\
8 & 59\,84 \\
7561 & 1\,16\,00 \\
1 & 75\,61
\end{array}
$$

$\sqrt{14.3} \approx 3.78$

EXERCISES 8.2

Find the following square roots correct to the nearest tenth.

1. $\sqrt{700}$ 2. $\sqrt{8200}$ 3. $\sqrt{6900}$ 4. $\sqrt{6085}$

5. $\sqrt{17.56}$ 6. $\sqrt{28.49}$ 7. $\sqrt{.06}$ 8. $\sqrt{5.4}$

9. $\sqrt{270,000}$ 10. $\sqrt{962,400}$

Find the following square roots correct to the nearest hundredth.

11. $\sqrt{5}$ 12. $\sqrt{3}$ 13. $\sqrt{14}$ 14. $\sqrt{.0215}$

15. $\sqrt{.0657}$ 16. $\sqrt{1.55}$ 17. $\sqrt{.333}$ 18. $\sqrt{75,421}$

19. $\sqrt{50,892}$ 20. $\sqrt{144,000}$

Find the following square roots correct to the nearest thousandth.

21. $\sqrt{5}$ 22. $\sqrt{2}$ 23. $\sqrt{.002468}$ 24. $\sqrt{1.035}$

25. $\sqrt{2.068}$

Find the square roots of the following perfect square numbers.

26. 454,276 27. 56,169 28. 236,196 29. 192,721 30. 79,524

31.   What number, to the nearest hundredth, has 76 as its square?

32.   If a number is multiplied by 4, and the square root of the new number is found, how will this square root compare with the square root of the original number?

8.3   THE DIVIDE-AND-AVERAGE METHOD

> **TO FIND THE SQUARE ROOT OF A NUMBER BY THE DIVIDE-AND-AVERAGE METHOD**
>
> 1. Estimate the square root.
>
> 2. Divide the number by the estimated square root.
>
> 3. Carry the quotient out to one more place than the estimate, and if the number in this place is even, leave it; if the number in this place is odd, replace it by the next larger even number.
>
> 4. Then average the estimate and the quotient.
>
> This average is the new estimate of the square root. The process of dividing and averaging is continued until the digits in the quotient and in the average are the same out to the desired decimal place.

An interesting fact about this process is that regardless of what number is used as the first estimate, eventually the same square root will result. This method is particularly adaptable for high-speed computers, but the square root algorithm will usually prove to be much faster for a person working only with pencil and paper.

EXAMPLE

Find $\sqrt{300}$ correct to the nearest hundredth by the divide-and-average method. Let the first estimate be 17 since $17^2 = 289$.

$$
\begin{array}{r}
17.6 \\
17\overline{)300.0} \\
\underline{17} \\
130 \\
\underline{119} \\
11\,0 \\
\underline{10\,2} \\
8
\end{array}
\qquad\qquad
\frac{17.6 + 17}{2} = \frac{34.6}{2} = 17.3
$$

$$
\begin{array}{r}
17.34 \\
17.3.\,\overline{)300.0.00} \\
\underline{173} \\
127\,0 \\
\underline{121\,1} \\
5\,9\,0 \\
\underline{5\,1\,9} \\
7\,10 \\
\underline{6\,92} \\
18
\end{array}
\qquad\qquad
\frac{17.34 + 17.3}{2} = \frac{34.64}{2} = 17.32
$$

$$\frac{17.322 + 17.32}{2} = \frac{34.642}{2} = 17.321$$

```
              17.321 use 17.322
    17.32. )300.00.0000
           173 2
           126 80
           121 24
             5 56 0
             5 19 6
               36 40
               34 64
                1 760
                1 732
                   28
```

$\sqrt{300} \approx 17.32$, to the nearest hundredth.

EXERCISES 8.3

Use the divide-and-average method and find the following square roots correct to the nearest hundredth.

1. $\sqrt{500}$ 2. $\sqrt{370}$ 3. $\sqrt{2}$ 4. $\sqrt{8}$ 5. $\sqrt{28,900}$

6. $\sqrt{22,500}$ 7. $\sqrt{250}$ 8. $\sqrt{425}$ 9. $\sqrt{95}$ 10. $\sqrt{70}$

11. $\sqrt{86}$ 12. $\sqrt{280}$ 13. $\sqrt{630}$ 14. $\sqrt{18}$ 15. $\sqrt{2500}$

16. If a number is multiplied by 9, and the square root of the new number is found, how will this square root compare with the square root of the original number? Do Problems 10 and 13 and Problems 3 and 14 confirm your answer?

8.4   THE PYTHAGOREAN THEOREM

The ancient Egyptians had quite a problem keeping track of property lines when the Nile River flooded each year. They used ropes with 12 equally spaced knots tied in them to set up right angles. The Egyptians knew that if they stretched the rope into a triangle so that the sides were of lengths 3, 4, and 5, the angle opposite the side 5 units long would be a right angle (90°). (See Figure 8.1.) They also knew that if the sides of a triangle were 5, 12, and 13, the angle opposite the side 13 units long would be a right angle. (See Figure 8.2.)

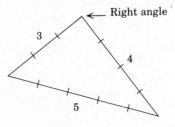

Figure 8.1

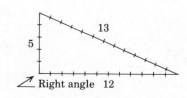

Figure 8.2

The longest side in a right triangle is called the **hypotenuse** and is always opposite the right angle. Pythagoras, a great mathematician, discovered that if he squared the hypotenuse and each side of the right triangles mentioned above, the sum of the square of the two sides was equal to the square of the hypotenuse.

$$3^2 + 4^2 = 5^2 \quad \text{or} \quad 9 + 16 = 25$$
$$5^2 + 12^2 = 13^2 \quad \text{or} \quad 25 + 144 = 169$$

The area of a square is found by squaring the length of one side; that is, multiplying the length of one side of a square by itself gives the number of square units in the area of the square. For example, a square of 4 units on a side contains 4^2, or 16, square units (see Figure 8.3).

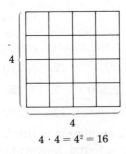

4

$$4 \cdot 4 = 4^2 = 16$$

Figure 8.3

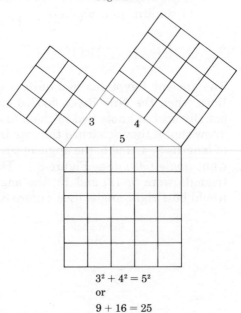

3   4

5

$$3^2 + 4^2 = 5^2$$
or
$$9 + 16 = 25$$

Figure 8.4

Thus, the Pythagorean relationship can be illustrated using the areas of squares whose sides have lengths the same as the sides of the right triangle (see Figure 8.4).

THE PYTHAGOREAN THEOREM
Let a, b, and c be the lengths of the sides of a triangle with c the largest number. Then the triangle is a right triangle if and only if

$$a^2 + b^2 = c^2$$

Figure 8.5 shows a right triangle where a, b, and c are the lengths of its sides, and where c is the hypotenuse.

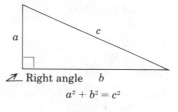

$$a^2 + b^2 = c^2$$

Figure 8.5

Pythagoras and his Society, a group of his students and fellow mathematicians, were living in an age when numbers were considered mystical. But the Pythagorean Society had discovered a new kind of number, the irrational number, and were so unpopular that they had to repress their knowledge of the new numbers or be punished and ridiculed by the citizens.

To understand how the Pythagorean Theorem leads to irrational numbers, consider a right triangle whose sides are both 1 unit long (Figure 8.6), and find the length of the hypotenuse.

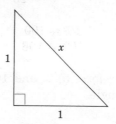

Figure 8.6

According to the theorem,

$$1^2 + 1^2 = x^2$$
$$1 + 1 = x^2$$
$$2 = x^2$$

So, $x = \sqrt{2}$

Thus, the hypotenuse is an irrational number.

EXAMPLES   1. Find the length of the hypotenuse (correct to the nearest hundredth) of a right triangle whose sides are 5 and 7.

Solution:

$5^2 + 7^2 = 25 + 49 = 74$

The length of the hypotenuse is $\sqrt{74}$.

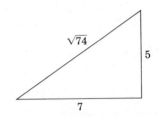

```
          8. 6  0  2  ≈ 8.60
      √74.00|00|00
          64
166     10 00
  6      9 96
1720        4 00
   0           0
  17202   4 00 00
         2  3 44 04
```

2. Are the numbers 5, 9, and 10 the lengths of the three sides of a right triangle?

Solution:

If 5, 9, and 10 are the lengths of the sides of a right triangle, then 10 must be the length of the hypotenuse (the longest side) and

$$5^2 + 9^2 = 10^2$$

But, $5^2 + 9^2 = 25 + 81 = 106$

and $10^2 = 100$
           $100 \neq 106$

Therefore, 5, 9, and 10 are *not* the lengths of the sides of a right triangle.

EXERCISES 8.4

The two given numbers represent the lengths of the sides of a right triangle. Find the length of the hypotenuse (correct to the nearest hundredth) in each case.

1. 1 and 2   2. 5 and 6   3. 6 and 8   4. 10 and 24

5. 9 and 12   6. 15 and 20   7. 3 and 3   8. 5 and 5

9. 10 and 25   10. 16 and 30

Decide if the three numbers in each problem could be the lengths of the sides and hypotenuse of a right triangle. The largest number would have to be the length of the hypotenuse.

11. 6, 8, 10   **12.** 26, 10, 24   **13.** 1, 2, $\sqrt{5}$ **14.** 1, $\sqrt{2}$, $\sqrt{3}$

15. 4, 5, 6   **16.** 7, 12, 16   **17.** 14, 20, 34   **18.** .6, .8, 1

19. .5, 1.2, 1.3   **20.** 300, 20, 320

21. Draw a right triangle with sides of 5, 12, and 13 units in length. Draw the squares adjacent to the triangle with sides corresponding to those of the triangle (similar to Figure 8.4) and count the square units to verify the Pythagorean Theorem for this triangle.

22. If two sides of a right triangle are 1 foot and $\sqrt{3}$ feet long, what is the length of the hypotenuse?

23. Draw a square that contains exactly 2 square centimeters. What are the lengths of the sides of this square?

24. Draw a square that contains exactly 3 square centimeters. What are the lengths of the sides of this square? What is the length of one of its diagonals?

25. If a right triangle has two equal sides and a hypotenuse that is $\sqrt{20}$ meters long, what is the length of one of the equal sides?

SUMMARY: CHAPTER 8

DEFINITION

The **square root** of a given number is the number whose square is the given number.

The **square root algorithm** is a process or pattern of steps to be used to find the square root of a number.

> TO FIND THE SQUARE ROOT OF A NUMBER BY THE DIVIDE-AND-AVERAGE METHOD
>
> 1. Estimate the square root.
>
> 2. Divide the number by the estimated square root.
>
> 3. Carry the quotient out to one more place than the estimate, and if the number in this place is even, leave it; if the number in this place is odd, replace it by the next larger even number.
>
> 4. Then average the estimate and the quotient.
>
> This average is the new estimate of the square root. The process of dividing and averaging is continued until the digits in the quotient and in the average are the same out to the desired decimal place.

THE PYTHAGOREAN
THEOREM

Let a, b, and c be the lengths of the sides of a triangle with c the largest number. Then the triangle is a right triangle if and only if

$$a^2 + b^2 = c^2$$

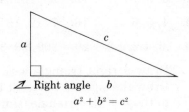

$$a^2 + b^2 = c^2$$

REVIEW QUESTIONS · CHAPTER 8

Which of the following are perfect square numbers?

1. 36   **2.** 75   **3.** 225   **4.** 144   **5.** 361   **6.** 200

Find the square of each of the following.

7. 14   **8.** 5   **9.** 3.4   **10.** .08   **11.** 6.1   **12.** .13

Use the table of square roots and a basic knowledge of fractions and decimals to find the following square roots.

13. $\sqrt{.25}$ **14.** $\sqrt{81}$ **15.** $\sqrt{.0081}$ **16.** $\sqrt{\dfrac{4}{25}}$ **17.** $\sqrt{\dfrac{289}{400}}$

18. $\sqrt{\dfrac{196}{121}}$

Use either the square root algorithm or the divide-and-average method to find the following square roots correct to the nearest hundredth.

19. $\sqrt{600}$ **20.** $\sqrt{6000}$ **21.** $\sqrt{.568}$ **22.** $\sqrt{10.87}$ **23.** $\sqrt{42,761}$

24. $\sqrt{.0961}$

The two numbers given represent the lengths of the sides of a right triangle. Find the length of the hypotenuse (correct to the nearest tenth) in each case.

25. 7 and 8   **26.** 5 and 12   **27.** 1 and $\dfrac{4}{3}$ **28.** 16 and 16

Decide if the three numbers in each problem could be the lengths of the sides and hypotenuse of a right triangle.

29. 15, 12, 9   **30.** 15, 39, 36   **31.** .05, .12, .13   **32.** .3, .4, .5

33. $\sqrt{7}$, $\sqrt{8}$, 3   **34.** 200, 35, 235

NEGATIVE NUMBERS

9

9.1   VECTORS AND NUMBER LINES

Vectors, in a technical sense, are quantities that have magnitude (or size) and direction. Thus, examples of vectors used in physical sciences are velocity, acceleration, and force. A wind blowing northeast at 3 miles per hour is a vector. An automobile traveling east at 80 kilometers per hour is a vector. For the purposes of this text, vector will be defined and used in an elementary geometric sense.

DEFINITION   A **geometric vector** (or **vector**) is a directed line segment.

Vectors are illustrated with half-headed arrows that have a **tail** (beginning point) and a **head** (tip). The length of the arrow indicates the vector's magnitude, and its direction is from the tail to the head (see Figure 9.1).

Figure 9.1

A point is a **zero vector.** It has zero magnitude (length) and any direction desired. *Two vectors are* **equal** *if they have the same magnitude and same direction. Two vectors are* **opposites** *if they have the same magnitude and opposite directions* (see Figure 9.2).

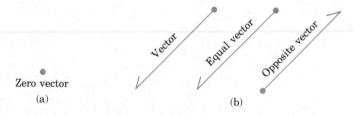

Zero vector

(a)   (b)

Figure 9.2

Now consider a straight line. Any straight line will do, but for convenience and consistency the lines will generally be horizontal in this section. Choose any point on the line and label it with the number 0. This point will be called the **origin** (see Figure 9.3).

Figure 9.3

Choose a point other than the origin, to the right of the origin, again only for convenience and consistency, and label it with the number 1. **A unit vector** has now been chosen and can be drawn with its tail above 0 and its head above 1 (see Figure 9.4).

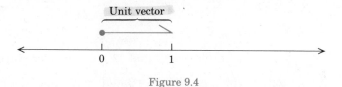

Figure 9.4

Using the unit length chosen, the points to the right of the origin can be labeled with the positive real numbers. There is a 1-to-1 correspondence between these points and the positive real numbers. That is, for each point there is a positive real number, and for each positive real number there is a point (see Figure 9.5).

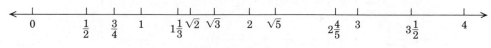

Figure 9.5

The numbers can be thought of as the points they label, or they can be thought of as vectors with length equal to the distance from 0 to the point and direction from 0 to the point. Several vectors corresponding to the number $1\frac{1}{2}$ are illustrated in Figure 9.6.

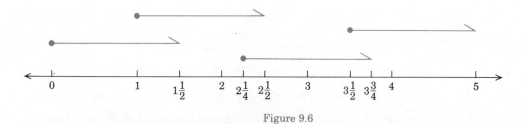

Figure 9.6

A vector is in **standard position** if its tail is directly above 0 and it points directly right (the **positive direction**) or left (the **negative direction**). **Positive numbers** (indicated by a $^+$ sign—that is, a *plus* superscript—or by no sign at all) are those corresponding to the vectors in standard position pointing to the right. **Negative numbers** (indicated by a ⁻ sign—that is, a *minus* superscript) are those corresponding to the vectors in standard position pointing to the left. The vectors ⁺2 and ⁻2 are illustrated in Figure 9.7.

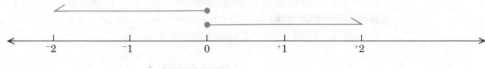

Figure 9.7

The numbers ⁺2 and ⁻2 are opposites since they correspond to opposite vectors. Similarly, ⁻6 and ⁺5 are opposites, $^+\frac{1}{2}$ and $^-\frac{1}{2}$ are opposites, and 15 and ⁻15 are opposites. *Zero (0) is neither positive nor negative* because the 0 vector can have any direction, and the opposite of 0 is 0.

Every point on the line is now labeled with a real number, either positive or negative, and the line is called **a real number line.** There is a 1-to-1 correspondence between the real numbers (positives, negatives, and 0) and the points on any line. Figure 9.8 illustrates a few points on a real number line.

Negative direction   Positive direction

Figure 9.8

EXERCISES 9.1

Define the following terms.

1. vector   2. equal vectors   3. opposite of a vector

Represent the following with vectors.

4. a plane flying north at 300 miles per hour

5. a rowboat moving east on a lake at 2 miles per hour

6. the opposite of driving southwest at 40 miles per hour

7. the opposite of walking southeast at 5 kilometers per hour

8. standing still

9.   a plane flying 50 miles per hour into a 50-mile-per-hour headwind

10. a man pushing a 200-pound rock with a force of 200 pounds

For each exercise, draw a real number line and indicate the vectors in standard position.

11. $^+3, ^+4, ^-\dfrac{1}{2}$ 12. $^+5, ^-4, 1\dfrac{1}{2}$ 13. $2\dfrac{1}{3}, ^+1, ^-1$

14. $^-\dfrac{1}{5}, ^-3, ^+2$ 15. $^-5, ^-6, 1\dfrac{1}{4}$ 16. $\sqrt{2}, \sqrt{3}, ^-\sqrt{5}$

Draw a real number line and indicate each of the following vectors in standard position; then draw three other equal vectors.

17. $^+3$ 18. $^-2$ 19. $^-\dfrac{3}{4}$ 20. $5\dfrac{1}{2}$

21. $^-4$ 22. $^+\dfrac{2}{3}$ 23. 1.6 24. $^-2.3$

9.2   ADDING VECTORS AND ABSOLUTE VALUE

Vectors, as used in this text, are not greatly important in themselves. They are to be used merely as visual aids in understanding addition and subtraction with negative numbers. Since vectors are treated as directed line segments, the sum of two vectors can be defined in the following way.

DEFINITION   Let **A** and **B** be vectors that have the same or opposite direction. Choose a vector equal to **B** with its tail at the head of **A**. The **sum** of **A** and **B** is the vector directed from the tail of **A** to the head of the chosen vector equal to **B**.

EXAMPLES   1. **Add the vectors 1 and 2**.

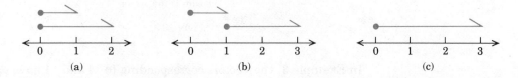

(a)   (b)   (c)

(a) Given vectors.
(b) Choose a vector equal to 2 with its tail at the head of the vector 1.
(c) The sum is the vector 3.

2.   Add the vectors 1 and ⁻5.

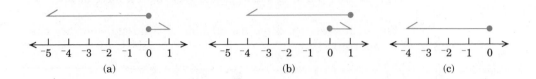

(a)   (b)   (c)

(a) Given vectors.
(b) Choose a vector equal to ⁻5 with its tail at the head of vector 1.
(c) The sum is the vector ⁻4.

3.   Add the vectors ⁺4 and ⁻4.

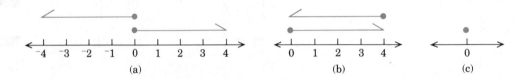

(a)   (b)   (c)

(a) Given vectors.
(b) Choose a vector equal to ⁻4 with its tail at the head of the vector 4.
(c) The sum is the zero vector.

The sum of any two opposite vectors will always be the zero vector.
    Vector addition is **commutative;** that is, if the order of the vectors being added is changed, the sum is the same vector. Consider Example 2 with the vectors being added as (⁻5) + (1) instead of (1) + (⁻5). The sum is still the vector ⁻4 (see Figure 9.9).

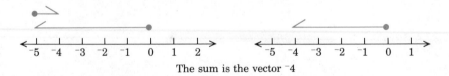

The sum is the vector ⁻4

Figure 9.9

    In Example 3, the vectors corresponding to ⁻4 and ⁺4 have opposite directions but the same length or magnitude. Although the numbers ⁻4 and ⁺4 are on opposite sides of 0, they are the same distance from 0. The ⁺ and ⁻ signs are used to indicate direction. How is magnitude indicated? The distance a number is from 0 on the number line *without regard to direction* is called its **absolute value.** Thus, ⁺4 and ⁻4 have the same absolute value. Similarly, $^+3$ and $^-3$ have the same absolute value. The absolute value of any number is either 0 or positive since it represents a distance or size only, without regard to direction. *Absolute value is symbolized with two vertical bars, one on each side of the number.*

EXAMPLES

1. $|^+3| = |^-3| = 3$ The absolute value of $^+3$ and the absolute value of $^-3$ are both equal to 3.

2. $|^-5| = |^+5| = 5$

3. $|0| = 0$ The absolute value of 0 is 0.

4. $\left|\dfrac{-3}{4}\right| = \left|\dfrac{+3}{4}\right| = \dfrac{3}{4}$

DEFINITION

If a is positive or zero, then $|a| = |^-a| = a$.

SELF QUIZ   **ANSWERS**

1. Find the sum of the vectors $^-4$ and 3.

      1.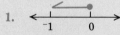

2. Find the absolute values of $^-7$ and $^+6$.

      2. $|^-7| = 7$
         $|^+6| = 6$

3. Which number has the larger absolute value, $^-9$ or $^+9$?

      3. Neither; both have the same absolute value.
         $|^-9| = |^+9|$

EXERCISES 9.2

Add each pair of vectors. Use three steps as in the examples.

1. 1, 2   2. 3, 4   3. $^-2, ^-3$ 4. $^-1, ^-5$

5. 6, $^-2$ 6. 2, $^-5$ 7. $^-6, 4$ 8. 6, $^-4$

9. 5, $^-6$ 10. 3, $^-4$ 11. $^-3, 3$ 12. $^-2\dfrac{1}{2}, 2\dfrac{1}{2}$

13. $\dfrac{1}{4}, \dfrac{-3}{4}$ 14. $\dfrac{1}{2}, 1\dfrac{3}{4}$ 15. $\dfrac{-1}{2}, ^-2\dfrac{1}{2}$ 16. 4.1, $^-3.6$

17. 5.6, $^-8.2$ 18. $^-1.6, ^-2.3$ 19. $^-.3, 1.4$ 20. $^-2, 3.7$

Find the absolute value of each number.

21. $^+7$ 22. $^+32$ 23. $^-6$ 24. $^-3.5$ 25. 16.7

26. -6.41   **27.** $+22\frac{1}{2}$ **28.** -10   **29.** $17\frac{3}{4}$ **30.** $+4\frac{1}{8}$

Which number in each of the following pairs has the larger absolute value?

31. 10, 13   **32.** -10, -13   **33.** -8, -6   **34.** -8, +6

35. $-\frac{3}{4}, +\frac{1}{2}$ **36.** -5, +6   **37.** -11, +11   **38.** $+\frac{5}{8}, -\frac{5}{8}$

39. $-\sqrt{2}, -\sqrt{3}$ **40.** $\sqrt{5}, -\sqrt{5}$

9.3   ADDING POSITIVE AND NEGATIVE NUMBERS

The format of the following discussion is to explain how to add first with vectors and then with numbers. Examples of all the possible combinations of positive and negative numbers are given. *Note carefully how absolute values are used in the discussions concerning the sums of numbers.*

Positive Plus Positive

(A) The sum of two positive vectors (both point to the right) is a positive vector whose length is the sum of the lengths of the two vectors, as shown in Figure 9.10.

(B) The sum of two positive numbers is a positive number that is the sum of the absolute values of the two numbers.

EXAMPLE

If a man deposits $300 in his savings account, then later deposits another $200, his account will be credited with total deposits of $500.

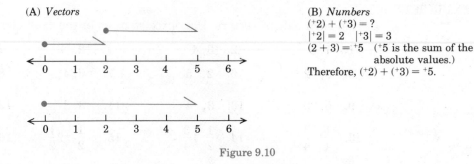

(A) *Vectors*

(B) *Numbers*
$(^+2) + (^+3) = ?$
$|^+2| = 2 \quad |^+3| = 3$
$(2 + 3) = {}^+5 \quad (^+5$ is the sum of the absolute values.)
Therefore, $(^+2) + (^+3) = {}^+5.$

Figure 9.10

[NOTE: The plus sign, +, indicating addition, and the positive sign, $^+$, indicating direction, are actually the same sign. And we must be careful, at least in the beginning, to distinguish between them by reading one "plus" and the other "positive." Thus, $(^+2) + (^+3) = (^+5)$ is read: "positive 2 plus positive 3 equals positive 5."]

Negative Plus Negative

(A) The sum of two negative vectors (both point to the left) is a negative vector whose length is the sum of the lengths of the two vectors, as shown in Figure 9.11.

(B) The sum of two negative numbers is a negative number that is the opposite of the sum of the absolute values of the two numbers.

EXAMPLE   If a man withdraws $100 from his savings account, then later withdraws another $300, his account will be charged for total withdrawals of $400.

(A) *Vectors*

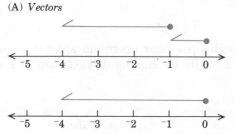

(B) *Numbers*
$(^-1) + (^-3) = ?$
$|^-1| = 1$ $|^-3| = 3$
$^-(1 + 3) = ^-4$ ($^-4$ is the opposite of the sum of the absolute values.)
Therefore, $(^-1) + (^-3) = ^-4$.

Figure 9.11

Positive Plus Negative

To add a positive and a negative vector or a positive and a negative number, there are three possible situations.

1.(A) If the *positive vector is longer than the negative vector,* their sum will be a positive vector whose length is the difference of the lengths of the two vectors (Figure 9.12).

(B) If the *positive number has a larger absolute value than the negative number,* their sum will be a positive number that is the difference of the absolute values of the two numbers.

EXAMPLE   A football player who gains 4 yards on one play and then loses 1 yard on the next play has a net gain of 3 yards. (See page 252.)

(A) *Vectors*

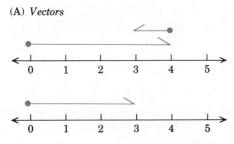

(B) *Numbers*
($^+$4) + ($^-$1) = ?
$|^+4| = 4$ $|^-1| = 1$
(4 − 1) = $^+$3   ($^+$3 is the difference of
the absolute values.)
Therefore, ($^+$4) + ($^-$1) = $^+$3.

Figure 9.12

2.(A) **If the *negative vector is longer than the positive vector*,** their sum will be a negative vector whose length is the difference of the lengths of the two vectors (Figure 9.13).

(B) If the *negative number has a larger absolute value than the positive number*, their sum will be a negative number that is the opposite of the difference of the absolute values of the two numbers.

EXAMPLE   If a man rows a boat 3 miles per hour upstream against a current of 5 miles per hour, he will actually be moving downstream at 2 miles per hour.

(A) *Vectors*

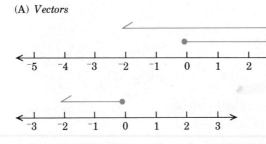

(B) *Numbers*
($^+$3) + ($^-$5) = ?
$|^+3| = 3$ $|^-5| = 5$
$^-$(5 − 3) = $^-$2   ($^-$2 is the opposite of
the difference of the
absolute values.)
Therefore, ($^+$3) + ($^-$5) = $^-$2.

Figure 9.13

3.(A) **If the *vectors are opposites*,** their sum is the zero vector (Figure 9.14).

(B) If the *numbers are opposites*, their sum is zero.

EXAMPLE   If a quarterback completes a pass for 5 yards on one play and then is tackled behind the line of scrimmage for a loss of 5 yards on the next play, he has a net gain of 0 yards.

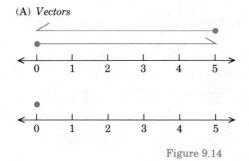

(B) *Numbers*
$(^+5) + (^-5) = ?$
$(^+5) + (^-5) = 0.$

Figure 9.14

If more than two numbers are to be added, the **associative law** may be used since it is true for all real numbers.

EXAMPLES   1. Find the sum $(^+2) + (^-6) + (^-3)$.

$$|^+2| = 2 \qquad |^-6| = 6 \qquad |^-3| = 3$$

We could have

$$(^+2) + (^-6) + (^-3) = [(^+2) + (^-6)] + (^-3) = (^-4) + (^-3) = \,^-7$$

or $\qquad (^+2) + (^-6) + (^-3) = (^+2) + [(^-6) + (^-3)] = (^+2) + (^-9) = \,^-7$

If the numbers are written vertically, the same rules apply, although sometimes adding all positive numbers and all negative numbers separately and then adding these two sums is easier.

2. Add:
$$\begin{array}{cc} ^+2 & ^+2 \\ ^-6 & ^-9 \\ \underline{^-3} & \underline{^-7} \\ ^-7 & \end{array}$$

3. Add:
$$\begin{array}{cc} ^+3 & ^+13 \\ ^-8 & \underline{^-12} \\ ^+10 & ^+1 \\ \underline{^-4} & \\ ^+1 & \end{array}$$

4. Add:
$$\begin{array}{cc} ^+2 & ^+13 \\ ^+5 & ^-7 \\ ^-4 & \underline{^+6} \\ ^+6 & \\ \underline{^-3} & \\ ^+6 & \end{array}$$

SELF QUIZ	Find the following sums.	ANSWERS
	1. $(^+13) + (^-8)$	**1.** 5
	2. $(^-13) + (^+8)$	**2.** $^-5$
	3. $(^+6) + (^+7) + (^-10) + (^-2)$	**3.** $^+1$
	4. $\begin{array}{c} ^+27 \\ ^-36 \\ \underline{^+5} \end{array}$	**4.** $^-4$

EXERCISES 9.3

Find the following sums. Use vectors as aids if necessary.

1. $(6) + (^-4)$ **2.** $(8) + (^-7)$ **3.** $(4) + (^+6)$

4. $(5) + (^-8)$ **5.** $(16) + (^+3)$ **6.** $(^-8) + (^-2)$

7. $(^-3) + (^-6)$ **8.** $(^-2) + (^+2)$ **9.** $(^+4) + (^-4)$

10. $(13) + (12)$ **11.** $(6) + (^-10)$ **12.** $(14) + (^-17)$

13. $(^+5) + (^-3)$ **14.** $(^+15) + (^-18)$ **15.** $(^-4) + (^-12)$

16. $(^-7) + (^-7)$ **17.** $(^+2) + (^-6) + (^-8)$ **18.** $(^-2) + (^-3) + (^-4)$

19. $(^-16) + (^+3) + (^+13)$ **20.** $(^-5) + (^+5) + (13)$

21. $(^-1.4) + (^-2.6) + (^+7)$ **22.** $(^+3.5) + (^-4) + \left(^-5\frac{1}{2}\right)$

23. $(^+6) + (^+3) + (^+5) + (^-4)$ **24.** $(^-18) + (^-3) + (^+5) + (^-7)$

25. $(^-1.5) + (^+2.33) + (^-4.8) + (^+1.61)$

26.	**27.**	**28.**	**29.**	**30.**	**31.**
+2	+8	+10	-16	-15	-4
-5	+3	-4	-8	-20	-17
-3	+1	+2	+12	-6	+11

32.	**33.**	**34.**	**35.**	**36.**	**37.**
+18	-8	+14	+13	-35	+66
+4	-5	-14	-5	+17	+34
-35	-13	+37	+17	-8	-81
+3	-22	-37	-25	-3	-16
				+7	-55

38.	**39.**	**40.**	**41.**	**42.**
$+21\frac{1}{2}$	$10\frac{5}{8}$	$36\frac{1}{5}$	-76	+17
$-13\frac{3}{4}$	$-3\frac{9}{10}$	$-11\frac{1}{10}$	-22	+25
$+15\frac{1}{4}$	$2\frac{1}{8}$	$-16\frac{1}{10}$	-87	+30
		-9	-24	+51
			-13	+18

9.4   SUBTRACTING POSITIVE AND NEGATIVE NUMBERS

Subtraction with negative numbers is defined in terms of addition just as it was with positive numbers; that is, the reason $8 - 5 = 3$ is that $8 = 3 + 5$. Deciding what to add to 5 to get 8 is easy because we are generally quite familiar with the addition combinations with positive numbers. But, what about $8 - (^-5)$? What should be added to $^-5$ to get 8? The answer is 13. To understand how to find this result, vectors are again helpful.

DEFINITION

The **difference** of two vectors **A** and **B** is the sum of **A** and the opposite of **B**. Symbolically, **A** − **B** = **A** + (−**B**).

[NOTE that nothing is said about **B** being positive or negative. In either case, the *opposite* of **B** is *added* to **A**.]

EXAMPLES

1. Find the difference of the vectors ⁺5 and ⁺3.

$$(^+5) - (^+3) = (^+5) + (^-3) = {}^+2$$

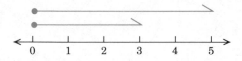 

2. Subtract the vector ⁻2 from the vector ⁺5.

$$(^+5) - (^-2) = (^+5) + (^+2) = {}^+7$$

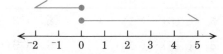 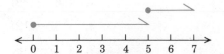

Restating the definition of the difference of two vectors with appropriate changes gives the definition of the difference of two real numbers, positive or negative.

DEFINITION

The **difference** of any two numbers a and b is the sum of a and the opposite of b. Symbolically, $a - b = a + (^-b)$.

The minus sign, −, means to subtract, and the negative sign, ⁻, implies direction. Thus, $a - b$ is read "a minus b," and $a + (^-b)$ is read "a plus the negative of b," or "a plus the opposite of b." It should be emphasized that *we now write* ⁻b *to mean the opposite of b whether b is positive or negative.* Thus, if b is positive, ⁻b is negative. And, if b is negative, ⁻b is positive.

EXAMPLES

1. Subtract ⁻6 from ⁺2.   The opposite of ⁻6 is ⁺6.

$$(^+2) - (^-6) = (^+2) + (^+6) = {}^+8$$

2. Subtract ⁻4 from ⁻1.   The opposite of ⁻4 is ⁺4.

$$(^-1) - (^-4) = (^-1) + (^+4) = {}^+3$$

3. Subtract: ⁻12   ⁻12   The opposite of ⁺5 is ⁻5.
                  +5   ⁻5
                ─────   ─────
                          ⁻17

SELF QUIZ	Find the following difference.	ANSWERS
1. $(^+16) - (^+5)$		**1.** ⁺11
2. $(^+16) - (^-5)$		**2.** ⁺21
3. $(^-18) - (^+8)$		**3.** ⁻26
4. $(^-20) - (^-6)$		**4.** ⁻14

EXERCISES 9.4

Find the following differences. Use vectors as aids if necessary.

1. $(^+5) - (^+2)$	**2.** $(^+16) - (^+3)$	**3.** $(^+8) - (^-3)$
4. $(^-5) - (^+2)$	**5.** $(^-10) - (^-1)$	**6.** $(^-3) - (^-7)$
7. $(^-4) - (^+6)$	**8.** $(^-13) - (^-14)$	**9.** $(^-9) - (^+9)$
10. $(^+15) - (^-2)$	**11.** $(^-17) - (^+14)$	**12.** $(^+3) - (^+8)$
13. $(^+6) - (^+17)$	**14.** $(^+7) - (^+12)$	**15.** $(11) - (13)$
16. $(17) - (21)$	**17.** $(^-8) - (^-6)$	**18.** $(^-35) - (^-18)$
19. $(^-70) - (^+61)$	**20.** $(^+68) - (^-32)$	

Subtract.

21. 18	**22.** 24	**23.** ⁻8	**24.** ⁻13	**25.** ⁻4
⁻12	16	⁻12	⁻18	+5

26. 32	**27.** ⁻6	**28.** ⁻25	**29.** ⁻45	**30.** 28
⁻48	⁻30	⁻13	⁻16	⁻15

9.5 ANOTHER NOTATION

The notation for adding and subtracting with negative numbers can be simplified because subtraction is related to addition. The addition sign and parentheses simply can be dropped, and all problems can be considered as *adding* positive and negative numbers. Also, the raised positive sign, ⁺, and the raised negative sign, ⁻, can be lowered. In a horizontal line of numbers being added, all positive and negative signs between numbers must be included. If the first number on the left is positive, the + sign may be eliminated, and the number will still be considered positive.

EXAMPLES

1. $(^+2) + (^+5) = 2 + 5 = 7$

2. $(^+7) + (^-3) = 7 - 3 = 4$

3. $(^-13) + (^+6) = -13 + 6 = -7$

4. $(^-4) + (^-8) = -4 - 8 = -12$

Thus, $9 - 12$ can be thought of as the sum of $^+9$ and $^-12$.

$$9 - 12 = (^+9) + (^-12) = -3$$
Or, simply
$$9 - 12 = -3$$

5. $+17 - 4 - 8 = 17 - 12 = 5$

6. $-14 + 30 - 5 + 1 = -19 + 31 = 12$

SELF QUIZ	Evaluate each of the following expressions.	ANSWERS
	1. $-26 - 6$	1. -32
	2. $-32 + 7 - 3$	2. -25
	3. $14 + 8 - 2 + 1$	3. 21
	4. $-6 - 23 - 20 + 16$	4. -33

EXERCISES 9.5

Evaluate each of the following expressions.

1. $6 + 2$ $\qquad$ 2. $7 - 1$ $\qquad$ 3. $4 + 6$ $\qquad$ 4. $-3 - 1$

5. $-2 - 6$ $\qquad$ 6. $8 - 1$ $\qquad$ 7. $9 - 3$ $\qquad$ 8. $12 - 6$

9. $-13 + 4$ $\qquad$ 10. $-9 + 5$ $\qquad$ 11. $-20 + 16$ $\qquad$ 12. $24 - 32$

ASSGN
ODD 1-29
WORK EVEN

13. $14 - 17$ **14.** $-2 - 8$ **15.** $-12 + 6$ **16.** $-15 + 18$

17. $-25 + 30$ **18.** $-72 - 30$ **19.** $-68 - 61$ **20.** $59 - 75$

21. $-4 + 16 - 8$ **22.** $-15 - 30 + 20$ **23.** $-6 - 8 - 13$

24. $-11 + 4 + 16$ **25.** $+50 + 12 - 18$ **26.** $-23 - 17 + 35 - 10$

27. $+75 + 15 - 36 + 4$ **28.** $-84 - 19 + 45 + 5 - 11$

29. $16 - 38 + 42 - 11 - 10$ **30.** $27 - 8 - 12 - 5 - 2$

9.6 MULTIPLICATION

In Chapter 3, the rational numbers were defined, and in Chapter 4, the real numbers were defined. Those definitions, however, were incomplete because negative numbers had not yet been introduced. The complete definitions or descriptions and symbolisms of important sets of numbers are given here for easy reference. The symbol ϵ, the Greek letter epsilon, is read "is a member of" or "is an element of."

1. **Natural Numbers** $N = \{1, 2, 3, 4, 5, 6, \ldots\}$

2. **Whole Numbers** $W = \{0, 1, 2, 3, 4, 5, \ldots\}$

3. **Integers** $I = \{\ldots, -3, -2, -1, 0, 1, 2, 3, \ldots\}$
    The set of integers consists of the set of whole numbers and their opposites.

4. **Rational Numbers** $Q = \left\{\dfrac{a}{b} \,\middle|\, a, b \,\epsilon\, I \text{ and } b \neq 0\right\}$

    A rational number is a number that can be written in the form $\dfrac{a}{b}$, where a and b are integers and $b \neq 0$. If a rational number is written in decimal form, it is an infinite repeating decimal.

5. **Irrational Numbers** $H = \{x \,|\, x$ can be written as an infinite nonrepeating decimal$\}$
    If an irrational number is written in decimal form, it is an infinite nonrepeating decimal.

6. **Real Numbers** $R = \{x \,|\, x \,\epsilon\, Q \text{ or } x \,\epsilon\, H\}$
    A real number is a number that is either a rational number or an irrational number. Real numbers can be written as infinite decimals (repeating or nonrepeating).

The set I (integers) is of particular importance because Q (rationals), H (irrationals), and R (reals) all depend on the definition of I; and because I contains negative numbers, Q, H, and R also contain negative numbers.

The discussion of multiplication with positive and negative numbers will be based on work with the integers. After the techniques for multiplying with integers are developed, these techniques will be used with all types of numbers without further justification.

First consider multiplication as repeated addition.

$$4 \cdot 7 = 7 + 7 + 7 + 7 = 28$$
$$6 \cdot 3 = 3 + 3 + 3 + 3 + 3 + 3 = 18$$

But suppose we want 4 times −7 and 6 times −3. Then the repeated addition yields the following results.

$$4(-7) = \underbrace{(-7) + (-7) + (-7) + (-7)}_{\text{add four −7's}} = -28$$

$$6(-3) = \underbrace{(-3) + (-3) + (-3) + (-3) + (-3) + (-3)}_{\text{add six −3's}} = -18$$

Repeated addition of negative numbers yields negative numbers. Thus, *the product of a positive integer and a negative integer will be a negative integer.*

EXAMPLES

1. $5(-2) = -10$

2. $7(-8) = -56$

3. $-9(5) = -45$

4. $-22(3) = -66$

5. $11(-11) = -121$

Now consider the following patterns of products and try to decide what the missing products should be.

−5	−5	−5	−5	−5	−5
+3	+2	+1	0	−1	−2
−15	−10	−5	0	?	?

−2	−2	−2	−2	−2	−2
+3	+2	+1	0	−1	−2
−6	−4	−2	0	?	?

Looking at the pattern of products −15, −10, −5, 0, ?, ?, you have probably figured out that $(-5)(-1) = +5$ and $(-5)(-2) = +10$. Also, looking at the pattern of products −6, −4, −2, 0, ?, ?, your intuition tells you that $(-2)(-1) = +2$ and $(-2)(-2) = +4$. You may make up similar examples using other numbers, and the results should lead you to believe that *the product of two negative integers is a positive integer.*

EXAMPLES

1. $(-6)(-4) = +24$

2. $(-7)(-9) = +63$

3. $-2(-3) = 6$

4. $-8(-20) = 160$

5. $-43(-5) = 215$

SUMMARY OF RULES FOR MULTIPLYING TWO INTEGERS

1. $(+ \text{integer}) \cdot (+ \text{integer}) = + \text{integer}$

2. $(+ \text{integer}) \cdot (- \text{integer}) = - \text{integer}$

3. $(- \text{integer}) \cdot (- \text{integer}) = + \text{integer}$

Or,

---

for $a, b \in N$:

1. $a \cdot b = ab$ 2. $a(-b) = -ab$ 3. $(-a)(-b) = ab$

Rules 2 and 3 can be proved using the distributive property of multiplication over addition. However, they will be accepted here without further justification.

Although the rules are stated only for integers, they also apply to all real numbers.

EXAMPLES

1. $(-\sqrt{2})(-\sqrt{2}) = +2$

2. $\left(-\dfrac{3}{4}\right)\left(+\dfrac{1}{5}\right) = -\dfrac{3}{20}$

3. $(-1.3)(-.4) = +.52$

SELF QUIZ	Find the following products.	ANSWERS
	1. $(-5)(5)$	**1.** -25
	2. $(-3)(-3)(0)$	**2.** 0
	3. $(-2.6)(-1)$	**3.** 2.6
	4. $(-2)(-2)(-2)$	**4.** -8

EXERCISES 9.6

1. $A = \left\{-3, -1, -\frac{1}{2}, -\sqrt{2}, 0, 1, \frac{5}{8}, 5\right\}$

   In set A, which of the numbers are (a) natural numbers? (b) whole numbers? (c) integers? (d) rational numbers? (e) irrational numbers? (f) real numbers?

2. Illustrate the product $5(-4)$ as repeated addition.

3. Give three examples of numbers in each of the sets $N, W, I, Q, H,$ and R.

Find the following products.

4. $(-7)(3)$	**5.** $6(-8)$	**6.** $(-15)(-4)$
7. $(-8)(-2)$	**8.** $\left(22\right)\left(-\frac{1}{2}\right)$	**9.** $(-2) \cdot 2 \cdot 3$
10. $(-\sqrt{2})(+\sqrt{2})$	**11.** $(\sqrt{3})(-\sqrt{3})$	**12.** $(-98)(-7)$
13. $9(21)$	**14.** $(-7)(-2)(-3)$	**15.** $(-6)(0)(-2)$
16. $(-3)(-3)(-3)$	**17.** $(-4)^2$	**18.** $(-2)^3$
19. $(-.5)(+.13)$	**20.** $\left(+\frac{3}{4}\right)\left(-\frac{2}{21}\right)$	**21.** $\left(-\frac{3}{5}\right)^2$
22. $(-1.4)(-3.2)$	**23.** $(-5)(-7)(0)$	**24.** $(-5)^3$
25. $(-1)(-1)(-1)(-1)$	**26.** $(-1)(-1)(-1)$	**27.** $\begin{array}{r} -72 \\ \underline{-16} \end{array}$
28. $\begin{array}{r} -36 \\ \underline{+19} \end{array}$	**29.** $\begin{array}{r} +83 \\ \underline{-25} \end{array}$	**30.** $\begin{array}{r} -17 \\ \underline{-35} \end{array}$
31. $\begin{array}{r} -23 \\ \underline{-91} \end{array}$	**32.** $\begin{array}{r} -63 \\ \underline{-15} \end{array}$	**33.** $\begin{array}{r} -106 \\ \underline{+112} \end{array}$
34. $|-6| \cdot |-5|$	**35.** $|-15| \cdot |+7|$	**36.** $|+25| \cdot |-4|$

ASSIGN
ODD 1-39

37. $|-5|^3$

38. $|3| \cdot |-7| \cdot |+6|$

39. $\left(\dfrac{1}{2}\right)\left(-\dfrac{3}{4}\right)\left(-\dfrac{3}{5}\right)$

40. $(-7)(+9)(-1)(-5)$

9.7 DIVISION

Divide 42 by 6. Why is the answer 7? Come on, think! Why does $42 \div 6 = 7$? One reason is that division can be thought of as the reverse of multiplication; that is,

$$\frac{42}{6} = 7 \quad \text{because} \quad 42 = 7 \cdot 6$$

DEFINITION   For $a, b \in I, b \neq 0, \dfrac{a}{b} = x$ means that $a = b \cdot x$.

[NOTE: The definition allows a and b to be integers. Thus, a can be positive or negative, and b can be positive or negative. The sign, $+$ or $-$, for the number x depends on the rules for multiplication. If $a = 0$, then of course $\dfrac{a}{b} = 0$.]

Consider the following examples carefully.

EXAMPLES   1. $\dfrac{+20}{+5} = x$

So $+20 = (+5) \cdot (x)$

In order for the product $(+5) \cdot (x)$ to be positive, x must be a positive number. Therefore

$$\frac{+20}{+5} = +4$$

2. $\dfrac{-36}{+9} = x$

So $-36 = (+9) \cdot (x)$

In order for the product $(+9) \cdot (x)$ to be negative, x must be a negative number. Therefore,

$$\frac{-36}{+9} = -4$$

3. $\dfrac{+56}{-7} = x$

So $+56 = (-7) \cdot (x)$

In order for the product $(-7) \cdot (x)$ to be positive, x must be a negative number. Therefore,

$$\frac{+56}{-7} = -8$$

4. $\dfrac{-22}{-2} = x$

So $-22 = (-2) \cdot (x)$

In order for the product $(-2) \cdot (x)$ to be negative, x must be a positive number. Therefore,

$$\frac{-22}{-2} = +11$$

These four examples illustrate all the possibilities when dividing with positive and negative integers. The rules for division with integers may be summarized in these manners similar to the summarizations for the rules of multiplication.

SUMMARY OF RULES FOR DIVISION WITH INTEGERS

1. $\dfrac{+ \text{ integer}}{+ \text{ integer}} = + \text{ number}$ 2. $\dfrac{- \text{ integer}}{+ \text{ integer}} = - \text{ number}$

3. $\dfrac{+ \text{ integer}}{- \text{ integer}} = - \text{ number}$ 4. $\dfrac{- \text{ integer}}{- \text{ integer}} = + \text{ number}$

Or,

---

for $a, b \in N$:

1. $\dfrac{+a}{+b} = +\dfrac{a}{b}$ 2. $\dfrac{-a}{+b} = -\dfrac{a}{b}$ 3. $\dfrac{+a}{-b} = -\dfrac{a}{b}$ 4. $\dfrac{-a}{-b} = +\dfrac{a}{b}$

[NOTE: The quotients need not be integers, although by the definition of rational numbers, the quotients will be rational numbers. The rules for finding the sign of the quotient still hold.]

EXAMPLES   1. $\dfrac{+16}{+10} = +1.6$

2. $\dfrac{-3}{+8} = -.375$

3. $\dfrac{+20}{-7} = -2\dfrac{6}{7}$

4. $\dfrac{-5}{-4} = +\dfrac{5}{4}$ or $+1.25$

Also, just as with multiplication, the rules for finding the sign of the quotient hold for all real numbers. The numerator and denominator need not be integers.

An important relationship that is very useful in simplifying expressions in algebra and in solving equations is the relationship indicated by rules 2 and 3.

$$\frac{-a}{+b} = \frac{+a}{-b} = -\frac{a}{b}$$

Thus, if a fraction has a negative sign, the negative sign may be with the numerator, or the denominator, or in front of the fraction without changing its value.

EXAMPLES   1. $\dfrac{-35}{+7} = \dfrac{+35}{-7} = -\dfrac{35}{7} = -5$

2. $\dfrac{-40}{4} = \dfrac{40}{-4} = -\dfrac{40}{4} = -10$

SELF QUIZ	Find the following quotients.	ANSWERS
	1. $\dfrac{-28}{14}$	1. -2
	2. $\dfrac{-72}{9}$	2. -8
	3. $\dfrac{-22}{-2}$	3. 11
	4. Evaluate the expression $4 \cdot 6 - 5 \cdot 8$	4. -16

EXERCISES 9.7

Find the following quotients.

1. $\dfrac{12}{3}$ 2. $\dfrac{-18}{3}$ 3. $\dfrac{22}{-11}$ 4. $\dfrac{-62}{31}$ 5. $\dfrac{-99}{9}$

6. $\dfrac{-46}{-2}$ 7. $\dfrac{-84}{-4}$ 8. $\dfrac{14}{-7}$ 9. $\dfrac{0}{-8}$ 10. $\dfrac{-200}{-50}$

11. $\dfrac{-8}{5}$ 12. $\dfrac{-4}{-5}$ 13. $\dfrac{-54}{-9}$ 14. $\dfrac{-18}{-18}$ 15. $\dfrac{72}{12}$

16. $\dfrac{-156}{6}$ 17. $\dfrac{105}{-7}$ 18. $\dfrac{-324}{-18}$ 19. $\dfrac{-289}{+17}$ 20. $\dfrac{-1.69}{-1.3}$

Evaluate the following expressions (remember the rules for the order of operations).

21. $4 \cdot 3 - 5 \cdot 7$

22. $18 \cdot 2 + \dfrac{6}{-2}$

23. $\dfrac{15}{-3} + 4(-8)$

24. $\dfrac{-18}{-9} + \dfrac{-35}{+7} + \dfrac{-10}{-2}$

25. $16 \div (-4) \cdot 2$

26. $(+12)(-6) \div 3 \cdot 2$

27. $-5(6 + 3) - \dfrac{12}{-6}$

28. $(16 - 25)(32 - 21)$

29. $8 \div 4 \cdot 3 - 2 - 16$

30. $15 \div (-3) - 2(-8)$

Fill in the blanks.

31. The product of two negative numbers is a _____ number.

32. The quotient of a positive and a negative number is a _____ number.

33. If x is a positive number, then x^2 is a _____ number.

34. If x is a negative number, and y is a positive number, then $x \cdot y$ is a _____ number.

35. If x is a negative number, then x^2 is a _____ number.

9.8 GRAPHING SETS OF REAL NUMBERS

The graph of a set of numbers consists of shaded dots over the corresponding points on a number line.

EXAMPLES   1. Graph $\left\{-\dfrac{1}{2}, 0, 1\right\}$.

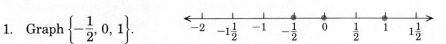

If an infinite set of distinct numbers is to be graphed, three dots, . . . , above the number line indicate that the pattern of graphing is to continue without end.

2. Graph {1, 2, 3, . . .}.

3. Graph {. . . , −4, −3, −2}.

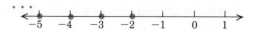

If a large finite set of numbers is to be graphed, the line may be "broken" to show that similar numbers between the numbers actually graphed are to be considered graphed.

4. Graph $\left\{-1\frac{1}{2}, -1, -\frac{1}{2}, \ldots, 37\frac{1}{2}, 38\right\}$.

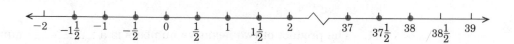

Suppose two numbers are to be compared. What can be said? One way of comparing numbers is to say that they are equal or they are not equal. That is, either $8 = 11$ or $8 \neq 11$. ($\neq$ is read "is not equal to.") Another comparison is to determine which number is smaller (or larger); for example,

$8 < 11$ is read "8 is less than 11" *or* "11 is greater than 8," depending on whether it is read from left to right or from right to left.

Similarly,

$11 > 8$ may be read "11 is greater than 8" or "8 is less than 11."

The symbols $\neq$, $<$, and $>$ are called symbols of **inequality.**

On a number line, the smaller of two numbers is always the one to the left. The following examples illustrate this fact.

EXAMPLES

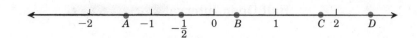

Refer to the number line above.

1. $-1 < 2$ 2. $-2 < -1$ 3. $C > A$ 4. $B > 0$

In the expression $x \in \{-1, 3, 5\}$, the letter x is used as a **variable** and represents any of the numbers in the set.

DEFINITION

A **variable** is a symbol that represents any element in a set that contains more than one element.

The set is called the **replacement set** of the variable. Thus, if $y \in \{4, 5, 6, 7\}$, then $\{4, 5, 6, 7\}$ is the replacement set for y. If the replacement set has only one element, then the symbol is called a constant. So, if $b \in \{-3\}$, then b is a constant. In this sense, any real number is also a constant.

Inequalities, graphs, and variables are three concepts related to **intervals** of real numbers. Suppose a and b are two real numbers and that $a < b$. Also, consider the set notation $\{x| \quad \}$, which is read "*the set of all x such that.*" In particular, $\{x|2 < x < 3\}$ is read "*the set of all x such that x is less than 3 and greater than 2.*" Intervals are classified and graphed in the following manner. (x is understood to represent real numbers.)

Open Interval $\{x|a < x < b\}$

The interval consists of all the real numbers between a and b. a and b are not included.

Closed Interval $\{x|a \leq x \leq b\}$

a and b are included. The circles around the points a and b are shaded in. $\leq$ is read "*less than or equal to*" from left to right and "*greater than or equal to*" from right to left.

Half-Open Interval $\{x|a \leq x < b\}$

a is included but b is not.

Half-Open Interval $\{x|a < x \le b\}$

b is included but a is not.

Open Interval $\{x|x > a\}$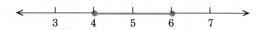

$\{x|x < a\}$

Half-Open Interval $\{x|x \ge b\}$

$\{x|x \le b\}$

EXAMPLES

1. Graph the closed interval $\{x|4 \le x \le 6\}$.

2. Represent the graph $\longleftrightarrow$ using set notation and tell what kind of interval it represents.

   Answer: $\{x|\ -2 < x < 0\}$ open interval

3. What members are included in the set $\{x|x < 3\}$? Is $2\frac{3}{4}$ in this set? Graph the set.

   Answer: The numbers included are all the real numbers less than 3. $2\frac{3}{4}$ is in the set since $2\frac{3}{4} < 3$.

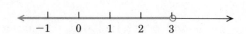

EXERCISES 9.8

Graph the following sets of numbers.

1. $\{2, 4, 6\}$ 2. $\{1, 3, 5\}$ 3. $\{4, 5, 6\}$ 4. $\{-1, 0, 2\}$

5. $\{-2, -1, 0\}$ 6. $\{-3, -1, 0, 1\}$ 7. $\{-2, 0, 2\}$ 8. $\{1, \sqrt{2}, 2, 3\}$

9. $\{\sqrt{2}, \sqrt{3}, 4\}$ **10.** $\{-\sqrt{2}, -1, 0, 1, \sqrt{2}\}$

11. $\{-\sqrt{3}, -1, 1, \sqrt{3}\}$ **12.** $\{-5, -4, -2, -1\}$

13. $\left\{-\dfrac{1}{2}, 0, \dfrac{1}{2}, 1, 1\dfrac{1}{2}\right\}$ **14.** $\left\{-\dfrac{2}{3}, -\dfrac{1}{3}, 0, \dfrac{1}{3}\right\}$

15. $\{64, 65, 66\}$ **16.** $\{71, 72, 73, 74\}$

17. $\{58, 59, 60, 61, 62\}$ **18.** $\{1, 2, 3, \ldots, 30\}$

19. $\{2, 4, 6, 8, \ldots, 30\}$ **20.** $\{-15, -14, -13, \ldots, 0\}$

21. $\{-1, 0, 1, 2, \ldots\}$ **22.** $\{-4, -3, -2, -1, \ldots\}$

23. $\left\{\dfrac{1}{2}, \dfrac{3}{4}, 1, 1\dfrac{1}{4}, 1\dfrac{1}{2}, \ldots\right\}$ **24.** $\{\ldots, -2, -1, 0\}$

25. $\{\ldots, 0, 1, 2\}$

State whether each of the following inequalities is true or false.

26. $3 \neq -3$ **27.** $-5 < -2$ **28.** $-13 > -1$ **29.** $|-7| \neq |+7|$

30. $|-7| < |+7|$ **31.** $|-4| > |+3|$ **32.** $-\dfrac{1}{2} < -\dfrac{3}{4}$ **33.** $\sqrt{2} > \sqrt{3}$

34. $-4 < -6$ **35.** $|-6| > 0$

Represent each of the following graphs using set notation and tell what kind of interval it represents.

36.

37.

38.

39.

40.

Graph each of the following sets of numbers and tell what type of interval each set represents.

41. $\{x | 7 < x < 10\}$ **42.** $\{x | -2 < x < 0\}$ **43.** $\{y | 1 \leq y \leq 4\}$

44. $\{y|-3 \le y \le 5\}$ **45.** $\{y|-1 < y \le 2\}$ **46.** $\{x|4 \le x < 7\}$

47. $\{x|68 \le x \le 72\}$ **48.** $\{x|105 < x \le 108\}$ **49.** $\{z|-12 < z < -7\}$

50. $\{z|14 < z < 20\}$ **51.** $\{x|x > -3\}$ **52.** $\{x|x \ge 0\}$

53. $\{x|x \ge \sqrt{2}\}$ **54.** $\{x|x < -\sqrt{3}\}$ **55.** $\left\{x|x \le \dfrac{2}{3}\right\}$

SUMMARY: CHAPTER 9

DEFINITION

A **geometric vector** (or **vector**) is a directed line segment.

A point is a **zero vector**.

Equal vectors have the same magnitude and the same direction.

Opposite vectors have the same magnitude and opposite direction.

Positive numbers correspond to vectors in standard position pointing to the right.

Negative numbers correspond to vectors in standard position pointing to the left.

Zero is neither positive nor negative.

DEFINITION

Let **A** and **B** be vectors that have the same or opposite direction. Choose a vector equal to **B** with its tail at the head of **A**. The **sum** of **A** and **B** is the vector directed from the tail of **A** to the head of the chosen vector equal to **B**.

The sum of any two opposite vectors is the zero vector.

DEFINITION

If a is positive or zero, then $|a| = |\bar{}a| = a$.

The sum of two positive numbers is positive.
The sum of two negative numbers is negative.
The sum of a positive number and a negative number is
   (a) *Positive* if the positive number has the larger absolute value.
   (b) *Negative* if the negative number has the larger absolute value.
   (c) *Zero* if the two numbers have the same absolute value.

DEFINITION

The **difference** of two vectors **A** and **B** is the sum of **A** and the opposite of **B**. Symbolically, $\mathbf{A} - \mathbf{B} = \mathbf{A} + (^-\mathbf{B})$.

DEFINITION

The **difference** of any two numbers a and b is the sum of a and the opposite of b. Symbolically, $a - b = a + (^-b)$.

Types of Numbers.

1. **Natural Numbers** $N = \{1, 2, 3, 4, 5, 6, \ldots\}$

2. **Whole Numbers** $W = \{0, 1, 2, 3, 4, 5, \ldots\}$

3. **Integers** $I = \{\ldots, -3, -2, -1, 0, 1, 2, 3, \ldots\}$
    The set of integers consists of the set of whole numbers and their opposites.

4. **Rational Numbers** $Q = \left\{\frac{a}{b} \middle| a, b \in I \text{ and } b \neq 0\right\}$
    A rational number is a number that can be written in the form $\frac{a}{b}$ where a and b are integers and $b \neq 0$. If a rational number is written in decimal form, it is an infinite repeating decimal.

5. **Irrational Numbers** $H = \{x | x \text{ can be written as an infinite nonrepeating decimal}\}$
    If an irrational number is written in decimal form, it is an infinite nonrepeating decimal.

6. **Real Numbers** $R = \{x | x \in Q \text{ or } x \in H\}$
    A real number is a number that is either a rational number or an irrational number. Real numbers can be written as infinite decimals (repeating or nonrepeating).

SUMMARY OF RULES FOR MULTIPLYING TWO INTEGERS

1. $(+ \text{ integer}) \cdot (+ \text{ integer}) = + \text{ integer}$

2. $(+ \text{ integer}) \cdot (- \text{ integer}) = - \text{ integer}$

3. $(- \text{ integer}) \cdot (- \text{ integer}) = + \text{ integer}$

Or,

---

for $a, b \in N$:

1. $a \cdot b = ab$ 2. $a(-b) = -ab$ 3. $(-a)(-b) = ab$

DEFINITION   For $a, b \in I$, $b \neq 0$, $\frac{a}{b} = x$ means that $a = b \cdot x$.

SUMMARY OF RULES FOR DIVISION WITH INTEGERS

1. $\dfrac{+ \text{ integer}}{+ \text{ integer}} = + \text{ number}$ 2. $\dfrac{- \text{ integer}}{+ \text{ integer}} = - \text{ number}$

3. $\dfrac{+ \text{ integer}}{- \text{ integer}} = - \text{ number}$ 4. $\dfrac{- \text{ integer}}{- \text{ integer}} = + \text{ number}$

Or,

for $a, b \in N$:

1. $\dfrac{+a}{+b} = +\dfrac{a}{b}$ 2. $\dfrac{-a}{+b} = -\dfrac{a}{b}$ 3. $\dfrac{+a}{-b} = -\dfrac{a}{b}$ 4. $\dfrac{-a}{-b} = +\dfrac{a}{b}$

$$\dfrac{-a}{+b} = \dfrac{+a}{-b} = -\dfrac{a}{b}$$

REVIEW QUESTIONS · CHAPTER 9

1. A geometric vector is a _____ line segment.

2. Two vectors are equal if they have the same _____ and the same
   _____ .

3. The opposite of a vector is a vector of the same _____ with the
   _____ _____ .

4. The _____ _____ of a number corresponds to the magnitude of a vector.

5. Define *variable*

Indicate whether the following inequalities are true or false.

6. $-10 < -11$ 7. $-10 > -11$ 8. $-10 \neq -11$

Add or subtract as indicated. Use vectors as aids if necessary.

9. $10 + (^-3)$ 10. $(^+14) + (^+5)$ 11. $(^-51) + (^-5)$

12. $(^-2) + (^+8)$ 13. $(^-2.5) + (^+3.6)$ 14. $(^+6) + (^+7) + (^-4)$

15. $(17) - (^-2)$ 16. $(4) - (5)$ 17. $(^-18) - (^-15)$

18. $(^-4) - (^-4)$ 19. $(72) + (^-72)$ 20. $(^-41) - (41)$

Evaluate each of the following.

21. $7 + 4$ 22. $16 - 5$ 23. $36 - 2 + 3$

24. $-15 - 8$ 25. $-9 - 7 + 16$ 26. $35 - 10 - 8 - 11$

27. $76 + 4 + 30 - 100$ 28. $-55 - 18 + 4 + 13$

29. $-63 + 72 - 3 - 1$

Graph each of the following sets of real numbers. If the set is an interval, tell what type of interval it is.

30. $\{-2, 1, 4\}$ **31.** $\{5, 6, \ldots, 75\}$ **32.** $\{-\sqrt{2}, 0, 1, \sqrt{3}\}$

33. {the odd counting numbers}

34. {the opposites of the odd counting numbers}

35. {all real numbers between -2 and $+2$}

36. $\left\{\text{all real numbers less than or equal to 7 and greater than } 4\frac{1}{2}\right\}$

37. Represent the graph in set notation and tell what type of interval it represents. Is $2\frac{1}{2}$ in the set?

Find the value of each of the following absolute values.

38. $|-6|$ **39.** $|-4 + 7 - 8|$ **40.** $|5 - 2 - 8 + 7|$

Determine whether each of the following is true or false. If a statement is false, give an example to show why.

41. The sum of two positive numbers must be a positive number.

42. The sum of two negative numbers must be a negative number.

43. The difference of two positive numbers must be a positive number.

44. The difference of two negative numbers must be a negative number.

45. The sum of a positive and a negative number must be positive.

46. The sum of a positive and a negative number must be negative.

47. The difference of a positive number and its opposite must be 0.

48. The sum of a positive number and its opposite must be 0.

49. The product of a positive number and its opposite must be negative.

50. The quotient of two negative numbers must be negative.

Evaluate the following expressions.

51. $(-8)(-2)(+3)$ **52.** $(-20)(+6)$

53. $\dfrac{-96}{-16}$ **54.** $\dfrac{25}{-5} + \dfrac{-36}{6}$

55. $(-32)(-3) + 14 \div (-7)$ **56.** $\left(-\dfrac{3}{4}\right)\left(-\dfrac{7}{6}\right) \div \left(-\dfrac{1}{2}\right)$

57. $-3(5 - 9) + 26 \div (-13)$ **58.** $6 \cdot 8(-2)(-5)(0)(-3)$

EQUATIONS
AND INEQUALITIES

10

10.1   EVALUATING EXPRESSIONS CONTAINING VARIABLES

What is the value of the expression $-15 + 4$? The answer is -11. Since the expression simply involves the sum of two integers, the rules for adding can be applied. But what is the value of $x + 3$ or $5x + 8x$ or $3a - 2$? Each expression contains a variable, and the value of the expression depends on the values the variable can take; that is, the value of the expression depends on the replacement set of the variable. Thus, if $x \in \{-2, 3, 5\}$, then the expression $x + 3$ could have the values

$$-2 + 3 = 1 \quad \text{or} \quad 3 + 3 = 6 \quad \text{or} \quad 5 + 3 = 8$$

> Any number in the replacement set of a variable may be substituted for the variable in an expression containing the variable.

EXAMPLES

1. Evaluate $x - 5$ if $x \in \{1, 2, 7\}$.
    Substituting gives

    $$1 - 5 = -4$$
    $$2 - 5 = -3$$
    $$7 - 5 = 2$$

    The possible values of $x - 5$ are -4, -3, and 2.

2. Evaluate $4 + 3y$ if $y \in \{-1, 0, 1\}$.
    Substituting gives

    $$4 + 3(-1) = 4 - 3 = 1$$
    $$4 + 3(0) = 4 + 0 = 4$$
    $$4 + 3(1) = 4 + 3 = 7$$

    The possible values of $4 + 3y$ are 1, 4, and 7. [NOTE: $3y$ means 3 times y, and in the expression $4 + 3(-1)$, the multiplication $+3(-1)$ comes first because of the rules for order of operations (see Section 1.9).]

    If a variable occurs more than once in an expression, the variable represents the same number from the replacement set in each occurrence. Another way of saying this is that, in evaluating an expression, only one number may be substituted for the variable each time the expression is evaluated.

EXAMPLE Evaluate $2x + 5x - 1$ if $x \in \{0, -1, 3\}$.
Substituting gives

$$2(0) + 5(0) - 1 = 0 + 0 - 1 = -1$$
$$2(-1) + 5(-1) - 1 = -2 - 5 - 1 = -8$$
$$2(3) + 5(3) - 1 = 6 + 15 - 1 = 20$$

The possible values of $2x + 5x - 1$ are -1, -8, and 20.

In Section 1.7, the *distributive property of multiplication over addition* was introduced for whole numbers. It holds for all real numbers and is stated here for all R.

DISTRIBUTIVE
PROPERTY OF
MULTIPLICATION
OVER ADDITION

$$a(b + c) = ab + ac \quad \text{and} \quad (b + c)a = ba + ca^*$$

The distributive property can be used to simplify expressions involving variables. To simplify $(2 + 3)x$, add 2 and 3 to get

$$(2 + 3)x = 5x$$

Now simplify $2x + 3x$. The distributive property may be used in the following way.

$$2x + 3x = (2 + 3)x = 5x$$

Similarly,

$$-4x + 5x + 3x + \frac{1}{2} = (-4 + 5 + 3)x + \frac{1}{2} = 4x + \frac{1}{2}$$

Therefore, to evaluate $-4x + 5x + 3x + \frac{1}{2}$ for $x \in \{2\}$, we may (a) substitute directly:

$$-4(2) + 5(2) + 3(2) + \frac{1}{2} = -8 + 10 + 6 + \frac{1}{2} = 8\frac{1}{2}$$

or (b) simplify and then substitute:

---

*Actually, $(b + c)a = ba + ca$ follows from the commutative property of multiplication applied to $a(b + c)$. The second equation is included here for convenience.

$$4x + \frac{1}{2} = 4(2) + \frac{1}{2} = 8 + \frac{1}{2} = 8\frac{1}{2}$$

EXAMPLES

1. Now, before evaluating $2x + 5x - 1$ if $x \in \{0, -1, 3\}$, first simplify $2x + 5x - 1$.

$$2x + 5x - 1 = (2 + 5)x - 1 = 7x - 1$$

Substituting using $x \in \{0, -1, 3\}$ gives

$$7(0) - 1 = 0 - 1 = -1$$
$$7(-1) - 1 = -7 - 1 = -8$$
$$7(3) - 1 = 21 - 1 = 20$$

2. Simplify $-3y - 4y + \frac{3}{4}$, then evaluate if $y \in \left\{\frac{1}{2}, -\frac{1}{2}\right\}$:

$$-3y - 4y + \frac{3}{4} = (-3 - 4)y + \frac{3}{4} = -7y + \frac{3}{4}$$

$$-7\left(\frac{1}{2}\right) + \frac{3}{4} = -\frac{7}{2} + \frac{3}{4} = -\frac{14}{4} + \frac{3}{4} = -\frac{11}{4}$$

$$-7\left(-\frac{1}{2}\right) + \frac{3}{4} = +\frac{7}{2} + \frac{3}{4} = \frac{14}{4} + \frac{3}{4} = \frac{17}{4}$$

The possible values of $-3y - 4y + \frac{3}{4}$ are $-\frac{11}{4}$ and $\frac{17}{4}$.

SELF QUIZ	Evaluate the following expressions for $x = -2$, $y = 3$, and $z = -1$.	ANSWERS
	1. $2x + y$	1. -1
	2. $3x - y + z$	2. -10
	3. $2x - 5x + 6x - 1$	3. -3

EXERCISES 10.1

Simplify the following expressions.

1. $6x + 2x$ 2. $4x - 3x$ 3. $5x + x$ 4. $7x - 3x$

5. $-10z + 2z$ 6. $-11y + 4y$ 7. $3x - 5x + 12x$ 8. $2a + 14a - 25a$

9. $32y + 16y - 50y - 4y$ 10. $40p - 15p - 6p - 8p$

11. $8x - 18x + 5$ **12.** $3y + 4y - 7y - 7$

Evaluate each of the following expressions if $x \in \{-5, -6, 7, 8\}$, $y \in \{2, 3\}$, $z \in \{-1\}$, and $a \in \{2\}$. Simplify first, if possible.

13. $x - 2$ **14.** $y - 2$ **15.** $2x + z$ **16.** $2y + a$ **17.** $3x + 2a$

18. $x - 2y$ **19.** $10 - x$ **20.** $20 - 2x$ **21.** $x + y$ **22.** $x - y$

23. $5x - 1$ **24.** $4y - z$ **25.** $3x + \dfrac{1}{2}$ **26.** $\dfrac{1}{4} + \dfrac{1}{2}x$ **27.** $\dfrac{2}{3} - \dfrac{2}{3}y$

28. $2x + 5 - x$ **29.** $y + 2 + 2y$ **30.** $x - 1 + 3x$

31. $2a + 4a - a$ **32.** $3x + a - x$ **33.** $2x - 4x + 2x + 3$

34. $5y - 2y - 3y - 4$ **35.** $z - a$

10.2   STATEMENTS AND OPEN SENTENCES

Is the equation $7 + 6 = 13$ true or false? Is the inequality $-7 < 0$ true or false? Is the equation $6 + 3 = -9$ true or false? Is the equation $x + 5 = -8$ true or false?

The answers to the four questions, in order, are *true, true, false,* and *don't know.*

The sentences $7 + 6 = 13$, $-7 < 0$, and $6 + 3 = -9$ are examples of **statements** because they can be judged as true or false.

The sentence $x + 5 = -8$ is an example of an **open sentence** because it cannot be judged as true or false without more knowledge about x.

DEFINITION   A **statement** is a sentence that can be judged to be true or false.

DEFINITION   An **open sentence** is a sentence that contains variables and becomes a statement when the variables are replaced by elements from their replacement sets.

Open sentences cannot be judged to be true or false.

EXAMPLES   1. $15 + 3 = 18$ (true statement)

2. $25 - 12 = -13$ (false statement)

3. $-18 < 0$ (true statement)

4. $5x + 2 = 7$ (open sentence)

5. $3x < 21$ (open sentence)

Now consider the open sentence $2x - 3 = 7$, where $x \in \{1, 2, 3, 4, 5\}$. Replacing x by each number in its replacement set gives the following five statements, only one of which is true.

$$2(1) - 3 = 7 \text{ (false)}$$
$$2(2) - 3 = 7 \text{ (false)}$$
$$2(3) - 3 = 7 \text{ (false)}$$
$$2(4) - 3 = 7 \text{ (false)}$$
$$2(5) - 3 = 7 \text{ (true)}$$

Thus, if x is replaced by 5 (that is, $x = 5$), then the open sentence $2x - 3 = 7$ becomes a *true* statement. The set $\{5\}$ is called the **solution set** of the open sentence $2x - 3 = 7$.

DEFINITION   The **solution set** of an open sentence with one variable is that set of numbers from the replacement set of the variable that makes the sentence a *true* statement when substituted for the variable.

EXAMPLES   1. $5x - 11 = 4, x \in \{1, 2, 3, 4, 5\}$
Substituting numbers from the replacement set yields only one true statement: $5(3) - 11 = 4$.
Therefore, $\{3\}$ is the *solution set* (abbreviated ss).

2. $4x < 10, x \in \{0, 1, 2\}$
Substituting gives

$$4(0) < 10$$
$$4(1) < 10$$
$$4(2) < 10$$

Since all three inequalities are true statements, **ss: $\{0, 1, 2\}$**.

3. $-15 = 3y - 3, y \in \{-6, -5, -4, -3\}$
Substituting numbers from the replacement set of y gives only one true statement: $-15 = 3(-4) - 3$.
ss: $\{-4\}$

SELF QUIZ   If $x \in \{-1, -\frac{1}{2}, 0, \frac{1}{2}, 1\}$, find the solution set   ANSWERS
for each of the following open sentences.

1. $3x + 4 = 1$

2. $2(x + 1) = 3x + x + 1$

3. $3x < x + 1$

1. ss: $\{-1\}$

2. ss: $\{\frac{1}{2}\}$

3. ss: $\left\{-1, -\frac{1}{2}, 0\right\}$

EXERCISES 10.2

Define the following terms.

1. statement   2. open sentence   3. solution set

Determine whether each of the following expressions is a true statement, a false statement, or an open sentence.

4. $x + 3 = 4$ 5. $6 + 8 = 2(3 + 4)$ 6. $-17 < -18$

7. $2x + 4 = 6x - 3$ 8. $5x < 3x + 5$ 9. $-5 = 19 - 14$

10. $3 + x = 2(x + 1)$ 11. $13x = 7x$ 12. $9 + 10 = 18 + 2$

If $x \in \{0, 1, 2, 3\}$, $y \in \{-2, -1, 0, 1, 2\}$, and $z \in \left\{-\frac{1}{3}, \frac{1}{3}, -\frac{1}{2}, \frac{1}{2}, -\frac{2}{3}, \frac{2}{3}\right\}$, find the solution set for each of the following open sentences.

13. $3x = 3$ 14. $-1 + y = -3$ 15. $4x - 1 = 7$

16. $3z - 1 = -2$ 17. $2z + 5 = 6$ 18. $x - 8 = -8$

19. $2y \leq 0$ 20. $2y \geq 0$ 21. $2x + 1 > 4$

22. $2z = -\frac{4}{3}$ 23. $z + \frac{1}{3} = -\frac{1}{3}$ 24. $2y + y = 6$

25. $-7 = 2x - 9$ 26. $z < 0$ 27. $5y - 6 = 2y$

28. $22 = 11x + 22$ 29. $15y - 2 = -32$ 30. $3z - 1 = 0$

10.3   SOLVING EQUATIONS

In the last two sections, the replacement sets for the various variables were generally given as some finite set of numbers. In the next two sections, the replacement set for each variable will be understood to be R, the set of all real numbers.

Inequalities have been discussed in some detail in Section 9.8, and the techniques for finding the solution sets to open sentences that are inequalities will be developed in Section 10.7. The discussion here will be restricted to solving open sentences that are **equations.**

DEFINITION An **equation** is either a statement or an open sentence using the equal sign, =, to indicate that two expressions represent the same real number.

The term *solving an equation* means to find the solution set of the equation. Consider the two equations

$$2x + 4 = 18 \quad \text{and} \quad x - 10 = -3$$

Both equations have the same solution set, $\{7\}$, and, therefore, the equations are **equivalent.**

DEFINITION Two equations are **equivalent** if they have the same solution set.

EXAMPLE $2x + 5 = x + 7$ is equivalent to $x = 2$.
Both have the solution set $\{2\}$.
The solution to $x = 2$ is *obvious*.
The solution to $2x + 5 = x + 7$ is *not obvious*.

The approach to solving an equation is to find an equivalent equation whose solution set is obvious. Two basic laws, or axioms, are necessary.

THE ADDITION AXIOM FOR EQUATIONS If the same number is added to both the left and right sides of an equation, the new equation is equivalent to the original equation.

This axiom covers subtraction also since adding a negative number is the same as subtracting a positive number.

THE MULTIPLICATION AXIOM FOR EQUATIONS If both sides of an equation are multiplied (or divided) by the same number (except 0), the new equation is equivalent to the original equation.

EXAMPLES 1. Solve $x + 4 = 10$.

$$x + 4 = 10$$

$$x + 4 + (-4) = 10 + (-4)$$ STEP 1. Add the composite of +4 to both sides of the equation.

$$x + 0 = 6$$ STEP 2. Simplify both sides of the equation using the property of additive inverses $(+4 - 4 = 0)$ and the identity property of addition $(x + 0 = x)$.

$$x = 6$$ STEP 3. The solution set is obviously $\{6\}$.

ss: {6}

Check: STEP 4. To be sure no mistakes were made, substitute the number in the original equation to see if the statement is true.

$$6 + 4 = 10$$

2. Find the solution set of the equation $x - 5 = -8$.

$$x - 5 = -8$$

$$x - 5 + 5 = -8 + 5$$ STEP 1. Add the opposite of −5 to both sides of the equation.

$$x + 0 = -3$$ STEP 2. Simplify both sides of the equation.

$$x = -3$$ STEP 3. This equation is equivalent to the original equation and has the obvious solution set $\{-3\}$.

ss: {−3}

Check: STEP 4. Substitution yields a true statement, so no mistakes were made.

$$-3 - 5 = -8$$

Because the equations in this chapter will have only one number in their solution sets, the notation **ss** will be omitted. Equations such as $x = 6$ in Example 1 and $x = -3$ in Example 2 are so simple that writing the solution set is an unnecessary step.

3. Solve the equation $6y = 42$.

$$6y = 42$$

$$\frac{6y}{6} = \frac{42}{6}$$ STEP 1. Divide both sides of the equation by 6.

$$y = 7$$ STEP 2. Simplify.

Check:

$$6 \cdot 7 = 42$$

4. Solve the equation $\frac{2}{3}x = -14$.

$$\frac{2}{3}x = -14$$

$$\frac{3}{2}\left(\frac{2}{3}x\right) = \frac{3}{2}(-14) \qquad \text{STEP 1. Multiply both sides of the equation by the reciprocal of } \frac{2}{3}.$$

$$\left(\frac{3}{2}\cdot\frac{2}{3}\right)x = -21 \qquad \text{STEP 2. Simplify, again using the associative and identity properties of}$$
$$x = -21 \qquad \qquad \text{multiplication.}$$

In summary, to obtain simplified equations equivalent to the original equation:

(a) If a constant is added to a variable, add its opposite to both sides of the equation.

(b) If a constant is multiplied by a variable, multiply both sides of the equation by its reciprocal, or divide both sides by the constant.

Solving equations using combinations of these procedures will be discussed in the next section.

SELF QUIZ	Solve the following equations. R is the replacement set for each variable.	ANSWERS
	1. $x + 5 = 14$	1. $x = 9$
	2. $y - 13 = 21$	2. $y = 34$
	3. $4x = 6$	3. $x = \frac{3}{2}$
	4. $-\frac{2}{3}y = 18$	4. $y = -27$

EXERCISES 10.3

Define the following terms.

1. equation

2. equivalent equations

3. State the Addition Axiom and the Multiplication Axiom.

Tell whether each of the following pairs of equations are equivalent equations.

4. $x + 4 = 9$, $x = 5$

5. $2y + 1 = 6$, $y = 2\frac{1}{2}$

6. $z - 7 = 4$, $z = 12$

7. $p = 13$, $2p - 10 = 14$

8. $p = 0$, $-8 + p = -8$

9. $2x = 1$, $2x + 4 = 5$

10. $\frac{1}{3} a = 5$, $\frac{a}{3} = 5$

11. $\frac{3}{4} x = 12$, $\frac{3x}{4} = 12$

12. $2x + 37 = 90$, $90 = 37 + 2x$

13. $7y - 8 = 62$, $62 = 7y - 8$

14. $3x = 8$, $\frac{8}{3} = x$

15. $5y = 10$, $y = 3$

Solve the following equations. The replacement set for each variable is the set of real numbers.

16. $x + 4 = 10$

17. $x + 13 = 20$

18. $y - 5 = 17$

19. $y - 12 = 4$

20. $-15 = x + 6$

21. $-35 = y - 10$

22. $3z = 15$

23. $4x = 24$

24. $-39 = 13x$

25. $48 = -16y$

26. $a + 5 = -2$

27. $b + 6 = -8$

28. $-7 = p + 2$

29. $-18 = -3 + p$

30. $4s = 5$

31. $6s = 40$

32. $25 = 12t$

33. $17 = -5t$

34. $x - \frac{1}{4} = -\frac{1}{5}$

35. $x + \frac{2}{3} = \frac{1}{10}$

36. $\frac{3}{4} x = 25$

37. $\frac{1}{5} y = 18$

38. $.7y = .35$

39. $.3x = -4.5$

40. $.75 = 2.5x$

41. $-.6x = -3$

42. $-4y = -20$

43. $-\frac{3}{4} x = -15$

44. $-\frac{5}{8} y = -\frac{2}{3}$

45. $-\frac{7}{16} x = -\frac{1}{2}$

10.4   MORE ON SOLVING EQUATIONS

Only the most basic types of equations were discussed in Section 10.3. In many cases, the solutions to the original equations were just as obvious as the solutions to the derived equivalent equations. In more complicated situations, a combination of the techniques used in Section 10.3 and the rules for simplifying expressions are used to find simpler equivalent equations with obvious solutions.

The following examples illustrate the appropriate procedures.

EXAMPLES   1. Solve the equation $4x + 3 = 11$.

$$4x + 3 = 11$$
$$4x + 3 - 3 = 11 - 3$$
$$4x = 8$$

Add the opposite of $+3$ to both sides to get one equivalent equation.

$$\frac{4x}{4} = \frac{8}{4}$$

Divide both sides of the new equation by 4.

$$x = 2$$

Check:
$$4 \cdot 2 + 3 \overset{?}{=} 11$$
$$8 + 3 = 11$$

2. Solve the equation $2x - 4 + 3x = 26$.

$$2x - 4 + 3x = 26$$
$$5x - 4 = 26$$

Simplify the left-hand side.

$$5x - 4 + 4 = 26 + 4$$
$$5x = 30$$

$$\frac{5x}{5} = \frac{30}{5}$$

$$x = 6$$

Check:
$$2 \cdot 6 - 4 + 3 \cdot 6 \overset{?}{=} 26$$
$$12 - 4 + 18 \overset{?}{=} 26$$
$$26 = 26$$

3. Solve the equation $5x + 2 = 3x - 8$. [NOTE: Variables are on both sides of the equation.]

$$5x + 2 = 3x - 8$$
$$5x + 2 - 3x = 3x - 8 - 3x$$

$-3x$ is added to both sides first so that there will not be variables on both sides of the new equation.

$$5x - 3x + 2 = 3x - 3x - 8$$
$$2x + 2 = -8$$
$$2x + 2 - 2 = -8 - 2$$
$$2x = -10$$

$$\frac{2x}{2} = \frac{-10}{2}$$

$$x = -5$$

Check:
$$5(-5) + 2 \overset{?}{=} 3(-5) - 8$$
$$-25 + 2 \overset{?}{=} -15 - 8$$
$$-23 = -23$$

4. Solve the equation $3(x + 2) = 18 - x$.

$$3(x + 2) = 18 - x$$
$$3x + 6 = 18 - x$$
$$3x + 6 + x = 18 - x + x$$
$$4x + 6 = 18$$
$$4x + 6 - 6 = 18 - 6$$
$$4x = 12$$

$$\frac{4x}{4} = \frac{12}{4}$$

$$x = 3$$

Check:
$$3(3 + 2) \overset{?}{=} 18 - 3$$
$$9 + 6 = 15$$

SELF QUIZ	Solve each of the following equations.	ANSWERS
	1. $10x + 4 = 7$	1. $x = \dfrac{3}{10}$
	2. $x + 5 - 2x = 7 + x$	2. $x = -1$
	3. $2(x - 7) = 5(x + 2)$	3. $x = -8$
	4. $\dfrac{3}{4}x + 6 = -3$	4. $x = -12$

EXERCISES 10.4

Solve each of the following equations.

1. $2x + 3 = 5$ 2. $3x - 4 = 8$ 3. $4y + 1 = 9$

4. $3x - 10 = 11$ 5. $6x + 4 = -14$ 6. $7y - 8 = -1$

7. $3 + 6y = 15$ 8. $6 + 5y = 9$ 9. $2x + 3 = -8$

10. $3x - 1 = -4$ 11. $5z + 12 = -3$ 12. $10p + 3 = 7$

13. $15 = 2x - 3$ 14. $20 = 3x - 1$ 15. $-17 = 5z - 2$

16. $30 = 4z + 6$ 17. $4 = 5x + 9$ 18. $8 = 12x - 2$

19. $-14 = 22x - 3$ **20.** $25y - 4 = 96$ **21.** $3x = x - 10$

22. $5y = 2y + 12$ **23.** $7z = 6z + 5$ **24.** $6p = 2p + 20$

25. $5x = 2x$ **26.** $4x + 3 = 2x + 9$ **27.** $5y - 2 = 4y - 6$

28. $7a + 14 = 10a + 5$ **29.** $5x + 20 = 8x - 4$

30. $17y + 32 = 25y - 8$ **31.** $\frac{1}{2}x + \frac{3}{4}x = -15$

32. $\frac{2}{3}y - 5 = \frac{1}{3}y + 20$ **33.** $\frac{3}{5}x - 4 = \frac{1}{5}x - 5$

34. $\frac{5}{8}y - \frac{1}{4} = \frac{2}{5}y + \frac{1}{3}$ **35.** $\frac{x}{8} + \frac{1}{6} = \frac{x}{10} + 2$

36. $5(x - 2) = 3(x - 8)$ **37.** $2(y + 1) = 3(y - 1)$

38. $4(x - 1) = 3x - 4$ **39.** $6y - 3 = 3(y - 1)$

40. $7y - 6y + 12 = 4(y + 3)$

10.5   WRITING MATHEMATICAL PHRASES

Word problems are made up of many phrases. To use algebraic techniques to solve such a problem, each phrase must be translated to mathematical symbols. Then, the symbols are related by an equation according to the conditions stated in the problem, and the equation is solved.

Developing the skills for translating English phrases into mathematical phrases and vice versa is the purpose of this section. Setting up equations and solving them will be discussed in the next section.

Observing key words in a phrase is important in translating English phrases into mathematical phrases. A list of key words is provided in Table 10.1.

Three more important terms are **consecutive integers, consecutive even integers,** and **consecutive odd integers.** 75, 76, and 77 are three **consecutive integers** because when they are written in order each is *one more* than the previous integer. Similarly, −3 and −2 are

TABLE 10.1   LIST OF KEY WORDS

ADDITION	SUBTRACTION	MULTIPLICATION	DIVISION
sum	difference	product	quotient
plus	minus	multiply	divide
add	subtract	times	
total	less than	of	
more than	decreased by		
increased by			

consecutive integers because −2 is one more than −3. Symbolically, if n represents an integer, then $n + 1$ is the next consecutive integer. Four consecutive integers can be represented as

$$n, n + 1, n + 2, \text{ and } n + 3$$

The numbers 12, 14, and 16 are **consecutive even integers** because they are even, and when written in order each is *two more* than the previous even integer. If n is an even integer, then $n + 2$ is the next consecutive even integer. Four consecutive even integers can be represented as

$$n, n + 2, n + 4, \text{ and } n + 6$$

The numbers 7, 9, and 11 are **consecutive odd integers** because they are odd, and when written in order each is *two more* than the previous odd integer. If n is an odd integer, then $n + 2$ is the next consecutive odd integer. Note that this "representation" is the same as for consecutive even integers. The difference is that n is odd instead of even. Four consecutive odd integers can be represented as

$$n, n + 2, n + 4, \text{ and } n + 6$$

EXAMPLES

1. The sum of a number and 7: $n + 7$

2. Twice a number increased by 3: $2n + 3$

3. The sum of three consecutive integers: $n + (n + 1) + (n + 2)$

4. The sum of three consecutive odd integers: $n + (n + 2) + (n + 4)$

5. The product of a number and 12 decreased by 5: $12n - 5$

6. The product of 4 and the sum of a number and 2: $4(n + 2)$

7. The quotient of a number and 2: $\dfrac{n}{2}$

Be careful to note that the examples are only phrases and not sentences. No equations are involved.

One mistake frequently made in translating English phrases into mathematical phrases is interpreting the word *and* to mean addition. *And* is merely a conjunction in most cases. It appears in Examples 1, 5, 6, and 7, but the key words are those indicating the operations *sum*, *product*, *decreased*, and *quotient*.

EXERCISES 10.5

Write the following English phrases as mathematical phrases. Use any letter to represent the unknown number in each phrase.

1. 12 subtracted from a number

2. a number divided by 8

3. the total of 16 and twice a number

4. the sum of two consecutive integers

5. three times a number decreased by 1

6. twice the sum of 13 and a number

7. ten minus half a certain number

8. two thirds decreased by one sixth a number

9. four less than eight times a certain number

10. twelve more than three times a number

11. the product of seven and a number

12. a number plus sixteen

13. seventeen increased by twice a number

14. five less than a number

15. a certain number less than twenty

16. thirty multiplied by a number

17. the sum of two consecutive even integers

18. the quotient of a number and eight

19. the difference of six and a number

20. Divide fourteen by a number.

21. the quotient of a number and eleven

22. Subtract a number from fifty.

23. three fourths of a number

24. thirteen added to a number

25. the sum of three consecutive odd integers

26. the sum of three consecutive even integers

27. the sum of three consecutive integers

28. one third of a number plus twice the same number

29. four times a number minus the product of the number and three

30. six less than the product of fifteen and a number

Change each of the following mathematical phrases to an English phrase. There may be several correct English phrases for each mathematical phrase.

31. $n + 6$ **32.** $n - 7$ **33.** $5n$ **34.** $\dfrac{1}{2} n$ **35.** $\dfrac{2}{3} n$

36. $\dfrac{n}{2}$ **37.** $\dfrac{5}{n}$ **38.** $\dfrac{14}{n}$ **39.** $\dfrac{n}{8}$ **40.** $12 + n$

41. $20 - n$ **42.** $2n + 5$ **43.** $5n - 6$ **44.** $\dfrac{5}{6} n + 2$ **45.** $3(n + 11)$

10.6 SOLVING WORD PROBLEMS

STEPS IN SOLVING WORD PROBLEMS

1. Read the problem carefully. Read the problem a second time.

2. Decide what is unknown and represent it with a letter.

3. Translate the English phrases into mathematical phrases and form an equation indicated by the sentence.

4. Solve the equation.

5. Check to see that the solution of the equation makes sense in the problem.

EXAMPLES

1. Five times a number is increased by 3, and the result is 38. What is the number?

Solution:

Let $n =$ the number
Translate "five times a number is increased by 3" to "$5n + 3$."
Translate "the result is" to "$=$."
The equation to be solved is

$$5n + 3 = 38$$
$$5n + 3 - 3 = 38 - 3$$
$$5n = 35$$

$$\frac{5n}{5} = \frac{35}{5}$$

$$n = 7$$

Check:
5 times 7 increased by 3 is $5 \cdot 7 + 3$, and $5 \cdot 7 + 3 = 35 + 3 = 38$.

2. The sum of three consecutive odd integers is −57. What are the integers?

Solution:
Let $n =$ first odd integer
$n + 2 =$ second consecutive odd integer
$n + 4 =$ third consecutive odd integer
The equation to be solved is

$$n + (n + 2) + (n + 4) = -57$$
$$n + n + 2 + n + 4 = -57$$
$$3n + 6 = -57$$
$$3n + 6 - 6 = -57 - 6$$
$$3n = -63$$

$$\frac{3n}{3} = \frac{-63}{3}$$

$$n = -21$$
$$n + 2 = -19$$
$$n + 4 = -17$$

Check: $(-21) + (-19) + (-17) = -57$

3. Seven less than four times a number is equal to twice the number increased by five. Find the number.

Solution:
Let $n =$ the number
Translate "seven less than four times a number" to $4n - 7$.
Translate "twice the number increased by five" to $2n + 5$.
The equation to be solved is

$$4n - 7 = 2n + 5$$
$$4n - 7 - 2n = 2n + 5 - 2n$$
$$2n - 7 = +5$$
$$2n - 7 + 7 = 5 + 7$$
$$2n = 12$$

$$\frac{2n}{2} = \frac{12}{2}$$

$$n = 6$$

Check: $4(6) - 7 \overset{?}{=} 2(6) + 5$
$24 - 7 \overset{?}{=} 12 + 5$
$17 = 17$

EXERCISES 10.6

Solve each of the following problems.

1. Find a number whose product with 3 is 57.

2. Find a number that when multiplied by 7 gives 84.

3. The sum of a number and 32 is 86. What is the number?

4. The difference of a number and 16 is −48. What is the number?

5. If the product of a number and 4 is decreased by 10, the result is 50. Find the number.

6. If the quotient of a number and 5 is increased by 2, the result is 7. Find the number.

7. If a number is divided by 8, and this quotient is increased by $\frac{1}{3}$, the result is 3. What is the number?

8. The sum of a number and 2 is equal to three times the number. What is the number?

9. If twice a number is decreased by 4, the result is the number. Find the number.

10. Three times the sum of a number and 4 is −60. What is the number?

11. One half the sum of a number and 5 is 15. What is the number?

12. If $\frac{1}{3}$ is added to a number, the result is four times the number. Find the number.

13. Twenty plus a number is equal to twice the number plus three times the same number. What is the number?

14. The sum of two consecutive integers is 37. Find the two integers.

15. The sum of three consecutive integers is −42. Find the three integers.

16. The sum of three consecutive odd integers is 27. Find the three integers.

17. Find three consecutive odd integers whose sum is 81.

18. Find three consecutive even integers whose sum is 30.

19. Find four consecutive even integers whose sum is 54 more than the smallest one.

20. If the product of four and the sum of a number and 3 is diminished by 6, the result is 26. Find the number.

21. What is the number whose sum with 5 is eleven times the number?

22. What is the number whose product with 7 decreased by four times the number is equal to the sum of the number and 10?

23. What is the number whose product with 6 increased by 8 is equal to twice the number increased by 9?

24. The difference of a number and 3 is equal to the difference of five times the number and 15. What is the number?

25. If the sum of two consecutive integers is multiplied by three, the result is −15. What are the two integers?

26. If the sum of two consecutive even integers is multiplied by 5, the result is 90. What are the two integers?

27. If one third of a number is increased by one half, the result is four fifths. What is the number?

28. The sum of a number, twice the number, and six times the number is equal to four times the number increased by 40. Find the number.

29. Find the number whose product with five is equal to one half the number plus 27.

30. If two thirds of a number is decreased by one fourth of the number, the result is 20. Find the number.

10.7   SOLVING INEQUALITIES

Open sentences such as $2x + 3 < 7$ and $x - 5 \geq 4$ can be solved with basically the same procedures used in solving equations. The solution sets will contain an infinite number of numbers and can be graphed as intervals on number lines.

An important difference between the techniques for solving equations and those for solving inequalities is in the multiplication axiom. Before we state the axioms, consider the following explanation. We know, for example that

$$12 < 20$$

Now, add and multiply both sides of this inequality by various numbers.

Add 3	*Add −5*
$12 < 20$	$12 < 20$
$12 + 3 < 20 + 3$	$12 - 5 < 20 - 5$
$15 < 23$	$7 < 15$

Multiply by 2	*Multiply by* $-\frac{1}{2}$
$12 < 20$	$12 < 20$
$2(12) < 2(20)$	$-\frac{1}{2}(12) > -\frac{1}{2}(20)$
$24 < 40$	$-6 > -10$

Figure 10.1 illustrates the effects of multiplying (or dividing) by a negative number. We see that multiplying both sides of the inequality $-1 < 2$ by -3 **reverses the sense** of the inequality. That is, $<$ is changed to $>$.

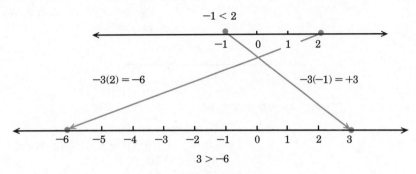

Figure 10.1

The following axioms are used to solve open sentences that are inequalities.

THE ADDITION AXIOM FOR INEQUALITIES	If the same number is added to both the left and right sides of an inequality, the new inequality is equivalent to the original inequality.
THE MULTIPLI-CATION AXIOM FOR INEQUALITIES	A. If both sides of an inequality are multiplied (or divided) by the same **positive** number, the new inequality is equivalent to the original and **has the same sense.** B. If both sides of an inequality are multiplied (or divided) by the same **negative** number, the new inequality is equivalent to the original but **has the opposite sense.**

EXAMPLES

Solve the following inequalities and graph each solution set on a number line.

1. $x - 3 < 2$

$$x - 3 < 2$$
$$x - 3 + 3 < 2 + 3 \qquad \text{Add 3 to both sides.}$$
$$x < 5$$

5

2. $-2x + 6 \geq 4$

$$-2x + 6 \geq 4$$
$$-2x + 6 - 6 \geq 4 - 6 \qquad \text{Add } -6 \text{ to both sides.}$$
$$-2x \geq -2$$

$$\frac{-2x}{-2} \leq \frac{-2}{-2} \qquad \text{Divide both sides by } -2 \text{ and } \textit{reverse the sense.}$$

$$x \leq 1$$

3. $7y - 8 > y + 10$

$$7y - 8 > y + 10$$
$$7y - 8 - y > y + 10 - y \qquad \text{Add } -y \text{ to both sides.}$$
$$6y - 8 > 10$$
$$6y - 8 + 8 > 10 + 8 \qquad \text{Add 8 to both sides.}$$
$$6y > 18$$

$$\frac{6y}{6} > \frac{18}{6} \qquad \begin{array}{l}\text{Divide both sides by 6.}\\ \text{The sense of the inequality}\\ \text{is unchanged.}\end{array}$$

$$y > 3$$

Sometimes conditions are placed on variables so that they must satisfy two inequalities at the same time. For example, suppose

$$5 < 2x + 3 \quad \text{and} \quad 2x + 3 < 10$$

We can solve each inequality separately, then choose the numbers that satisfy *both* conditions. Or, we can combine both inequalities into one expression, such as

$$5 < 2x + 3 < 10$$

and solve both inequalities at the same time.

EXAMPLES   1.   Find the values of x that satisfy both $5 < 2x + 3$ and $2x + 3 < 10$. Graph the solution set.

$$5 < 2x + 3 < 10$$
$$5 - 3 < 2x + 3 - 3 < 10 - 3 \qquad \text{Add } -3 \text{ to each member.}$$
$$2 < 2x < 7$$

$$\frac{2}{2} < \frac{2x}{2} < \frac{7}{2}$$ Divide each member by 2.

$$1 < x < \frac{7}{2}$$

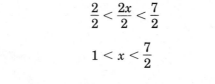

The solution set consists of all real numbers between 1 and $\frac{7}{2}$. This is an open interval.

2. Find and graph the solution set for $16 \le -3x + 1 \le 22$.

$$16 \le -3x + 1 \le 22$$
$$16 - 1 \le -3x + 1 - 1 \le 22 - 1$$ Add -1 to each member.
$$15 \le -3x \le 21$$

$$\frac{15}{-3} \ge \frac{-3x}{-3} \ge \frac{21}{-3}$$ Divide each member by -3 and reverse the sense of each inequality.

$$-5 \ge x \ge -7$$

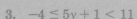

The solution set is the closed interval $\{x \mid -5 \ge x \ge -7\}$.

SELF QUIZ	Solve and graph the solution set for each of the following inequalities.	ANSWERS
	1. $2x - 1 > 7$	1. $x > 4$
	2. $3x + 2 \le 5x + 1$	2. $\frac{1}{2} \le x$
	3. $-4 \le 5y + 1 < 11$	3. $-1 \le y < 2$

EXERCISES 10.7

Solve each inequality and graph the solution set on a number line.

1. $x + 4 < 7$ 2. $x - 6 > -2$

3. $y - 3 \ge -1$ 4. $y + 5 \le 2$

5. $2y \le 3$ **6.** $5y > -6$

7. $6y + 1 > 5$ **8.** $7x - 2 < 9$

9. $x + 2 < 3x + 2$ **10.** $x - 4 > 2x + 1$

11. $2x - 5 \ge x + 2$ **12.** $3x - 8 \le x + 2$

13. $\frac{1}{2}x - 4 < 2$ **14.** $\frac{1}{3}x + 1 > -1$

15. $-x + 3 < -2$ **16.** $-x - 5 \ge -4$

17. $7y - 1 \ge 5y + 1$ **18.** $6x + 3 > x - 2$

19. $-4x - 4 \ge -4 + x$ **20.** $3x + 15 < x + 5$

21. $4 \le x + 7 \le 5$ **22.** $-2 \le x - 3 \le 1$

23. $-1 \le 2y + 1 \le 0$ **24.** $0 \le 3x - 1 \le 5$

25. $-3 < 4x + 1 \le 5$ **26.** $-5 \le y - 2 \le -1$

27. $7 \le -2x - 3 < 9$ **28.** $14 < 5x - 1 \le 24$

29. $-7 \le -2y - 3 \le 9$ **30.** $-2 \le 6x + 3 \le 0$

SUMMARY: CHAPTER 10

> Any number in the replacement set of a variable may be substituted for the variable in an expression containing the variable.

DEFINITION   A **statement** is a sentence that can be judged to be true or false.

DEFINITION   An **open sentence** is a sentence that contains variables and becomes a statement when the variables are replaced by elements from their replacement sets.

DEFINITION   The **solution set** of an open sentence with one variable is that set of numbers from the replacement set of the variable that makes the sentence a *true* statement when substituted for the variable.

DEFINITION   An **equation** is either a statement or an open sentence using the equal sign, =, to indicate that two expressions represent the same real number.

DEFINITION   Two equations are **equivalent** if they have the same solution set.

THE ADDITION AXIOM FOR EQUATIONS   If the same number is added to both the left and right sides of an equation, the new equation is equivalent to the original equation.

THE MULTIPLICATION AXIOM FOR EQUATIONS   If both sides of an equation are multiplied (or divided) by the same number (except 0), the new equation is equivalent to the original equation.

TABLE 10.1   LIST OF KEY WORDS

ADDITION	SUBTRACTION	MULTIPLICATION	DIVISION
sum	difference	product	quotient
plus	minus	multiply	divide
add	subtract	times	
total	less than	of	
more than	decreased by		
increased by			

STEPS IN SOLVING WORD PROBLEMS

1. Read the problem carefully. Read the problem a second time.
2. Decide what is unknown and represent it with a letter.
3. Translate the English phrases into mathematical phrases and form an equation indicated by the sentence.
4. Solve the equation.
5. Check to see that the solution of the equation makes sense in the problem.

THE ADDITION AXIOM FOR INEQUALITIES   If the same number is added to both the left and right sides of an inequality, the new inequality is equivalent to the original inequality.

THE MULTIPLICATION AXIOM FOR INEQUALITIES

A. If both sides of an inequality are multiplied (or divided) by the same **positive** number, the new inequality is equivalent to the original and **has the same sense.**

B. If both sides of an inequality are multiplied (or divided) by the same **negative** number, the new inequality is equivalent to the original but **has the opposite sense.**

REVIEW QUESTIONS · CHAPTER 10

1. A statement is a sentence that can be judged _____ or _____.

2. An open sentence is a sentence that contains variables and becomes a _____ when the variables are replaced by elements from their _____ _____.

3. The _____ _____ of an open sentence with one variable is the set of numbers from the replacement set of the variable that makes the sentence a true _____ when substituted for the variable.

4. Define *equation*.

5. Two equations are equivalent if they have the same _____ _____.

If $x \in \{0, 1, 2\}$, $y \in \{5\}$, and $z \in \{-1, -2\}$, find all possible values for each of the following expressions.

6. $x + 3$ 7. $x + y$ 8. $x + z$ 9. $x - y$ 10. $2x - z$

11. $\frac{2}{3} + \frac{1}{3}x$ 12. $\frac{5}{6} - \frac{1}{6}y$ 13. $3y + z$ 14. $2y - z$ 15. $2y + z - y$

Determine whether each of the following is a statement or an open sentence.

16. $4 + 6 = 7 + 3$ 17. $x - 5 = 17$

18. Is the sentence $3x - 1 = 5$ true or false? Explain your answer.

Simplify the following expressions using the distributive property.

19. $7x + 3x$ 20. $8x - 4x - 3x$ 21. $-17y - 5y - 2$

22. $13z - 13z + 6$ 23. $42p - 32p - p + 8p$

Tell which of the following pairs of equations are equivalent.

24. $y + 3 = 7$, $y = 4$ 25. $4 = 5x + 1$, $x = \frac{3}{5}$

26. $7x - 7 = 0$, $2x = 14$ 27. $3x = -1$, $6x = -2$

Solve each of the following equations. The replacement set for each variable is R.

28. $5x + 2 = 12$ 29. $5x - 2 = -12$ 30. $4y - 3 = 17$

31. $-42 = 7x + 14$ 32. $4y + 5 = 10y - 1$ 33. $5z + 16 = -40$

34. $6a - 5 = 2a - 1$ **35.** $3a = \dfrac{3}{5}$ **36.** $\dfrac{x}{4} + \dfrac{1}{16} = \dfrac{21}{16}$

37. $\dfrac{2x}{7} + 1 = \dfrac{5x}{14} - 8$ **38.** $3y + 5 - 4y = 16$

39. $23x - 11x + 2 = 3x - 1$ **40.** $2y + 8 + y = 22 - 4y$

Write each English phrase as a mathematical phrase. Use any convenient letter to represent the unknown number.

41. three times a number

42. the product of a number and three

43. the difference of a number and thirteen

44. the difference of 42 and twice a number

45. the sum of five times a number, seven, and twice the same number

46. the sum of two consecutive even integers

47. the quotient of 28 and a number

48. fifteen decreased by the product of a number and three

49. the quotient of twice a number and 8, increased by three eighths

50. twice the difference of a number and five, plus four times the sum of the number and seven

Solve each of the following problems.

51. If the product of a number and 5 is increased by 7, the sum is 47. What is the number?

52. What is the number whose product with 6 is the sum of the number and 8?

53. What number has its product with 15 equal to its product with 3?

54. One third of a number minus $\dfrac{5}{8}$ is equal to $\dfrac{5}{12}$. What is the number?

55. Find three consecutive integers whose sum is −66.

56. Find three consecutive odd integers whose sum is −123.

57. Seventeen is equal to 3 plus seven times a number. What is the number?

58. A number plus the quotient of the number and eight equals the number increased by one sixteenth. What is the number?

59. What number plus five equals the sum of five and four times the number?

60. Multiply a number by two; divide the product by five; increase the quotient by fourteen. The result is negative fifty-six. What is the number?

Solve each of the following inequalities and graph the solution sets on number lines.

61. $x + 4 \leq -5$ **62.** $2x + 14 \geq 8$ **63.** $5x - 2 < x + 10$

64. $-7x + 3 \geq -11$ **65.** $-3 \leq 2x + 3 \leq 3$

ANCIENT
NUMERATION SYSTEMS

APPENDIX

I

I.1

The number systems used by ancient peoples are interesting from a historical point of view, but from a mathematical point of view, they are difficult to work with. One of the many things that determined the progress of any civilization was its system of numeration. Mankind has made its most rapid progress since the invention of the zero and the place value system (which we will discuss in the next section) by the Hindu-Arabic peoples about A.D. 800.

Egyptian Numerals (Hieroglyphics)

The ancient Egyptians used a set of symbols called hieroglyphics as early as 3500 B.C. (see Table I.1).

TABLE I.1 EGYPTIAN HIEROGLYPHIC NUMERALS

SYMBOL	NAME	VALUE	
|	Staff (vertical stroke)	1	one
∩	Heel bone (arch)	10	ten
?	Coil of rope (scroll)	100	one hundred
?	Lotus flower	1000	one thousand
?	Pointing finger	10,000	ten thousand
?	Bourbot (tadpole)	100,000	one hundred thousand
?	Astonished man	1,000,000	one million

To write the numeral for a number, the Egyptians wrote the symbols next to each other from left to right, and the number represented was the sum of the values of the symbols. The most times any symbol was used was nine. Instead of using a symbol ten times, they used the symbol for the next higher number. They also grouped the symbols in threes or fours.

EXAMPLE

? ? ? ? ∩ ∩ | | | |
? ? ? ? ∩ ∩ | | |

represents the number one thousand, six hundred twenty-seven, or
$1000 + 600 + 20 + 7 = 1627$

Mayan System

The Mayans used a system of dots and bars (for numbers from 1 to 19) combined with a place value system. A dot represented one and a bar represented five. They had a symbol, ⬭ , for zero and based their system, with one exception, on twenty. (See Table I.2.) The symbols were arranged vertically, smaller values starting at the bottom. The value of the third place up was 360 (18 times the value of the second place), but all other places were 20 times the value of the previous place.

TABLE I.2 MAYAN NUMERALS

SYMBOL	VALUE	
.	1	one
—	5	five
⬭	0	zero

EXAMPLES

1. ⋯ $(3 + 5 = 8)$

2. ☰☰ $(3 \cdot 5 + 4 = 19)$

3. ⋯ 3 20's
   ⬭ 0 units
   $(3 \cdot 20 + 0 = 60)$

4. .. 2 7200's
   ⬭ 0 360's
   . 6 20's
   .. 7 units
   $(2 \cdot 7200 + 0 \cdot 360 + 6 \cdot 20 + 7 = 14{,}527)$

[Note: ⬭ is used as a place holder.]

Attic Greek System

The Greeks used two numeration systems, the Attic (see Table I.3) and the Alexandrian. (See Section I.2 for information on the Alexandrian system.) In the Attic system, no numeral was used more than four times. When a symbol was needed five or more times, the symbol for five was used as shown in the examples.

EXAMPLES

1. X X ⊬ H H ⊬ I I I I $(2 \cdot 1000 + 7 \cdot 100 + 5 \cdot 10 + 4 = 2754)$

2. ⊬ H H H H Δ Δ Γ $(5 \cdot 1000 + 4 \cdot 100 + 2 \cdot 10 + 5 = 5425)$

TABLE I.3   ATTIC GREEK NUMERALS

SYMBOL	VALUE	
I	1	one
Γ	5	five
Δ	10	ten
H	100	one hundred
X	1000	one thousand
M	10,000	ten thousand

Roman System

The Romans used a system (Table I.4) that we still see in evidence as hours on clocks and dates on buildings.

TABLE I.4   ROMAN NUMERALS

SYMBOL	VALUE	
I	1	one
V	5	five
X	10	ten
L	50	fifty
C	100	one hundred
D	500	five hundred
M	1000	one thousand

The symbols were written largest to smallest, from left to right. The value of the numeral was the sum of the values of the individual symbols. Each symbol was used as many times as necessary with the following exceptions: when the Romans got to 4, 9, 40, 90, 400, or 900, they used a system of subtraction.

$$IV = 5 - 1 = 4 \qquad XL = 50 - 10 = 40 \qquad CD = 500 - 100 = 400$$
$$IX = 10 - 1 = 9 \qquad XC = 100 - 10 = 90 \qquad CM = 1000 - 100 = 900$$

EXAMPLES

1. VII   represents 7

2. DXLIV   represents 544

3. MCCCXXVIII   represents 1328

EXERCISES I.1

Find the values of the following ancient numerals.

1. ꟼ ꟼ ∩∩∩ ∩∩ III

2. ◠ ℓℓℓ ℓℓℓ ⚇⚇⚇ ∩∩∩∩∩

3. ℓℓ ⚇ ꟼ ꟼ ꟼ ∩∩ IIII III

4. ⦙⦙⦙

5.

6.

7. Ɣ I I I

8. Ϝ H H Δ Δ Δ Ɣ I

9. X X H H H Ϝ Δ I I

10. XCVII

11. DCCXLIV

12. MMMCDLXV

13. CMLXXVIII

14. Write 64
    (a) as an Egyptian numeral   (b) as a Mayan numeral
    (c) as an Attic Greek numeral   (d) as a Roman numeral

15. Follow the instructions for Problem 14, using 532 in place of 64.

16. Follow the same instructions, using 1969.

17. Follow the same instructions, using 846.

I.2

Babylonian System (Cuneiform Numerals)

The Babylonians (about 3500 B.C.) used a place value system based on the number sixty, called a sexagesimal system. They had only two symbols, V and < . (See Table I.5.) These wedge shapes are called cuneiform numerals since *cuneus* means *wedge* in Latin.

TABLE I.5   CUNEIFORM NUMERALS

SYMBOL	VALUE	
V	1	one
<	10	ten

The symbol for one was used as many as nine times, and the symbol for ten as many as five times; however, since there was no symbol for zero, many Babylonian numbers could be read several ways. For our purposes, we will group the symbols to avoid some of the ambiguities inherent in the system.

EXAMPLES

1. $\text{VVV} \quad << \begin{matrix}\text{VVV}\\\text{VV}\end{matrix} \quad <<< \text{VV}$

$$(3 \cdot 60^2) + (25 \cdot 60^1) + (32 \cdot 1) = (3 \cdot 3600) + (25 \cdot 60) + 32$$
$$= 10{,}800 + 1500 + 32$$
$$= 12{,}332$$

2. $\text{V} \quad <<<< \begin{matrix}\text{VVV}\\\text{VVV}\end{matrix} \quad < \begin{matrix}\text{VVVV}\\\text{VVV}\end{matrix}$

$$(1 \cdot 60^2) + (46 \cdot 60^1) + (17 \cdot 1) = 3600 + 2760 + 17 = 6377$$

Alexandrian Greek System

The Greeks used two numeration systems, the Attic and the Alexandrian. We discussed the Attic Greek system in Section I.1.

In the Alexandrian system (Table I.6), the letters were written next to each other, largest to smallest, from left to right. Since the numerals were also part of the Greek alphabet, an accent mark or bar was sometimes used above a letter to indicate that it represented a number. Multiples of 1000 were indicated by strikes in front of the unit symbols, and multiples of 10,000 were indicated by placing the unit symbols above the symbol M.

TABLE I.6 ALEXANDRIAN GREEK SYMBOLS

SYMBOL	NAME		VALUE	SYMBOL	NAME		VALUE
A	Alpha	1	one	Ξ	Xi	60	sixty
B	Beta	2	two	O	Omicron	70	seventy
Γ	Gamma	3	three	Π	Pi	80	eighty
Δ	Delta	4	four	Ϙ	Koppa	90	ninety
E	Epsilon	5	five	P	Rho	100	one hundred
F	Digamma (or Vau)	6	six	Σ	Sigma	200	two hundred
Z	Zeta	7	seven	T	Tau	300	three hundred
H	Eta	8	eight	Y	Upsilon	400	four hundred
Θ	Theta	9	nine	Φ	Phi	500	five hundred
I	Iota	10	ten	X	Chi	600	six hundred
K	Kappa	20	twenty	Ψ	Psi	700	seven hundred
Λ	Lambda	30	thirty	Ω	Omega	800	eight hundred
M	Mu	40	forty	ϡ	Sampi	900	nine hundred
N	Nu	50	fifty				

EXAMPLES   1. $\overline{\Phi\ \Xi\ Z}$ $(500 + 60 + 7 = 567)$

2. $\overline{\text{B}}$
   M T N Δ $(20{,}000 + 300 + 50 + 4 = 20{,}354)$

Chinese-Japanese System

The Chinese-Japanese system (Table I.7) uses a different numeral for each of the digits up to ten, then a symbol for each power of ten. A digit written above a power of ten is to be multiplied by that power, and all such results are to be added to find the value of the numeral.

TABLE I.7   CHINESE-JAPANESE NUMERALS

SYMBOL	VALUE		SYMBOL	VALUE	
―	1	one	七	7	seven
ニ	2	two	八	8	eight
三	3	three	九	9	nine
四	4	four	十	10	ten
五	5	five	百	100	one hundred
六	6	six	千	1000	one thousand

EXAMPLES

1. 三 } 30
   十

   九   9

   $(30 + 9 = 39)$

2. 五 千 } 5000

   四 百 } 400

   八 十 } 80

   ニ } 2

   $(5000 + 400 + 80 + 2 = 5482)$

EXERCISES I.2

Find the value of each of the following ancient numerals.

1. V <<VVV VVV   2. <<<VV   3. <<< VV

4. $\overline{Y\ N\ E}$

5. $\overline{\Delta}$
   M /Z π

6. $\overline{\Sigma\ K\ B}$

7. 四
   十
   六

8. 五
   千
   一
   百
   八

9. 九
   百
   九
   十
   九

Write each of the following numbers as (a) Babylonian numerals, (b) Alexandrian Greek numerals, and (c) Chinese-Japanese numerals.

10. 472 11. 596 12. 5047 13. 3665 14. 7293 15. 10,852

BASE TWO AND BASE FIVE

APPENDIX

II

II.1   THE BINARY SYSTEM (BASE TWO)

In the decimal system, ten is the base. You might ask if another number could be chosen as the base in a place value system. And, if so, would the system be any better or more useful than the decimal system? The fact is that computers do operate under a place value system whose base is two. In the **binary system** (or base two system), only two digits are needed, 0 and 1. These two digits correspond to the two possible conditions of an electric current, either *on* or *off*.

Any number can be represented in base two or in base ten. However, base ten has a definite advantage when large numbers are involved, as you will see. The advantage of base two over base ten is that for base two only two digits are needed, while ten digits are needed for base ten.

If the base of a place value system were not ten but two, then the beginning point would not be a decimal point but a **binary point.** The value of each place would be a power of two, as shown in Figure II.1.

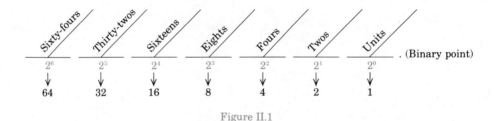

Figure II.1

TO WRITE NUMBERS IN THE BASE TWO SYSTEM,
REMEMBER THREE THINGS

1.   {0, 1} is the set of digits we can use.

2.   The value of each place from the binary point is in powers of two.

3.   The symbol 2 does not exist in the binary system, just as there is no digit for ten in the decimal system.

To avoid confusion with the base ten numerals, we will write $_{(2)}$ to the lower right of each base two numeral. We could write $_{(10)}$ to the lower right of each base ten numeral, but this would not be practical since most of the numerals we work with are in base ten. Therefore, *if no base is indicated, the numeral will be understood to be in base ten.*

EXAMPLES   1.   Find the value of the numeral $1101._{(2)}$.
Writing the value of each place under the digit gives

$$\frac{1}{2^3} \quad \frac{1}{2^2} \quad \frac{0}{2^1} \quad \frac{1}{2^0} \quad \cdot_{(2)}$$

In expanded notation,

$$1101._{(2)} = 1(2^3) + 1(2^2) + 0(2^1) + 1(2^0)$$
$$= 1(8) + 1(4) + 0(2) + 1(1)$$
$$= 8 + 4 + 0 + 1$$
$$= 13$$

Thus, to a computer, the symbol $1101._{(2)}$ means "thirteen."

2. $1._{(2)} = 1$

3. $10._{(2)} = 1(2) + 0 = 2$

4. $11._{(2)} = 1(2) + 1 = 2 + 1 = 3$

5. $100._{(2)} = 1(2^2) + 0(2) + 0(1) = 4 + 0 + 0 = 4$

6. $101._{(2)} = 1(2^2) + 0(2) + 1(1) = 4 + 0 + 1 = 5$

7. $110._{(2)} = 1(2^2) + 1(2) + 0(1) = 4 + 2 + 0 = 6$

8. $111._{(2)} = 1(2^2) + 1(2) + 1(1) = 4 + 2 + 1 = 7$

9. $1000._{(2)} = 1(2^3) + 0(2^2) + 0(2) + 0(1) = 8 + 0 + 0 + 0 = 8$

10. $1001._{(2)} = 1(2^3) + 0(2^2) + 0(2) + 1(1) = 8 + 0 + 0 + 1 = 9$

11. $1010._{(2)} = 1(2^3) + 0(2^2) + 1(2) + 0(1) = 8 + 0 + 2 + 0 = 10$

Do *not* read $100_{(2)}$ as "one hundred" because the 1 is not in the hundreds place. The 1 is in the fours place. So, $100_{(2)}$ is read "four" or "one, zero, zero—base two." Similarly, $111_{(2)}$ is read "seven" or "one, one, one—base two."

EXERCISES II.1

Write the following base ten numerals in expanded form using components.

Example: $273 = 2(10^2) + 7(10^1) + 3(10^0)$

1. 35   **2.** 761   **3.** 8469   **4.** 500   **5.** 62,322

Write the following base two numerals in expanded form and find the value of each numeral.

Example: $110_{(2)} = 1(2^2) + 1(2^1) + 0(2^0)$
$= 1(4) + 1(2) + 0(1)$
$= 4 + 2 + 0$
$= 6$

6. $11_{(2)}$ **7.** $101_{(2)}$ **8.** $111_{(2)}$ **9.** $1011_{(2)}$

10. $1101_{(2)}$ **11.** $110,111_{(2)}$ **12.** $11,110_{(2)}$ **13.** $101,011_{(2)}$

14. $11,010_{(2)}$ **15.** $1000_{(2)}$ **16.** $1,000,010_{(2)}$ **17.** $11,101_{(2)}$

18. $10,110_{(2)}$ **19.** $111,111_{(2)}$ **20.** $1111_{(2)}$

21. If a computer is directed to place some information in memory space number $1,101,111_{(2)}$, what is the number of this memory space in base ten?

II.2   THE QUINARY SYSTEM (BASE FIVE)

Many numbers may be used as bases for place value systems. To illustrate this point and to emphasize the concept of place value, one more base system, base five, will be discussed. Interested students may want to try writing numerals in base three or base eight or base eleven.

Again, the system relies on powers of the base and a set of digits. In the **quinary system** (base five system), the powers of 5 are $5^0, 5^1, 5^2,$ $5^3, 5^4$, and so on, and the digits to be used are $\{0, 1, 2, 3, 4\}$. The **quinary point** is the beginning point, as shown in Figure II.2.

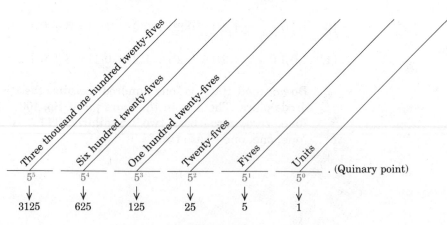

Figure II.2

EXAMPLES   **1.** $\underline{1}._{(5)} = 1$

            2. $\underline{2}._{(5)} = 2$

3. $\underline{3}._{(5)} = 3$

4. $\underline{4}._{(5)} = 4$

5. $\underline{1}\,\underline{0}._{(5)} = 1(5) + 0(1) = 5 + 0 = 5$

6. $\underline{1}\,\underline{1}._{(5)} = 1(5) + 1(1) = 5 + 1 = 6$

7. $\underline{1}\,\underline{2}._{(5)} = 1(5) + 2(1) = 5 + 2 = 7$

8. $\underline{1}\,\underline{3}._{(5)} = 1(5) + 3(1) = 5 + 3 = 8$

9. $\underline{1}\,\underline{4}._{(5)} = 1(5) + 4(1) = 5 + 4 = 9$

10. $\underline{2}\,\underline{0}._{(5)} = 2(5) + 0(1) = 10 + 0 = 10$

11. $\underline{2}\,\underline{1}._{(5)} = 2(5) + 1(1) = 10 + 1 = 11$

12. $\underline{2}\,\underline{2}._{(5)} = 2(5) + 2(1) = 10 + 2 = 12$

13. $\begin{aligned}\underline{3}\,\underline{2}\,\underline{4}._{(5)} &= 3(5^2) + 2(5^1) + 4(5^0) \\ &= 3(25) + 2(5) + 4(1) \\ &= 75 + 10 + 4 \\ &= 89\end{aligned}$

EXERCISES II.2

Write the following base five numerals in expanded form and find the value of each.

1. $24_{(5)}$ 2. $13_{(5)}$ 3. $10_{(5)}$ 4. $43_{(5)}$ 5. $104_{(5)}$

6. $312_{(5)}$ 7. $32_{(5)}$ 8. $230_{(5)}$ 9. $423_{(5)}$ 10. $444_{(5)}$

11. $1034_{(5)}$ 12. $4124_{(5)}$ 13. $244_{(5)}$ 14. $3204_{(5)}$ 15. $13{,}042_{(5)}$

16. Do the numerals $101_{(2)}$ and $10_{(5)}$ represent the same number? If so, what is the number?

17. Answer the questions in Problem 16 about the numerals $1101_{(2)}$ and $23_{(5)}$.

18. Answer the questions in Problem 16 about the numerals $11{,}100_{(2)}$ and $103_{(5)}$.

19. What set of digits do you think would be used in a base eight system?

20. What set of digits do you think would be used in a base twelve system? [HINT: New symbols for some new digits must be introduced.]

II.3 ADDITION AND MULTIPLICATION IN BASE TWO AND BASE FIVE

Now that we have two new numeration systems, base two and base five, a natural question to ask is. How are addition and multiplication* performed in these systems? The basic techniques are the same as for base ten since place value is involved. However, because different bases are involved, the numerals will be different. For example, to add five plus seven in base ten, we write $5 + 7 = 12$. In base two, this same sum is written $101_{(2)} + 111_{(2)} = 1100_{(2)}$.

Writing the numerals vertically (one under the other) gives

$$
\begin{array}{cc}
5 & 101_{(2)} \\
7 & 111_{(2)} \\
\hline
12 & 1100_{(2)}
\end{array}
$$

Now a step-by-step analysis of the sum in base two will be discussed.

EXAMPLES   1. (a) $101_{(2)}$
$111_{(2)}$

The numerals are written so that the digits of the same place value line up.

(b) $\overset{1}{101}_{(2)}$
$111_{(2)}$
$\overline{0_{(2)}}$

Adding $1 + 1$ in the units column gives "two," which is written $10_{(2)}$. 0 is written in the units column and 1 is "carried" to the "twos" column.

(c) $\overset{11}{101}_{(2)}$
$111_{(2)}$
$\overline{00_{(2)}}$

Now, in the twos column, $1 + 0 + 1$ is again "two," or $10_{(2)}$. Again 0 is written, and 1 is "carried" to the next column, the "fours" column.

(d) $\overset{11}{101}_{(2)}$
$111_{(2)}$
$\overline{1100_{(2)}}$

In the fours column (or third column), $1 + 1 + 1$ is "three," or $11_{(2)}$. Since there are no digits in the "eights" column (or fourth column), 11 is written, and the sum is $1100_{(2)}$.

Checking in Base Ten:

$$
\begin{array}{llll}
101_{(2)} = & 1(2^2) + 0(2^1) + 1(2^0) = & 4 + 0 + 1 = \ 5 & 5 \\
111_{(2)} = & 1(2^2) + 1(2^1) + 1(2^0) = & 4 + 2 + 1 = \ 7 & 7 \\
\hline
1100_{(2)} & 1(2^3) + 1(2^2) + 0(2^1) + 0(2^0) = 8 + 4 + 0 + 0 = 12 & 12
\end{array}
$$

*Subtraction and division may also be performed in base two and base five but will not be discussed here for reasons of time. Some students may want to investigate these operations on their own.

Addition in base five is similar. Although the thinking is done in base ten, which is familiar to us, the numerals written must be in base five.

2. (a) $143_{(5)}$
   $34_{(5)}$

The numerals are written so that the digits of the same place value line up.

   (b) 1
   $143_{(5)}$
   $34_{(5)}$
   $2_{(5)}$

Adding $3 + 4$ in the units column gives "seven," which is written $12_{(5)}$. 2 is written in the units column, and 1 is carried to the fives column.

   (c) 11
   $143_{(5)}$
   $34_{(5)}$
   $32_{(5)}$

In the fives column, $1 + 4 + 3$ gives "eight," which is $13_{(5)}$. The 3 is written, and 1 is carried to the next column (the twenty-fives column).

   (d) 11
   $143_{(5)}$
   $34_{(5)}$
   $232_{(5)}$

In the third column, $1 + 1$ gives 2. The sum is $232_{(5)}$.

Checking in Base Ten:

$$143_{(5)} = 1(5^2) + 4(5^1) + 3(5^0) = 25 + 20 + 3 = 48 \qquad 48$$
$$34_{(5)} = 3(5^1) + 4(5^0) = 15 + 4 = 19 \qquad 19$$
$$232_{(2)} = 2(5^2) + 3(5^1) + 2(5^0) = 50 + 15 + 2 = 67 \qquad 67$$

Multiplication in each base is performed and checked in the same manner as addition. Of course, the difference is that you multiply instead of add. When you multiply, be sure to write the correct symbol for the number in the base being used. Also remember to add in the correct base.

3. $101_{(2)}$
   $111_{(2)}$
   ${}_1 101$
   ${}_1\,101$
   101
   $100011_{(2)}$

Multiplication in base two is easy since we are multiplying by only 1's or 0's. The adding must be done in base two.

Checking gives:

$101_{(2)} = 5$
$111_{(2)} = 7$
$\phantom{111_{(2)} =}\,101 \qquad 35$
$\phantom{111_{(2)} =}\,101$
$\phantom{111_{(2)} =}\,101$
$\phantom{111_{(2)} =}100011_{(2)}$

Remember to *multiply* the checking numbers.

$100,011_{(2)} = 1(2^5) + 0(2^4) + 0(2^3) + 0(2^2)$
$\phantom{100,011_{(2)} =} + 1(2) + 1(1)$
$\phantom{100,011_{(2)} =} = 32 + 2 + 1 = 35$

4.　　　　$\overset{2}{}$
　　　$\overset{1}{3}4_{(5)}$
　　　　$23_{(5)}$
　　　─────
　　　　212
　　　　123
　　　─────
　　　$1442_{(5)}$

Multiplying 3×4 gives "twelve," which is $22_{(5)}$. Write 2 and carry 2 just as in regular multiplication. Then, 3×3 is "nine," and "nine" plus 2 is "eleven"; but in base five, "eleven" is $21_{(5)}$. Similarly, 2×4 is "eight," or $13_{(5)}$. Write the 3, carry the 1. 2×3 is "six," and "six" plus 1 is "seven," or $12_{(5)}$.

Checking gives:

$$34_{(5)} = 19 \Big\}$$
$$23_{(5)} = 13 \Big\}$$
　　　─────
　　212　　57
　　123　　19
　　─────
$1442_{(5)}$　247

Remember to *multiply* the checking numbers.
$1442_{(5)} = 1(5^3) + 4(5^2) + 4(5) + 2(1)$
$\qquad = 125 + 100 + 20 + 2 = 247$

EXERCISES II.3

Add in the base indicated and check your work in base ten.

1. $101_{(2)}$ $\underline{11_{(2)}}$	**2.** $43_{(5)}$ $\underline{213_{(5)}}$	**3.** $1101_{(2)}$ $\underline{1011_{(2)}}$	**4.** $111_{(2)}$ $\underline{1010_{(2)}}$	**5.** $134_{(5)}$ $\underline{243_{(5)}}$

6. $11_{(2)}$ $10_{(2)}$ $\underline{11_{(2)}}$	**7.** $11_{(2)}$ $11_{(2)}$ $\underline{101_{(2)}}$	**8.** $214_{(5)}$ $\underline{343_{(5)}}$	**9.** $14_{(5)}$ $321_{(5)}$ $\underline{43_{(5)}}$	**10.** $431_{(5)}$ $214_{(5)}$ $\underline{102_{(5)}}$

11. $11_{(2)}$ $101_{(2)}$ $111_{(2)}$ $\underline{101_{(2)}}$	**12.** $111_{(2)}$ $11_{(2)}$ $110_{(2)}$ $\underline{111_{(2)}}$	**13.** $101_{(2)}$ $101_{(2)}$ $101_{(2)}$ $\underline{101_{(2)}}$	**14.** $23_{(5)}$ $103_{(5)}$ $214_{(5)}$ $\underline{322_{(5)}}$	**15.** $414_{(5)}$ $211_{(5)}$ $334_{(5)}$ $\underline{222_{(5)}}$

Multiply in the base indicated and check your work in base ten.

16. $1101_{(2)}$ $\underline{111_{(2)}}$	**17.** $1011_{(2)}$ $\underline{101_{(2)}}$	**18.** $423_{(5)}$ $\underline{30_{(5)}}$	**19.** $104_{(5)}$ $\underline{23_{(5)}}$	**20.** $223_{(5)}$ $\underline{44_{(5)}}$

21. $423_{(5)}$ $\underline{32_{(5)}}$	**22.** $1111_{(2)}$ $\underline{111_{(2)}}$	**23.** $111_{(2)}$ $\underline{111_{(2)}}$	**24.** $2212_{(5)}$ $\underline{43_{(5)}}$	**25.** $10{,}111_{(2)}$ $\underline{110_{(2)}}$

ANSWER KEY

CHAPTER 1

EXERCISES 1.1   PAGE 4

	Exponent	Base	Power		Exponent	Base	Power
1.	3	2	8	2.	5	2	32
3.	2	5	25	5.	0	7	1
6.	2	11	121	7.	4	1	1
9.	0	4	1	10.	6	3	729
11.	2	3	9	13.	0	5	1
14.	50	1	1	15.	1	62	62
17.	2	10	100	18.	3	10	1000
19.	2	4	16	21.	4	10	10,000
22.	3	5	125	23.	3	6	216
25.	0	9	1				

26. 2^2 27. 5^2 29. 3^3 30. 2^5 31. 11^2

33. 2^3 34. 3^2 35. 6^2 37. 9^2 or 3^4

38. 8^2 or 4^3 or 2^6 39. 10^2 41. 10^4 42. 6^3

43. 12^2 45. 3^5 46. 1, 2, 4, 5, 10, 20   47. 1, 2, 7, 14

49. 1, 17   50. 1, 3, 17, 51   51. 1, 3, 5, 15, 25, 75

53. 1, 5, 25   54. 1, 29   55. 1, 2, 4, 8, 16   57. 1, 2, 31, 62

58. 1, 2, 3, 4, 6, 8, 12, 24   59. 1, 79

EXERCISES 1.2   PAGE 7

1. $3(10^1) + 7(10^0)$
   thirty-seven

2. $8(10^1) + 4(10^0)$
   eighty-four

3. $9(10^1) + 8(10^0)$
   ninety-eight

5. $1(10^2) + 2(10^1) + 2(10^0)$
   one hundred twenty-two

6. $4(10^2) + 9(10^1) + 3(10^0)$
   four hundred ninety-three

7. $8(10^2) + 2(10^1) + 1(10^0)$
   eight hundred twenty-one

9. $8(10^2) + 2(10^1) + 8(10^0)$
   eight hundred twenty-eight

10. $5(10^3) + 4(10^2) + 9(10^1) + 6(10^0)$
    five thousand, four hundred ninety-six

11. $1(10^4) + 2(10^3) + 5(10^2) + 1(10^1) + 7(10^0)$
    twelve thousand, five hundred seventeen

13. $2(10^5) + 4(10^4) + 3(10^3) + 4(10^2) + 0(10^1) + 0(10^0)$
    two hundred forty-three thousand, four hundred

14. $8(10^5) + 9(10^4) + 1(10^3) + 5(10^2) + 4(10^1) + 0(10^0)$
    eight hundred ninety-one thousand, five hundred forty

15. $4(10^4) + 3(10^3) + 6(10^2) + 5(10^1) + 5(10^0)$
    forty-three thousand, six hundred fifty-five

17. $8(10^6) + 4(10^5) + 0(10^4) + 0(10^3) + 8(10^2) + 1(10^1) + 0(10^0)$
     eight million, four hundred thousand, eight hundred ten

18. $5(10^6) + 6(10^5) + 6(10^4) + 3(10^3) + 7(10^2) + 0(10^1) + 1(10^0)$
     five million, six hundred sixty-three thousand, seven hundred one

19. $1(10^7) + 6(10^6) + 3(10^5) + 0(10^4) + 2(10^3) + 5(10^2) + 9(10^1) + 0(10^0)$
     sixteen million, three hundred two thousand, five hundred ninety

21. $8(10^7) + 3(10^6) + 0(10^5) + 0(10^4) + 0(10^3) + 6(10^2) + 0(10^1) + 5(10^0)$
     eighty-three million, six hundred five

22. $1(10^8) + 5(10^7) + 2(10^6) + 4(10^5) + 0(10^4) + 3(10^3) + 6(10^2) + 7(10^1)$
     $+ 2(10^0)$
     one hundred fifty-two million, four hundred three thousand, six
     hundred seventy-two

23. $6(10^8) + 7(10^7) + 9(10^6) + 0(10^5) + 7(10^4) + 8(10^3) + 1(10^2) + 0(10^1)$
     $+0(10^0)$
     six hundred seventy-nine million, seventy-eight thousand, one
     hundred

25. $8(10^9) + 5(10^8) + 7(10^7) + 2(10^6) + 0(10^5) + 0(10^4) + 3(10^3) + 4(10^2)$
     $+2(10^1) + 5(10^0)$
     eight billion, five hundred seventy-two million, three thousand,
     four hundred twenty-five

26. 76	27. 132	29. 3842	30. 2005	
31. 192,151	33. 21,400	34. 33,333	35. 5,045,000	
37. 10,639,582	38. 281,300,501	39. 530,000,700	41. 90,090,090	
42. 82,700,000	43. 175,000,002	45. 757		

EXERCISES 1.3 PAGE 12

1. 16	2. 15	3. 13	5. 21	6. 20	7. 19	9. 12	
10. 17	11. 18	13. 12	14. 21	15. 17	17. 27	18. 18	
19. 21	21. commutative	22. commutative	23. associative				
25. associative	26. associative	27. identity					
29. identity	30. associative	31. 162					
33. 239	34. 835	35. 1298	37. 1236	38. 4168			
39. 6869	41. 1,603,426	42. 1,463,930	43. 2,610,667				
45. 2762 miles	46. $6,313,323	47. $18,463					
49. 1518 students	50. 33,830 appliances						

EXERCISES 1.4 PAGE 15

1. 3	2. 13	3. 0	5. 9	6. 17	7. 9	
9. 5	10. 6	11. 0	13. 13	14. 20	15. 20	
17. 5	18. 45	19. 13	21. 94	22. 126	23. 218	

25.	475	26.	376	27.	593	29.	188	30.	478	31.	1569

25. 475 26. 376 27. 593 29. 188 30. 478 31. 1569
33. 1568 34. 1531 35. 1493 37. 694 38. 5871 39. 2517
41. 2,806,644 42. 3,800,559 43. 1,006,958
45. 5,671,011 46. 222 47. 195
49. 44 points 50. $250,404 51. $934
53. $235,456 54. 1,909,096

EXERCISES 1.5 PAGE 20

1. 72 2. 54 3. 56 5. 0 6. 7 7. 32 9. 96
10. 63 11. 15 13. 100 14. 100 15. 0

25.

GIVEN NO.	ADD 5	DOUBLE	SUBTRACT 10
2	7	14	4
0	5	10	0
1	6	12	2
7	12	24	14
8	13	26	16
5	10	20	10

26. The numbers in the last column are three times the numbers in the first column. Subtracting 18 eliminates the effect of tripling 6 just as subtracting 10 eliminates the effect of doubling 5 in Problem 25.

EXERCISES 1.6 PAGE 23

1. 250 2. 7600 3. 47,000 5. 72 6. 13
7. 3000 9. 400 10. 3600 11. 1200 13. 6300
14. 9000 15. 4000 17. 15,000 18. 3600 19. 5200
21. 16,000 22. 18,000 23. 10,000 25. 60,000
26. 25,000 27. 12,000 29. 240,000 30. 800,000
31. 48,000 33. 600,000 34. 16,000,000 35. 630,000

EXERCISES 1.7 PAGE 28

1. 224 2. 162 3. 432 5. 344 6. 432
7. 455 9. 252 10. 760 11. 2352 13. 330
14. 1189 15. 2412 17. 960 18. 2790 19. 7055
21. 544 22. 2548 23. 880 25. 375 26. 4371
27. 2064 29. 156 30. 2916 31. 5166 33. 2850
34. 7632 35. 29,601 37. 9800 38. 174,045 39. 125,178
41. 31,200 42. 66,960 43. 380,000 45. 496,400
46. 1,504,000 47. 217,300 49. 249,600 50. 897,000
51. 583,128 53. $456,055 54. $1320 55. 2602; 2004

EXERCISES 1.8   PAGE 32

1. 30	2. 10	3. 21	5. 12	6. 30
7. 5	9. 6 R4	10. 7 R2	11. 24	13. 10 R11
14. 14	15. 11 R7	17. 41	18. 45 R6	19. 47 R8
21. 234 R9	22. 135 R14	23. 212 R7	25. 304	26. 5 R2
27. 2 R3	29. 7	30. 6	31. 12	33. 38
34. 32 R2	35. 3 R10	37. 38 R6	38. 8	39. 15 R5
41. 28 R7	42. 42 R3	43. 50	45. 20	46. 30
47. 400 R3	49. 400 R6	50. 301 R4	51. 2 R2	53. 20 R13
54. 61 R15	55. 54 R3		57. 2	58. 22 R74
59. 4 R192	61. 201 R62	62. 196 R370		63. 221 R308
65. 302	66. 461 R242	67. 305		69. 693
70. 611	71. 98			

73. 14; 0 R7; no; division is not commutative

74. 55 years old

EXERCISES 1.9   PAGE 36

1. 3	2. 15	3. 7	5. 22	6. 31
7. 3	9. 5	10. 10	11. 5	13. 5
14. 0	15. 3	17. 7	18. 0	19. 0
21. 6	22. 3	23. 3	25. 27	26. 0
27. 26	29. 0	30. 5	31. 69	33. 68
34. 80	35. 140	37. for example, $21 \div 7 \neq 7 \div 21$		

38. for example, $48 \div (12 \div 4) = 48 \div 3 = 16$.
     but $(48 \div 12) \div 4 = 4 \div 4 = 1$

39. $10 = 7 + 3$; only if the division is on the right; division is not commutative, and $5 \div (35 + 15) \neq 5 \div 35 + 5 \div 15$

41. 485	42. 3000	43. 586
45. six thousand, one hundred eighty-four		46. 1973
47. $665; $2495	49. 85	50. $4.59

REVIEW QUESTIONS · CHAPTER 1   PAGE 39

1. base; exponent   2. factors   3. 2^7 4. 13^2

5. $4(10^2) + 9(10^1) + 6(10^0)$
    four hundred ninety-six

6. $7(10^3) + 8(10^2) + 4(10^1) + 2(10^0)$
    seven thousand, eight hundred forty-two

7. $8(10^6) + 0(10^5) + 0(10^4) + 0(10^3) + 5(10^2) + 7(10^1) + 0(10^0)$
    eight million, five hundred seventy

8. 4856   9. 15,032,197   10. 672,340,083

11. (a) 1, 2, 3, 6, 34, 51, 102   (b) 1, 3, 13, 39   (c) 1, 2, 4, 38, 76
12. commutative prop. of add.   13. associative prop. of mult.
14. associative prop. of add.   15. commutative prop. of mult.
16. $32 \div 8 = 4$; $2 \div 2 = 1$; division is not associative

17. 15   18. 46   19. 35   20. 13   21. 2   22. 2
23. 10,541   24. 1674   25. 2400   26. 0   27. 0
28. 508   29. 2384   30. 2112   31. 5952   32. 3,822,498
33. 14,388,000   34. 292 R2   35. 135 R81   36. 703 R7
37. 1059   38. 35   39. 9   40. $1485; $99   41. 83

42. If a is a whole number, then there is a unique whole number 0 with
     the property that $a + 0 = a$. If a is a whole number, then there is a
     unique whole number 1 with the property that $a \cdot 1 = a$.

43. 70

44.

GIVEN NO.	ADD 100	DOUBLE	SUBTRACT 200
3	103	206	6
20	120	240	40
15	115	230	30
8	108	216	16

45. 20 times

CHAPTER 2

EXERCISES 2.1   PAGE 45

1. 2   2. 2   3. 3, 5, 15   5. 3   6. 2, 5, 10
7. None   9. 3   10. 5   11. 3, 5, 15
13. 2, 3, 5, 6, 9, 10, 15 (all)   14. 3, 5, 15   15. 2, 3, 6   17. None
18. 5   19. 2   21. 3, 5, 15   22. 2, 3, 6   23. 3, 9
25. 2, 3, 6, 9   26. 2, 3, 5, 6, 9, 10, 15 (all)   27. 2, 5, 10   29. 3, 9
30. 2, 3, 6, 9   31. 2, 3, 5, 6, 10, 15   33. 2
34. 2, 3, 5, 6, 9, 10, 15 (all)   35. 2, 3, 5, 6, 10, 15   37. 3, 5, 15
38. 3, 5, 9, 15   39. None   41. 2, 3, 6, 9   42. 3, 5, 15
43. 2, 3, 6   45. 2, 3, 6, 9   46. 2, 3, 5, 6, 10, 15   47. 3, 9
49. 2, 3, 6   50. 5   51. Yes; 18, 36, 54, 72, 90

EXERCISES 2.2   PAGE 49

1. $1 \cdot 28$ 2. $1 \cdot 32$ 3. $1 \cdot 16$ 5. $1 \cdot 9$ 6. $1 \cdot 105$
    $2 \cdot 14$ $2 \cdot 16$ $2 \cdot 8$ $3 \cdot 3$ $3 \cdot 35$
    $4 \cdot 7$ $4 \cdot 8$ $4 \cdot 4$ $5 \cdot 21$
                                                                                     $7 \cdot 15$

7. $1 \cdot 35$
$5 \cdot 7$

9. $1 \cdot 36$
$2 \cdot 18$
$3 \cdot 12$
$4 \cdot 9$
$6 \cdot 6$

10. $1 \cdot 72$
$2 \cdot 36$
$3 \cdot 24$
$4 \cdot 18$
$6 \cdot 12$
$8 \cdot 9$

11. $1 \cdot 14$
$2 \cdot 7$

13. $1 \cdot 10$
$2 \cdot 5$

14. $1 \cdot 50$
$2 \cdot 25$
$5 \cdot 10$

15. $1 \cdot 24$
$2 \cdot 12$
$3 \cdot 8$
$4 \cdot 6$

17. $1 \cdot 100$
$2 \cdot 50$
$4 \cdot 25$
$5 \cdot 20$
$10 \cdot 10$

18. $1 \cdot 65$
$5 \cdot 13$

19. $1 \cdot 51$
$3 \cdot 17$

21. $\{3, 6, 9, 12, \ldots\}$
22. $\{7, 14, 21, 28, \ldots\}$
23. $\{6, 12, 18, 24, \ldots\}$
25. $\{20, 40, 60, 80, \ldots\}$
26. $\{16, 32, 48, 64, \ldots\}$
27. $\{17, 34, 51, 68, \ldots\}$
29. $\{25, 50, 75, 100, \ldots\}$
30. $\{21, 42, 63, 84, \ldots\}$

31.

1	②	③	4	⑤	6	⑦	8	9	10
⑪	12	⑬	14	15	16	⑰	18	⑲	20
21	22	㉓	24	25	26	27	28	㉙	30
㉛	32	33	34	35	36	㊲	38	39	40
㊶	42	㊸	44	45	46	㊼	48	49	50
51	52	㊾	54	55	56	57	58	㊾	60
㊶	62	63	64	65	66	㊻	68	69	70
㊹	72	㊻	74	75	76	77	78	㊾	80
81	82	㊸	84	85	86	87	88	㊾	90
91	92	93	94	95	96	㊾	98	99	100

33. 3, 4
34. 8, 2
35. 12, 1
37. 25, 2
38. 5, 4
39. 8, 3
41. 12, 3
42. 7, 1
43. 21, 3
45. 5, 5
46. 4, 4
47. 12, 5
49. 9, 3
50. 18, 4
51. $\{2\}$
53. $1 \cdot 52$ 52 1 26 2 13 4
$2 \cdot 26$ $1\overline{)52}$ $52\overline{)52}$ $2\overline{)52}$ $26\overline{)52}$ $4\overline{)52}$ $13\overline{)52}$
$4 \cdot 13$

EXERCISES 2.3   PAGE 53

1. $2^3 \cdot 3$
2. $2^2 \cdot 7$
3. 3^3
5. $2^2 \cdot 3^2$
6. $2^2 \cdot 3 \cdot 5$
7. $2^3 \cdot 3^2$
9. 3^4
10. $3 \cdot 5 \cdot 7$
11. 5^3
13. $3 \cdot 5^2$
14. $2 \cdot 3 \cdot 5^2$
15. $2 \cdot 3 \cdot 5 \cdot 7$
17. $2 \cdot 5^3$
18. $3 \cdot 31$
19. $2^3 \cdot 3 \cdot 7$
21. $2 \cdot 3^2 \cdot 7$
22. $2^4 \cdot 3$
23. 17
25. $3 \cdot 17$
26. $2^4 \cdot 3^2$
27. 11^2
29. $3^2 \cdot 5^2$
30. $2^2 \cdot 13$
31. 2^5
33. $2^2 \cdot 3^3$
34. 103
35. 101
37. $2 \cdot 3 \cdot 13$

38. $2^2 \cdot 5^3$ **39.** $2^4 \cdot 5^4$

41. {1, 2, 3, 4, 6, 12} **42.** {1, 2, 3, 6, 9, 18}

43. {1, 2, 4, 7, 14, 28} **45.** {1, 11, 121}

46. {1, 3, 5, 9, 15, 45} **47.** {1, 3, 5, 7, 15, 21, 35, 105}

49. {1, 97} **50.** {1, 2, 4, 6, 8, 9, 12, 16, 18, 24, 36, 48, 72, 144}

EXERCISES 2.4 PAGE 56

1. 4 **2.** 4 **3.** 17 **5.** 10 **6.** 6 **7.** 3 **9.** 6

10. 11 **11.** 2 **13.** 8 **14.** 11 **15.** 4 **17.** 12 **18.** 11

19. 25 **21.** 1 **22.** 20 **23.** 1 **25.** 35

Problems that have relative prime pairs are: 26, 27, 29, 31, 32, 33, and 35.

EXERCISES 2.5 PAGE 59

1. 24 **2.** 105 **3.** 36 **5.** 432 **6.** 200 **7.** 600

9. 216 **10.** 210 **11.** 338 **13.** 1560 **14.** 2250 **15.** 918

17. 120 24 · 5 **18.** 120 8 · 15 **19.** 30 6 · 5
15 · 8　　　　　10 · 12　　　　　15 · 2
10 · 12　　　　　120 · 1　　　　　30 · 1

21. 1140 228 · 5 **22.** 4410 63 · 70 **23.** 1960 56 · 35
12 · 95　　　　　98 · 45　　　　　40 · 49
95 · 12　　　　　45 · 98　　　　　196 · 10

25. 14,157 99 · 143
363 · 39
143 · 99

26. GCD – 1 **27.** GCD – 1 **29.** GCD – 3
LCM – 45　　　　LCM – 110　　　　LCM – 60

30. GCD – 1 **31.** GCD – 5 **33.** GCD – 27
LCM – 1225　　　LCM – 2610　　　LCM – 324

34. GCD – 15 **35.** GCD – 13
LCM – 675　　　LCM – 41,405

36. 252 minutes **37.** 48 hours; 4 orbits and 3 orbits, respectively

38. Once every 30 days; once every 30 days

REVIEW QUESTIONS · CHAPTER 2 PAGE 62

1. natural; 1; divisors; 1; itself **2.** composite **3.** factors

4. composite; prime factorization **5.** relatively prime; 1

6. 3, 5, 9, 15 **7.** 2, 3, 6, 9 **8.** none

9. 2, 3, 5, 6, 9, 10, 15 (all) **10.** 2 **11.** 3

12. {14, 28, 42, 56, . . .}

13. {2, 3, 5, 7, 11, 13, 17, 19, 23, 29, 31, 37, 41, 43, 47, 53, 59, 61, 67}

14. 6 and 4   15. 5 and 12

16. (a) $2 \cdot 3 \cdot 5^2$ 17. (a) {1, 3, 5, 15, 25, 75}   18. (a) 10
     (b) $5 \cdot 13$ (b) {1, 2, 3, 4, 6, 9, 12, 18, 36}   (b) 4
     (c) $2^2 \cdot 3 \cdot 7$ (c) {1, 13, 169}   (c) 1

19. (a) 168   20. GCD -3 21. 210 seconds;
     (b) 1800   LCM -1638 14 laps and 12 laps

CHAPTER 3

EXERCISES 3.1   PAGE 72

1. A rational number is a number that can be written in the form $\dfrac{a}{b}$ where a is a whole number and b is a natural number.

3. $\dfrac{13}{52}$ 5. $\dfrac{1}{25}$ 6. $\dfrac{3}{12}$ 7. $\dfrac{1}{2}$ 9. $\dfrac{2}{6}$ 10. $\dfrac{14}{36}$ 11. $\dfrac{5}{40}$

13. $\dfrac{1}{4}$ 14. $\dfrac{1}{4}$ 15. $\dfrac{3}{16}$ 17. $\dfrac{11}{22}$ 18. no; yes

19. (a) $\left\{ \dfrac{5}{6}, \dfrac{10}{12}, \dfrac{15}{18}, \dfrac{20}{24}, \dfrac{25}{30}, \cdots \right\}$ (b) $\left\{ \dfrac{4}{3}, \dfrac{8}{6}, \dfrac{12}{9}, \dfrac{16}{12}, \dfrac{20}{15}, \cdots \right\}$

21. $\dfrac{2}{3} = \dfrac{18}{27}$ 22. equivalent   23. equivalent

25. not equivalent   26. equivalent   27. equivalent
29. not equivalent   30. equivalent   31. equivalent

33. $\dfrac{8}{3}$ 34. $\dfrac{12}{35}$ 35. $\dfrac{3}{32}$ 37. $\dfrac{1}{4}$ 38. $\dfrac{3}{112}$

39. $\dfrac{0}{24}$ 41. $\dfrac{6}{385}$ 42. $\dfrac{64}{117}$ 43. $\dfrac{48}{455}$ 45. $\dfrac{1}{360}$

46. $\dfrac{27}{100}$ 47. 840   49. $\dfrac{297}{16,000}$ 50. $\dfrac{1}{252,000}$

51.   53.   54.

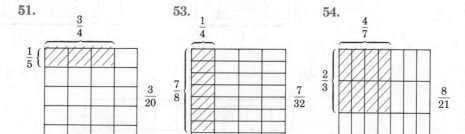

55. $\dfrac{2}{3}$

$\dfrac{2}{9}$ {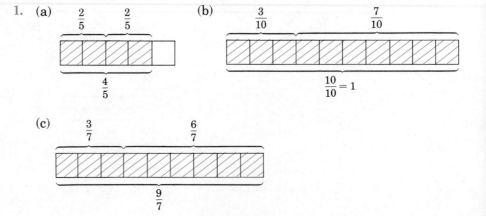

$\dfrac{4}{27}$

57. Commutative property of multiplication
58. 0

EXERCISES 3.2   PAGE 79

1. 21	2. 4	3. 10	5. 45	6. 20	7. 16
9. 42	10. 60	11. 54	13. 40	14. 15	15. 44
17. 10	18. 32	19. 45	21. 80	22. 240	23. 18
25. 2	26. 28	27. 35	29. 30	30. 700	31. 190

33. $\dfrac{1}{3}$ 34. $\dfrac{2}{3}$ 35. $\dfrac{3}{4}$ 37. $\dfrac{2}{5}$ 38. $\dfrac{4}{5}$ 39. $\dfrac{7}{18}$

41. $\dfrac{0}{25}=0$ 42. $\dfrac{3}{4}$ 43. $\dfrac{2}{5}$ 45. $\dfrac{5}{6}$ 46. $\dfrac{1}{4}$ 47. $\dfrac{2}{3}$

49. $\dfrac{2}{3}$ 50. $\dfrac{3}{4}$ 51. $\dfrac{6}{25}$ 53. $\dfrac{2}{3}$ 54. $\dfrac{2}{3}$ 55. $\dfrac{12}{35}$

57. $\dfrac{2}{9}$ 58. $\dfrac{3}{7}$ 59. $\dfrac{25}{76}$ 61. $\dfrac{1}{2}$ 62. $\dfrac{4}{1}=4$ 63. $\dfrac{6}{1}=6$

65. $\dfrac{2}{17}$ 66. $\dfrac{3}{8}$ 67. $\dfrac{9}{20}$ 69. $\dfrac{10}{9}$ 70. $\dfrac{11}{15}$ 71. $\dfrac{5}{4}$

73. $\dfrac{5}{18}$ 74. $\dfrac{1}{6}$ 75. $\dfrac{1}{4}$ 77. $\dfrac{21}{4}$ 78. $\dfrac{189}{52}$ 79. $\dfrac{77}{4}$

81. $\dfrac{8}{5}$ 82. $\dfrac{3}{10}$ 83. $\dfrac{2}{7}$

EXERCISES 3.3   PAGE 87

1. (a) $\dfrac{2}{5}$ $\dfrac{2}{5}$

$\dfrac{4}{5}$

(b) $\dfrac{3}{10}$ $\dfrac{7}{10}$

$\dfrac{10}{10}=1$

(c) $\dfrac{3}{7}$ $\dfrac{6}{7}$

$\dfrac{9}{7}$

3. $\left(\dfrac{3}{100}+\dfrac{7}{100}\right)+\dfrac{1}{10}=\left(\dfrac{10}{100}\right)+\dfrac{10}{100}=\dfrac{20}{100}=\dfrac{1}{5}$

$\dfrac{3}{100}+\left(\dfrac{7}{100}+\dfrac{1}{10}\right)=\dfrac{3}{100}+\left(\dfrac{7}{100}+\dfrac{10}{100}\right)=\dfrac{3}{100}+\left(\dfrac{17}{100}\right)=\dfrac{20}{100}=\dfrac{1}{5}$

5. For any rational number $\dfrac{a}{b}$, $\dfrac{a}{b}+0=\dfrac{a}{b}+\dfrac{0}{b}=\dfrac{a+0}{b}=\dfrac{a}{b}$

6. $\dfrac{10}{10}=1$ 7. $\dfrac{5}{14}$ 9. $\dfrac{3}{2}$ 10. $\dfrac{3}{2}$ 11. $\dfrac{10}{5}=2$

13. $\dfrac{5}{3}$ 14. $\dfrac{3}{5}$ 15. $\dfrac{13}{18}$ 17. $\dfrac{11}{16}$ 18. $\dfrac{23}{20}$

19. $\dfrac{3}{5}$ 21. $\dfrac{12}{12}=1$ 22. $\dfrac{11}{16}$ 23. $\dfrac{17}{20}$ 25. $\dfrac{17}{21}$

26. $\dfrac{3}{4}$ 27. $\dfrac{9}{13}$ 29. $\dfrac{23}{54}$ 30. $\dfrac{151}{140}$ 31. $\dfrac{23}{72}$

33. $\dfrac{106}{945}$ 34. $\dfrac{343}{432}$ 35. $\dfrac{31}{96}$ 37. $\dfrac{13}{9}$ 38. $\dfrac{73}{48}$

39. $\dfrac{627}{400}$ 41. $\dfrac{99}{70}$ 42. $\dfrac{1879}{1008}$ 43. $\dfrac{143}{147}$ 45. $\dfrac{1}{5}$

46. $\dfrac{7}{6}$ 47. $\dfrac{317}{1000}$ 49. $\dfrac{271}{10,000}$ 50. $\dfrac{631}{100}$ 51. $\dfrac{8191}{1000}$

53. $\dfrac{753}{1000}$ 54. $\dfrac{63}{50}$ 55. $\dfrac{89}{200}$ 57. $\dfrac{613}{1000}$

58. $\dfrac{27,683}{10,000}$ 59. $\dfrac{5134}{1000}$

EXERCISES 3.4 PAGE 90

1. (a) $\dfrac{2}{3}$ (b) $\dfrac{3}{4}=\dfrac{9}{12}$

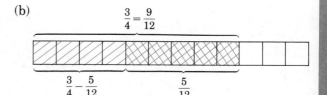

$\dfrac{2}{3}-\dfrac{1}{3}\quad\dfrac{1}{3}$ $\dfrac{3}{4}-\dfrac{5}{12}\qquad\dfrac{5}{12}$

(c) $2=\dfrac{8}{4}$

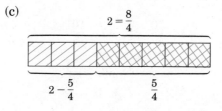

$2-\dfrac{5}{4}\qquad\dfrac{5}{4}$

2. $\dfrac{3}{4}$ by $\dfrac{1}{12}$ 3. $\dfrac{7}{8}$ by $\dfrac{1}{24}$ 5. $\dfrac{4}{10}$ by $\dfrac{1}{40}$ 6. $\dfrac{13}{20}$ by $\dfrac{1}{40}$

7. $\dfrac{21}{25}$ by $\dfrac{11}{400}$ 9. $\dfrac{7}{24}$ by $\dfrac{1}{72}$ 10. $\dfrac{11}{48}$ by $\dfrac{1}{60}$ 11. $\dfrac{37}{100}$ by $\dfrac{1}{20}$

13. $\dfrac{8}{9}, \dfrac{9}{10}, \dfrac{11}{12}$ 14. $\dfrac{11}{12}, \dfrac{19}{20}, \dfrac{7}{6}$ 15. $\dfrac{40}{36}, \dfrac{31}{24}, \dfrac{17}{12}$ 17. $\dfrac{13}{18}, \dfrac{31}{36}, \dfrac{7}{8}$

18. $\dfrac{20}{10,000}, \dfrac{3}{1000}, \dfrac{1}{100}$ 19. $\dfrac{298}{1000}, \dfrac{3}{10}, \dfrac{32}{100}, \dfrac{3,333}{10,000}$

21. $\dfrac{3}{7}$ 22. $\dfrac{6}{10} = \dfrac{3}{5}$ 23. $\dfrac{4}{8} = \dfrac{1}{2}$ 25. $\dfrac{9}{15} = \dfrac{3}{5}$ 26. $\dfrac{3}{6} = \dfrac{1}{2}$

27. $\dfrac{13}{30}$ 29. $\dfrac{9}{32}$ 30. $\dfrac{1}{40}$ 31. $\dfrac{7}{54}$ 33. $\dfrac{8}{15}$

34. $\dfrac{19}{180}$ 35. $\dfrac{31}{288}$ 37. $\dfrac{1}{6}$ 38. $\dfrac{29}{84}$ 39. $\dfrac{1}{7}$

41. $\dfrac{87}{100}$ 42. $\dfrac{59}{1000}$ 43. $\dfrac{3759}{5000}$ 45. $\dfrac{247}{240}$ 46. $\dfrac{11}{72}$

47. $\dfrac{3}{32}$ 49. $\dfrac{5}{12}$

EXERCISES 3.5   PAGE 96

1. A mixed number is the sum of a whole number and a rational number not a whole number, indicated by writing the numbers next to each other.

2. $\dfrac{5}{2}$ 3. $\dfrac{4}{3}$ 5. $\dfrac{3}{2}$ 6. $\dfrac{3}{2}$ 7. $\dfrac{7}{5}$ 9. $\dfrac{5}{4}$

10. $\dfrac{5}{4}$ 11. $\dfrac{25}{6}$ 13. $1\dfrac{1}{3}$ 14. $1\dfrac{1}{4}$ 15. $1\dfrac{1}{2}$ 17. $6\dfrac{1}{7}$

18. $2\dfrac{1}{8}$ 19. $7\dfrac{1}{2}$ 21. $3\dfrac{1}{9}$ 22. $2\dfrac{1}{15}$ 23. 3 25. $4\dfrac{1}{2}$

26. $2\dfrac{1}{17}$ 27. $\dfrac{3}{1} = 3$ 29. $1\dfrac{3}{4}$ 30. $1\dfrac{17}{20}$ 31. $\dfrac{37}{8}$ 33. $\dfrac{76}{15}$

34. $\dfrac{8}{5}$ 35. $\dfrac{46}{11}$ 37. $\dfrac{7}{3}$ 38. $\dfrac{34}{7}$ 39. $\dfrac{32}{3}$ 41. $\dfrac{34}{5}$

42. $\dfrac{71}{5}$ 43. $\dfrac{50}{3}$ 45. $\dfrac{101}{5}$ 46. $\dfrac{47}{5}$ 47. $\dfrac{92}{7}$

49. $\dfrac{17}{1} = 17$ 50. $\dfrac{151}{50}$

EXERCISES 3.6 PAGE 99

1. $7\frac{2}{3}$ 2. $10\frac{3}{8}$ 3. $42\frac{7}{20}$ 5. $10\frac{3}{4}$ 6. $7\frac{13}{35}$

7. $12\frac{7}{27}$ 9. $18\frac{23}{45}$ 10. $18\frac{1}{15}$ 11. $28\frac{1}{2}$ 13. $9\frac{29}{40}$

14. $12\frac{17}{24}$ 15. $12\frac{47}{60}$ 17. $63\frac{11}{24}$ 18. $26\frac{35}{48}$ 19. $53\frac{83}{192}$

21. $71\frac{19}{40}$ 22. $38\frac{13}{21}$ 23. $85\frac{23}{54}$ 25. $27\frac{5}{72}$ 26. $24\frac{26}{35}$

27. $110\frac{6}{25}$ 29. $136\frac{67}{390}$ 30. $213\frac{117}{125}$ 31. $35\frac{11}{40}$ kilometers

33. (a) no; $15\frac{1}{2} + 1\frac{3}{10}$ is less than $16\frac{9}{10}$
     (b) yes

34. $9\frac{1}{60}$ hours; 56 minutes

EXERCISES 3.7 PAGE 102

1. $3\frac{1}{2}$ 2. $4\frac{3}{5}$ 3. $3\frac{5}{8}$ 5. $3\frac{7}{8}$ 6. $10\frac{4}{5}$

7. $3\frac{29}{32}$ 9. $3\frac{1}{60}$ 10. $1\frac{7}{16}$ 11. $3\frac{9}{10}$ 13. $7\frac{4}{5}$

14. $\frac{5}{8}$ 15. $57\frac{1}{6}$ 17. $\frac{13}{15}$ 18. $1\frac{13}{16}$ 19. 17

21. $4\frac{11}{20}$ 22. $5\frac{23}{48}$ 23. $1\frac{5}{6}$ 25. $12\frac{1}{12}$ 26. $16\frac{1}{2}$

27. $3\frac{26}{27}$ 29. $13\frac{31}{72}$ 30. $29\frac{97}{102}$ 31. $\frac{3}{5}$ hour; 36 minutes

33. $7\frac{33}{80}$ inches   34. $\frac{43}{100}$

35. $4\frac{3}{5}$ centimeters; $\frac{3}{5}$ centimeters; $56\frac{1}{5}$ centimeters

 37. $8\frac{17}{20}$ 38. $56\frac{9}{16}$, $28\frac{7}{16}$, $14\frac{11}{16}$ 39. $219\frac{13}{16}$ pounds

EXERCISES 3.8 PAGE 106

1. $7\frac{7}{12}$ 2. $1\frac{13}{35}$ 3. $10\frac{1}{2}$ 5. $22\frac{1}{2}$ 6. 12

7. $31\frac{1}{6}$ 9. 8   10. 12   11. 30   13. $21\frac{3}{4}$

14. 1   15. 1   17. $27\frac{1}{2}$ 18. $10\frac{1}{5}$ 19. $19\frac{1}{4}$

21. $\frac{1}{7}$ 22. $\frac{1}{5}$ 23. $\frac{1}{10}$ 25. $\frac{36}{385}$ 26. $39\frac{3}{8}$

27. 11   29. 242   30. $15\frac{99}{160}$ 31. $11\frac{3}{8}$ 33. 17

34. $1\frac{3}{10}$ 35. $2\frac{1}{10}$ 37. $1827\frac{1}{10}$ 38. $1291\frac{5}{42}$

39. $3016\frac{26}{75}$ 41. $325\frac{34}{45}$ 42. $592\frac{13}{32}$ 43. $2861\frac{21}{100}$

45. $1204\frac{49}{288}$ 46. $20\frac{1}{4}$ inches   47. 177 miles

49. $22\frac{11}{80}$ feet above   50. $\frac{5}{12}$

$10\frac{1}{16}$ feet below

51. Two possibilities exist depending on the relative positions of the towns: $100\frac{2}{5}$ or $6\frac{3}{5}$ kilometers.

EXERCISES 3.9   PAGE 111

1. $\frac{8}{9}$ 2. $\frac{4}{15}$ 3. $\frac{5}{7}$ 5. $\frac{7}{5}=1\frac{2}{5}$

6. $\frac{11}{3}=3\frac{2}{3}$ 7. $\frac{1}{3}$ 9. $\frac{3}{4}$ 10. $\frac{7}{16}$

11. $\frac{4}{9}$ 13. $\frac{5}{8}$ 14. $\frac{24}{35}$ 15. $\frac{39}{32}=1\frac{7}{32}$

17. $\frac{9}{10}$ 18. $\frac{10}{9}=1\frac{1}{9}$ 19. 1   21. $\frac{32}{21}=1\frac{11}{21}$

22. $\frac{8}{3}=2\frac{2}{3}$ 23. $\frac{10}{3}=3\frac{1}{3}$ 25. $\frac{16}{21}$ 26. $\frac{10}{39}$

27. $\frac{7}{60}$ 29. $\frac{63}{50}=1\frac{13}{50}$ 30. $\frac{28}{17}=1\frac{11}{17}$ 31. $\frac{62}{39}=1\frac{23}{39}$

33. $\frac{41}{12}=3\frac{5}{12}$ 34. $\frac{7}{5}=1\frac{2}{5}$ 35. $\frac{41}{3}=13\frac{2}{3}$ 37. $\frac{37}{2}=18\frac{1}{2}$

38. $\dfrac{9}{32}$ 39. $\dfrac{20}{11} = 1\dfrac{9}{11}$ 41. $\dfrac{2}{3}$ 42. $\dfrac{1}{5}$

43. 0 45. $\dfrac{43}{450}$ 46. $\dfrac{172}{9} = 19\dfrac{1}{9}$ 47. $\dfrac{111}{20} = 5\dfrac{11}{20}$

49. $\dfrac{57}{38} = 1\dfrac{1}{2}$ 50. $\dfrac{3}{5}$ 51. $\dfrac{353}{630}$ 53. $\dfrac{50}{27} = 1\dfrac{23}{27}$

54. $\dfrac{62}{43} = 1\dfrac{19}{43}$ 55. $\dfrac{27}{46}$ 57. $48   58. $135,000

59. $375; $7\dfrac{7}{8}; $1\dfrac{7}{8}$; the profit, $375.

EXERCISES 3.10   PAGE 114

1. $\dfrac{15}{7}$ 2. $\dfrac{44}{25}$ 3. $\dfrac{11}{70}$ 5. $\dfrac{5}{7}$ 6. $\dfrac{12}{7}$ 7. $\dfrac{1}{5}$

9. $\dfrac{111}{31}$ 10. $\dfrac{5}{2}$ 11. $\dfrac{25}{21}$ 13. $\dfrac{33}{4}$ 14. $\dfrac{328}{133}$ 15. $\dfrac{76}{135}$

17. $\dfrac{429}{365}$ 18. $\dfrac{27}{40}$ 19. $\dfrac{63}{19}$ 21. 140 22. $\dfrac{69}{35}$ 23. $\dfrac{60}{167}$

25. $\dfrac{377}{373}$ 26. 1 27. 0 29. 1 30. $\dfrac{211}{1211}$

31. $\dfrac{245}{184}$ or $1\dfrac{61}{184}$ 33. $\dfrac{192}{11}$ or $17\dfrac{5}{11}$ 34. $\dfrac{215}{26}$ or $8\dfrac{7}{26}$

REVIEW QUESTIONS · CHAPTER 3   PAGE 118

1. $\dfrac{a}{b}$; a; b 2. 0; undefined; undefined 3. 1 4. lowest terms

5. 1 6. (a) associative property of addition
              (b) commutative property of multiplication

7.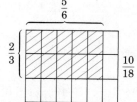

8. $\dfrac{25}{100}$ 9. $\dfrac{56}{72}$

10.

$$\overbrace{\frac{1}{2} = \frac{3}{6}}$$

$$\underbrace{\frac{1}{2} - \frac{1}{3}} \quad \underbrace{\frac{1}{3} = \frac{2}{6}}$$

11. $\dfrac{1}{30}$ 12. $\dfrac{3}{49}$ 13. $\dfrac{1}{12}$

14. 2   15. 60   16. 55   17. 0   18. $\dfrac{3}{4}$

19. $\dfrac{5}{4}$ 20. $\dfrac{9}{8}$ 21. 0   22. $\dfrac{6}{11}$ 23. $\dfrac{5}{7}$

24. $\dfrac{2}{3}$ 25. $\dfrac{1}{10}$ 26. $\dfrac{49}{72}$ 27. $\dfrac{1}{4}$ 28. $\dfrac{29}{66}$

29. $\dfrac{9}{10}$ 30. $\dfrac{1}{8}$ 31. $\dfrac{341}{270}$ 32. $7\dfrac{5}{6}$ 33. $6\dfrac{1}{4}$

34. $3\dfrac{21}{50}$ 35. $4\dfrac{34}{35}$ 36. $\dfrac{51}{10}$ 37. $\dfrac{182}{5}$ 38. $\dfrac{29}{4}$

39. $\dfrac{41}{3}$ 40. $30\dfrac{3}{4}$ 41. $6\dfrac{7}{12}$ 42. $43\dfrac{11}{60}$ 43. $14\dfrac{1}{10}$

44. $6\dfrac{47}{56}$ 45. $4\dfrac{5}{6}$ 46. $\dfrac{1}{5}$ 47. 104   48. $\dfrac{2}{9}$

49. $\dfrac{7}{12}$ 50. $\dfrac{3}{4}$ 51. $\dfrac{353}{480}$ 52. $\dfrac{25}{112}$ 53. $\dfrac{33}{4}$

54. $\dfrac{1737}{55}$, or $31\dfrac{32}{55}$ 55. $31\dfrac{7}{8}$ feet   56. $\dfrac{6}{55}$; $\dfrac{6}{11}$ 57. $33\dfrac{3}{5}$ cents

58. $10\dfrac{1}{7}$ 59. $\dfrac{39}{124}$ 60. $\dfrac{3}{4}$ hour; 45 minutes

CHAPTER 4

EXERCISES 4.1 PAGE 128

1. A decimal number is a rational number that has a power of ten as its denominator.

2. $3(10^1) + 7(10^0) + 4\left(\dfrac{1}{10}\right) + 9\left(\dfrac{1}{10^2}\right) + 8\left(\dfrac{1}{10^3}\right)$

3. $5(10^2) + 6(10^1) + 2(10^0) + 3\left(\dfrac{1}{10}\right)$

5. $9(10^2) + 4(10^1) + 6(10^0) + 3\left(\dfrac{1}{10}\right) + 4\left(\dfrac{1}{10^2}\right) + 6\left(\dfrac{1}{10^3}\right)$

6. $7(10^3) + 8(10^2) + 6(10^1) + 2(10^0) + 3\left(\dfrac{1}{10}\right) + 5\left(\dfrac{1}{10^2}\right) + 7\left(\dfrac{1}{10^3}\right) + 1\left(\dfrac{1}{10^4}\right)$

7. $8(10^2) + 9(10^1) + 4(10^0) + 6\left(\dfrac{1}{10}\right) + 7(\dfrac{1}{10^2}) + 7\left(\dfrac{1}{10^3}\right) + 1\left(\dfrac{1}{10^4}\right)$

9. $2(10^2) + 0(10^1) + 0(10^0) + 0\left(\dfrac{1}{10}\right) + 0\left(\dfrac{1}{10^2}\right) + 7\left(\dfrac{1}{10^3}\right) + 9\left(\dfrac{1}{10^4}\right) + 3\left(\dfrac{1}{10^5}\right)$

10. $4(10^1) + 0(10^0) + 0\left(\dfrac{1}{10}\right) + 0\left(\dfrac{1}{10^2}\right) + 0\left(\dfrac{1}{10^3}\right) + 5\left(\dfrac{1}{10^4}\right)$

11. $2(10^3) + 9(10^2) + 3(10^1) + 6(10^0) + 7\left(\dfrac{1}{10}\right) + 8\left(\dfrac{1}{10^2}\right) + 1\left(\dfrac{1}{10^3}\right) +$

$0\left(\dfrac{1}{10^4}\right) + 3\left(\dfrac{1}{10^5}\right)$

13. .014   **14.** .17   **15.** 6.28   **17.** 72.392   **18.** 850.0036
19. 700.77   **21.** 600,500.402   **22.** 322,476,000.084   **23.** 705.4293
25. 1,000,000.0037   **26.** five tenths   **27.** ninety-three hundredths
29. thirty-two and fifty-eight hundredths
30. seventy-one and six hundredths
31. thirty-five and seventy-eight thousandths
33. eighteen and one hundred two thousandths
34. fifty and eight thousandths
35. six hundred seven and six hundred seven thousandths
37. five hundred ninety-three and eighty-six hundredths (eight hundred sixty thousandths)
38. four thousand, seven hundred and six hundred seventeen thousandths
39. five thousand and five thousandths
41. nine hundred and four thousand, six hundred thirty-eight ten-thousandths
42. five thousand five and five hundred five thousandths
43. four and five thousand, six hundred seventy-one hundred-thousandths
45. eight hundred seventy-nine and fifty-eight thousand, nine hundred thirty-two hundred-thousandths
46. (a) $372.58   three hundred seventy-two and $\dfrac{58}{100}$ dollars

(b) $577.50   five hundred seventy-seven and $\dfrac{50}{100}$ dollars.

(c) $2405.37   two thousand, four hundred five and $\dfrac{37}{100}$ dollars.

(d) $1,476,324.75   one million, four hundred seventy-six thousand, three hundred twenty-four and $\dfrac{75}{100}$ dollars.

EXERCISES 4.2   PAGE 131

1.	4.8	2.	5.0	3.	76.3	5.	89.0	6.	7.6		
7.	18.0	9.	14.3	10.	0.0	11.	.39	13.	5.72		
14.	8.99	15.	7.00	17.	.08	18.	6.00	19.	5.71		
21.	.067	22.	.056	23.	.634	25.	32.479	26.	9.430		
27.	17.364	29.	.002	30.	20.770	31.	479	33.	18.		
34.	20.	35.	382.	37.	440.	38.	701.	39.	6333		
41.	5160.	42.	6480.	43.	500.	45.	1000.	46.	380.		
47.	5480.	49.	92,540.	50.	7010.	51.	7000.	53.	48,000		
54.	103,000	55.	217,000.			57.	380,000.	58.	4,501,000.		
59.	7,305,000.	61.	.00058			62.	.54	63.	470		
65.	500	66.	6000			67.	3.230	69.	80,000		
70.	78,420										

EXERCISES 4.3   PAGE 135

1.	2.3	2.	11.5	3.	7.55	5.	72.31
6.	4.6926	7.	276.096	9.	44.6516	10.	481.25
11.	118.333	13.	7.148	14.	7.4914	15.	93.877
17.	103.429	18.	46.943	19.	137.150	21.	1.44
22.	8.93	23.	15.89	25.	64.947	26.	4.895
27.	4.7974	29.	2.9434	30.	34.186	31.	1113.665
33.	155.073	34.	6.7855	35.	.91	37.	.253 inches
38.	12.05 inches			39.	3.0284; 95.08; 98.1084		

EXERCISES 4.4   PAGE 138

1.	.42	2.	7.5	3.	.28	5.	18.6	6.	.004
7.	.025	9.	.246	10.	.07	11.	.0006	13.	.094
14.	.0663	15.	10.79	17.	3.	18.	14.	19.	1.
21.	346.	22.	2057.	23.	782.	25.	4.35	26.	4178.2
27.	.38	29.	71,200	30.	251,480.	31.	.000045		
33.	.1484	34.	.00092481	35.	7.905642	37.	$197.08		
38.	422.4 miles	39.	37.125 feet	41.	211.464	42.	7.8; 8.0; no		

43. 119.4; 119.0; 8; 6; 119; 117. Rounding off before multiplying can give answers different from rounding off after multiplying. Also, either procedure may yield the larger product.

45. Three and sixty-two ten-thousandths; five hundred seventy-eight thousandths; three and five thousand, eight hundred forty-two ten-thousandths; one and seven million, three hundred seventy-five thousand, eight hundred thirty-six ten-millionths; two and four thousand, two hundred eighty-two ten-thousandths

EXERCISES 4.5   PAGE 143

1. 2.34	2. .57	3. 9.9
5. .08	6. .9	7. 2056
9. 20.	10. 5.	11. .002
13. .9428 ≈ .943	14. .13	15. .8444 ≈ .844
17. .1666 ≈ .167	18. 1.3	19. 3.03
21. .0706 ≈ .071	22. 12.6567 ≈ 12.657	23. 310.6
25. 20.9876 ≈ 20.988	26. 511.1111 ≈ 511.111	27. 7.3239 ≈ 7.324
29. 28.4	30. 4.2465 ≈ 4.247	31. .784
33. .0005036	34. .45621	35. 7.682
37. .000001826	38. 91.112	39. .0006122
41. 55.93	42. 34 cents	43. 309.6 miles
45. 57.695 miles per hour	46. 5.625¢	47. 60¢ a pound
49. 775 feet	50. $142.00	51. $11,458.33

53. $68,560

54. 1140.625, or $1140\frac{5}{8}$ miles

EXERCISES 4.6   PAGE 148

1. A real number is any number that can be written as an infinite decimal.

2. infinite repeating decimal

5. .20 . . .	6. .$\bar{6}$	7. .8750 . . .	9. .$\bar{3}$
10. .$\bar{1}$	11. .$\bar{4}$	13. .08$\bar{3}$	14. .093750 . . .
15. .18750 . . .	17. .$\bar{7}$	18. .$\overline{54}$	19. .2$\bar{3}$
21. 4.$\overline{142857}$	22. 3.10 . . .	23. 1.2$\overline{142857}$	25. 2.8$\bar{3}$

26. $\frac{1}{3}$	27. $\frac{7}{33}$	29. $\frac{5}{9}$	30. $\frac{2}{3}$
31. $\frac{7}{9}$	33. $\frac{1}{1} = 1$	34. $\frac{41}{333}$	35. $\frac{17}{33}$
37. $\frac{43}{333}$	38. $\frac{32}{111}$	39. $\frac{2611}{3333}$	

42. 5.717	43. 72.984	45. .774	46. 7.238
47. 97.186	49. .000	50. .566	

51. (a) $\frac{22}{7} = 3.142857142857 . . .$

$$3.142857142857 . . .$$
$$- 3.141592653589 . . .$$
$$.001263 . . .$$

(b) $3.141592653589 . . .$
$$- 3.14000 . . .$$
$$.00159 . . .$$

(c) $3.1416000 . . .$
$$- 3.141592653589 . . .$$
$$.0000173 . . .$$

Of the three numbers $\frac{22}{7}$, 3.14, and 3.1416, the number 3.1416 is closest to π.

REVIEW QUESTIONS · CHAPTER 4　　PAGE 151

1. rational number　　2. true　　3. false　　4. true

5. $5(10^1) + 6(10^0) + 4\left(\dfrac{1}{10^1}\right) + 9\left(\dfrac{1}{10^2}\right)$　　6. four and eight thousandths

7. nine hundred and five tenths
8. six and five thousand, seven hundred eighty-one ten-thousandths

9. 200.17	10. 84.075	11. 3003.003
12. 5900	13. 7.6	14. .039
15. 73.00	16. 26.82	17. 64.151
18. 93.418	19. 17.79	20. 9.02
21. 3.9623	22. 15.775	23. 126.72
24. 4568.1	25. 15.00	26. 2.50
27. .00	28. 365,280	29. .007641
30. 40.7	31. $.\overline{16}$	32. $.\overline{285714}$
33. $.6\overline{0}$	34. $.\overline{45}$	35. $1.\overline{307692}$

36. $\dfrac{1}{9}$　　37. $\dfrac{32}{99}$　　38. $\dfrac{505}{999}$

39. $\dfrac{27}{99} = \dfrac{3}{11}$　　40. $\dfrac{107}{333}$　　41. 7.66298

42. 149.8 miles　　43. 4.6855　　44. 166.85 miles
45. 21.6 cents　　46. $550; $2640; $3960

CHAPTER 5

EXERCISES 5.1　PAGE 154

1. $\dfrac{14}{13}$　　2. $\dfrac{6}{5}$　　3. $\dfrac{1}{1} = 1$　　5. $\dfrac{2}{5}$　　6. $\dfrac{6}{5}$　　7. $\dfrac{5}{6}$

9. $\dfrac{15}{7}$　　10. $\dfrac{8}{3}$　　11. $\dfrac{4}{1} = 4$　　13. $\dfrac{1}{1} = 1$　　14. $\dfrac{1}{1} = 1$　　15. $\dfrac{6}{5}$

17. $\dfrac{5}{7}$　　18. $\dfrac{4}{3}$　　19. $\dfrac{3}{8}$　　21. $\dfrac{7}{50}$　　22. $\dfrac{655}{917}$　　23. $\dfrac{1}{2}$

25. $\dfrac{1}{1} = 1$　26. $\dfrac{18}{1} = 18$　27. $\dfrac{1}{1} = 1$　29. $\dfrac{1}{1} = 1$　30. $\dfrac{125}{1} = 125$

EXERCISES 5.2　PAGE 157

1. T　　2. F　　3. T　　5. F　　6. T　　7. T　　9. T　　10. T
11. T　　13. T　　14. T　　15. T　　17. F　　18. F　　19. F

21. T; $\dfrac{6}{3} = \dfrac{8}{4}, \dfrac{8}{6} = \dfrac{4}{3}, \dfrac{3}{4} = \dfrac{6}{8}$　　22. T; $\dfrac{18}{12} = \dfrac{21}{14}, \dfrac{21}{18} = \dfrac{14}{12}, \dfrac{12}{14} = \dfrac{18}{21}$

23. F; $\dfrac{6}{5} \neq \dfrac{8}{7}, \dfrac{8}{6} \neq \dfrac{7}{5}, \dfrac{5}{7} \neq \dfrac{6}{8}$ 25. T; $\dfrac{3.1}{6.2} = \dfrac{5.1}{10.2}, \dfrac{5.1}{3.1} = \dfrac{10.2}{6.2}, \dfrac{6.2}{10.2} = \dfrac{3.1}{5.1}$

26. T; $\dfrac{2\frac{1}{3}}{8\frac{1}{2}} = \dfrac{1\frac{1}{6}}{4\frac{1}{4}}, \dfrac{1\frac{1}{6}}{2\frac{1}{3}} = \dfrac{4\frac{1}{4}}{8\frac{1}{2}}, \dfrac{8\frac{1}{2}}{4\frac{1}{4}} = \dfrac{2\frac{1}{3}}{1\frac{1}{6}}$ 27. T; $\dfrac{1\frac{1}{7}}{6\frac{1}{5}} = \dfrac{\frac{8}{14}}{3\frac{1}{10}}, \dfrac{\frac{8}{14}}{1\frac{1}{7}} = \dfrac{3\frac{1}{10}}{6\frac{1}{5}}, \dfrac{6\frac{1}{5}}{3\frac{1}{10}} = \dfrac{1\frac{1}{7}}{\frac{8}{14}}$

29. T; $\dfrac{16}{7} = \dfrac{8}{3\frac{1}{2}}, \dfrac{8}{16} = \dfrac{3\frac{1}{2}}{7}, \dfrac{7}{3\frac{1}{2}} = \dfrac{16}{8}$ 30. T; $\dfrac{17}{10} = \dfrac{8\frac{1}{2}}{5}, \dfrac{8\frac{1}{2}}{17} = \dfrac{5}{10}, \dfrac{10}{5} = \dfrac{17}{8\frac{1}{2}}$

EXERCISES 5.3   PAGE 160

1. $x = 12\frac{1}{2}$ 2. $x = 12$ 3. $y = 2$ 5. $z = 26\frac{2}{3}$

6. $x = 21$ 7. $y = 32$ 9. $x = 3$ 10. $y = 10\frac{1}{2}$

11. $x = 3\frac{1}{3}$ 13. $y = 10$ 14. $y = 37\frac{1}{2}$ 15. $x = 16\frac{2}{3}$

17. $x = 1$ 18. $x = 33\frac{1}{3}$ 19. $y = 40$ 21. $x = 50$

22. $z = 100$ 23. $y = \frac{1}{2}$ 25. $x = 48$ 26. $x = 20$

27. $x = 22$ 29. $z = 13\frac{1}{2}$ 30. $x = 5$ 31. $A = 3$

33. $R = 12\frac{1}{2}$ 34. $A = 10$ 35. $B = 5$ 37. $R = 125$

38. $B = 1300$ 39. $A = 45$ 41. $x = 18.6$ 42. $R = 37\frac{1}{2}$

43. $P = 43,000$ 45. $I = 180$ 46. $x = 3\pi$ 47. $C = 16\pi$

49. $y = \frac{1}{3}$ 50. $w = 180$

EXERCISES 5.4   PAGE 162

1. $41.67   2. $1.05   3. 930 gallons

5. $106,200; $3540   6. 63 men   7. $2\frac{1}{10}$ miles

9. $1\frac{1}{4}$ inches   10. 200 feet   11. $3\frac{1}{5}$ units

13. $437\frac{1}{2}$ grams   14. 4 quarts   15. 23,400 miles

17. 7 words; 35 words per minute   18. 12 weeks

19. $28\frac{4}{7}$ or 29 stores   21. \$420   22. \$25

23. 259,200 revolutions   25. $11\frac{1}{4}$ hours

26. Co. A, \$12,000; Co. B, \$16,000   27. \$325.80
29. 6:4; 9:6; 12:8; and so on   30. 7000 millimeters

31. 5300 grams   33. $17\frac{1}{7}$ pages; 35 hours

34. \$9.01   35. $133,333\frac{1}{3}$ miles

REVIEW QUESTIONS · CHAPTER 5   PAGE 165

1. (1) 5 of 8 equal parts;   (2) $5 \div 8$;   (3) Ratio 5:8

2. ratios; equal   3. four   4. $\frac{15}{16}$ 5. $\frac{5}{9}$ 6. $\frac{500}{5500} = \frac{1}{11}$

7. True; $\frac{21}{12} = \frac{7}{4}, \frac{4}{7} = \frac{12}{21}, \frac{21}{7} = \frac{12}{4}$ 8. False

9. True; $\frac{100}{5} = \frac{80}{4}, \frac{4}{80} = \frac{5}{100}, \frac{100}{80} = \frac{5}{4}$

10. True; $\frac{16}{32} = \frac{4\frac{1}{2}}{9}, \frac{9}{4\frac{1}{2}} = \frac{32}{16}, \frac{16}{4\frac{1}{2}} = \frac{32}{9}$

11. $x = 25$ 12. $y = 20$ 13. $x = 57$ 14. $z = 19\frac{1}{2}$ 15. 40,000

16. 45 mph   17. \$1.95   18. \$5.55   19. \$9120   20. 200 miles

CHAPTER 6

EXERCISES 6.1   PAGE 169

1. 30%   2. 16%   3. 96%   5. $2\frac{1}{2}\%$ 6. 53%

7. 1.8%   9. 73%   10. 12.5%   11. 37.5%   13. $5\frac{1}{3}\%$

14. .6%   15. .5%   17. $\frac{1}{2}\%$ 18. $\frac{1}{4}\%$ 19. $\frac{2}{3}\%$

21.	132%	22.	500%	23.	625%	25.	95%	26.	148%
27.	950%	29.	1000%	30.	1500%				

EXERCISES 6.2   PAGE 171

1.	2%	2.	9%	3.	10%	5.	36%	6.	70%
7.	50%	9.	5%	10.	1.6%	11.	.5%	13.	2.5%
14.	.25%	15.	17.5%	17.	47.6%	18.	4.5%	19.	125%
21.	100.2%	22.	210%	23.	560%	25.	450%	26.	700%
27.	2300%	29.	100%	30.	1000%	31.	.1	33.	.06
34.	.27	35.	.32	37.	.6	38.	.7	39.	.12
41.	.006	42.	.013	43.	.591	45.	.169	46.	1.75
47.	2	49.	1.252	50.	1	51.	.035	53.	.0725

54. .015   55. .02125   57. $.25\frac{2}{3}$ 58. $.50\frac{1}{6}$ 59. .35375

61. .166   62. .1005   63. .151   65. $.63\frac{1}{9}$ 66. 49

67. 51.32   69. .724   70. 65.1403

EXERCISES 6.3   PAGE 174

1.	3%	2.	16%	3.	7%	5.	70%	6.	87%
7.	83%	9.	63%	10.	97%	11.	102%	13.	257%
14.	200%	15.	300%	17.	125%	18.	250%	19.	10%

21. 50%   22. 25%   23. 75%   25. $66\frac{2}{3}\%$ 26. $12\frac{1}{2}\%$

27. $62\frac{1}{2}\%$ 29. $11\frac{1}{9}\%$ 30. $55\frac{5}{9}\%$ 31. 20%   33. 4%

34. 2%   35. 102%   37. 80%   38. 150%   39. 220%

41. 15%   42. 165%   43. 405%   45. $83\frac{1}{3}\%$ 46. $77\frac{7}{9}\%$

47. 110%   49. 107%   50. $6\frac{2}{3}\%$ 51. $8\frac{1}{3}\%$ 53. 125%

54. $28\frac{4}{7}\%$ 55. $113\frac{1}{3}\%$

EXERCISES 6.4   PAGE 175

1. $\frac{1}{10}$ 2. $\frac{1}{20}$ 3. $\frac{3}{20}$ 5. $\frac{1}{4}$ 6. $\frac{1}{2}$ 7. $\frac{3}{10}$

9. $\frac{4}{5}$ 10. $\frac{3}{4}$ 11. $\frac{2}{3}$ 13. $\frac{33}{100}$ 14. $\frac{1}{5}$ 15. $\frac{1}{8}$

17. $\dfrac{3}{5}$ 18. $\dfrac{9}{10}$ 19. $\dfrac{19}{20}$ 21. $\dfrac{1}{50}$ 22. $\dfrac{5}{8}$ 23. $\dfrac{9}{400}$

25. $\dfrac{1}{400}$ 26. $\dfrac{1}{200}$ 27. $\dfrac{1}{9}$ 29. $\dfrac{4}{9}$ 30. 1   31. 2

33. $1\dfrac{1}{5}$ 34. $1\dfrac{3}{4}$ 35. $\dfrac{11}{20}$ 37. 10   38. 52   39. $48\dfrac{1}{5}$

41. $\dfrac{11}{200}$ 42. $\dfrac{3}{400}$ 43. $2\dfrac{4}{25}$ 45. $\dfrac{18}{25}$ 46. $\dfrac{9}{25}$ 47. $\dfrac{9}{175}$

49. $\dfrac{6}{25}$ 50. $\dfrac{7}{12}$

EXERCISES 6.5   PAGE 179

1. (1) 7   2. (1) 3.1   3. (1) 9   5. (2) 150

6. (2) 85   7. (2) $466\dfrac{2}{3}$ 9. (3) 20%   10. (3) 50%

11. (3) 50%   13. (3) $33\dfrac{1}{3}$%   14. (3) 20%   15. (3) 40%

17. (2) $23\dfrac{17}{21}$ 18. (1) 57   19. (1) 69.6   21. (1) 12.5

22. (2) $254\dfrac{6}{11}$ 23. (2) $62\dfrac{38}{51}$ 25. (1) 2.109   26. (1) 4

27. (3) $84\dfrac{3}{8}$%   29. (2) $25\dfrac{1}{3}$ 30. (1) 150   31. (2) 16

33. (3) 900%   34. (1) 18   35. (3) 150%   37. (1) 6
38. (1) 21   39. (2) 320   41. 76   42. 30
43. 82   45. 75   46. 10   47. 12
49. 130%   50. 200%

EXERCISES 6.6   PAGE 183

1. $3480   2. $105; $3395   3. $10,000
5. $5.95 for sheets, $3.15 for towels; $4.30   6. $3500
7. 400 cavities   9. 20 problems   10. 40%; 60%

11. $750   13. $250; 20% discount; $33\dfrac{1}{3}$%; 25%   14. $66\dfrac{2}{3}$%

15. $1.82; $32.02   17. 20% rent; 40% food; 15% for taxes
18. 88%; 250 pages originally; 30 pages cut   19. $10.53; 60%

21. $240; $24; $12\frac{1}{2}$%; $11\frac{1}{9}$%   22. $750; $150; 20%

23.   10%; $11\frac{1}{9}$%; The percents are different because different bases were used; 200 pounds.

25. $656.40

EXERCISES 6.7   PAGE 188

1. $40   2. $42   3. $5.25   5. $11.25   6. $8.34
7. $36   9. $365   10. $1030   11. $463.50; $36.50
13. $37,500   14. 72 days   15. 10%

17.   (a) interest—$7.50   18. $1000; 270 days or $\frac{3}{4}$ year
      (b) time—60 days
      (c) rate—18%
      (d) principal—$100
19. 12%; $9000

EXERCISES 6.8   PAGE 192

1. $13,260; $13,795.70   2. $10,418.60
3. $306.82; $5306.82; $6.82   5. $370.80; $376.53; $379.48
6. $5610.21   7. $800; $1441.96
9. $19,965; No, the value will be only $17,395.40, and the difference is $2569.60; Yes
10. $129.28; $129.92; $.64
11. Almost double in 7 years

YEAR	PRINCIPAL	INTEREST	TOTAL
TABLE C INFLATION AT 10%			
1	$10,000.00	$1,000.00	$11,000.00
2	11,000.00	1,100.00	12,100.00
3	12,100.00	1,210.00	13,310.00
4	13,310.00	1,331.00	14,641.00
5	14,641.00	1,464.10	16,105.10
6	16,105.10	1,610.51	17,715.61
7	17,715.61	1,771.56	19,487.17
8	19,487.17	1,948.72	21,435.89

REVIEW QUESTIONS · CHAPTER 6   PAGE 195

1. 85%   2.   37%   3.   60%   4.   130%

5. 120%   6. 325%   7. 575%   8. $41\frac{2}{3}\%$

9. $\frac{14}{100} = \frac{7}{50}$ 10. $\frac{40}{100} = \frac{2}{5}$ 11. $\frac{66}{100} = \frac{33}{50}$ 12. 4

13. $\frac{2500}{100} = 25$ 14. $\frac{27}{100}$ 15. $\frac{16\frac{2}{3}}{100} = \frac{1}{6}$ 16. $\frac{1}{400}$

17. 6%   18. 30%   19. 67%   20. 2.7%
21. 459%   22. 500%   23. .35   24. .04   25. .0025
26. .0025   27. .1375   28. .071   29. 15.6   30. 2.55

31. $233\frac{1}{3}$ 32. $42\frac{6}{7}$ 33. 20%   34. $33\frac{1}{3}\%$ 35. 25%

36. .02   37. 50   38. $254\frac{6}{11}$ 39. 1   40. 200%

41. $4044.   42. $7.50   43. $25,400

44. $33\frac{1}{3}\%$; 25%   45. $30.   46. 60 days

47. (a) Interest—$12   (b) Time—$1\frac{1}{2}$ years.

     (c) Rate—8.5%   (d) Principal—$2000
48. $803.40; $815.81; $822.20

CHAPTER 7

EXERCISES 7.1   PAGE 205

5. 3 cm   6. 11 cm   7. 7 cm   9. 20 mm
10. 82 mm   11. 96 mm   13. 104 mm   14. 20 km
15. 20 km   17. same distance   18. 100 cm   19. 1200 cm
21. 4000 mm   22. 700 mm   23. 160 mm   25. 7.5 cm
26. 0.036 m   27. 0.025 m   29. 15 000 m   30. 1700 mm
31. 0.25 m   33. 1 000 000 mm   34. 3 km   35. 4 mm
37. 0.15 m   38. 0.0561 cm   39. 7.5 dm   41. 0.17 km
42. 1300 m   43. 75 000 m   45. 7860 mm   46. 342 000 m
47. 590 dm   49. 5.7 dm   50. 0.896 m

EXERCISES 7.2   PAGE 210

1. 7000 mg   2. 200 g   3. 0.0345 g
5. 4 t   6. 5.6 kg   7. 73 000 000 mg
9. 540 mg   10. 700 mg   11. 5000 kg
13. 2000 kg   14. 0.896 g   15. 896 000 mg

17. 75 kg   18. 3 g   19. 7 000 000 g
21. 0.00034 kg   22. 780 mg   23. 0.016 g
25. 3940 mg   26. 0.0923 kg   27. 5600 kg
29. 3.547 t   30. 2.963 t

EXERCISES 7.3   PAGE 215

1. 300 mm²   2. 560 mm²   3. 870 mm²   5. 6 cm²
6. 0.28 cm²   7. 14 cm²   9. 400 cm² = 40 000 mm²
10. 730 cm² = 73 000 mm²   11. 5700 cm² = 570 000 mm²
13. 1700 dm² = 170 000 cm² = 17 000 000 mm²
14. 290 dm² = 29 000 cm² = 2 900 000 mm²
15. 3 dm² = 300 cm² = 30 000 mm²   17. 1.42 cm²
18. 58 cm²   19. 2 m²   21. 780 m²   22. 30 000 m²
23. 4 m²   25. 869 a = 86 900 m²   26. 781 a = 78 100 m²
27. 16 a = 1600 m²   29. 0.01 ha   30. 0.15 ha
31. 50 000 a = 500 ha   33. 3000 a = 30 ha
34. 5320 a = 53.2 ha   35. 875 cm²   37. 20 mm²
38. 78.5 m²   39. 7.065 cm²   41. 38.5 mm²
45. 5 cm² or 500 mm²   43. 6 cm²   45. 57.12 mm²
46. 21.195 m²   47. 32.28 dm²   49. 75.36 a   50. 1580 a

EXERCISES 7.4   PAGE 221

1. 1000 mm³   2. 10 cm   3. 10 dm
   1000 cm³   100 mm   100 cm
   1000 dm³   100 cm²   100 dm²
   1 000 000 000 m³   10 000 mm²   10 000 cm²
                        1000 cm³   1000 dm³
                        1 000 000 mm³   1 000 000 cm³
5. 73 000 ℓ   6. 900 ℓ   7. 0.4 ℓ   9. 63 000 ml
10. 8700 ml   11. 0.5 kl   13. 19 cm³   14. 5000 mm³
15. 2000 cm³   17. 5300 ml   18. 30 ml   19. 0.03 ℓ
21. 48 000 ℓ   22. 72 kl   23. 0.32 hl   25. 0.29 kl
26. 0.569 ℓ   27. 7 200 000 ml 29. 9500 ℓ   30. 0.72 hl
31. 70 dm³   33. 381.51 cm³ 34. 904.32 dm³
35. 12.56 dm³ 37. 224 cm³   38. 9106 dm³   39. 282.6 cm³

REVIEW QUESTIONS · CHAPTER 7   PAGE 228

1. 1500 cm   2. 3500 dm   3. 3700 mm²
4. 1700 cm²   5. 300 a   6. 30 000 m²
7. 5000 cm³   8. 36 000 ml   9. 13 000 cm³
10. 68 000 mm³   11. 5000 g   12. 3400 mg
13. 6710 kg   14. 0.019 g   15. 8000 g

16. 4.29 kg   17. 5 cm²   18. 2280 mm²
19. 44.13 m²   20. 70 mm²   21. 294 ℓ
22. 0.02931 ℓ (rounded off)

CHAPTER 8

EXERCISES 8.1 PAGE 232

1. 9   2. 49   3. 64   5. 256   6. 289

7. 324   9. 400   10. 625   11. $\frac{1}{4}$ 13. $\frac{4}{9}$

14. $\frac{25}{36}$ 15. $\frac{100}{121}$ 17. $\frac{625}{961}$ 18. .01   19. .16

21. .81   22. 1.69   23. 2.25   25. 5.76   26. 12.96
27. 37.21   29. .0025   30. .0064   31. .0144   33. .0361
34. 4.0401   35. 12.3904   37. 5   38. 9   39. 13
41. 18   42. 20   43. 1.3   45. 1.7   46. .2
47. .02   49. .03   50. .1   51. .18   53. .12

54. .08   55. .06   57. $\frac{4}{5}$ 58. $\frac{7}{6}$ 59. $\frac{3}{7}$

61. $\frac{10}{11}$ 62. $\frac{18}{13}$ 63. $\frac{20}{7}$ 65. $\frac{16}{19}$ 66. 1.73

67. 2.24   69. $\sqrt{3.24} = 1.8$ 70. 3026

EXERCISES 8.2 PAGE 235

1. 26.5   2. 90.6   3. 83.1   5. 4.2   6. 5.3
7. .2   9. 519.6   10. 981.0   11. 2.24   13. 3.74
14. .15   15. .26   17. .58   18. 274.63   19. 225.59
21. 2.236   22. 1.414   23. .050   25. 1.438   26. 674.
27. 237.   29. 439.   30. 282.   31. 8.72

EXERCISES 8.3 PAGE 237

1. 22.36   2. 19.24   3. 1.41   5. 170.00   6. 150.00
7. 15.81   9. 9.75   10. 8.37   11. 9.27   13. 25.10
14. 4.24   15. 50

EXERCISES 8.4 PAGE 240

1. 2.24   2. 7.81   3. 10   5. 15   6. 25   7. 4.24
9. 26.93   10. 34   11. yes 13. yes   14. yes   15. no
17. no   18. yes   19. yes 22. 2 feet

23. $\sqrt{2}$ centimeters   25. $\sqrt{10}$ meters

REVIEW QUESTIONS · CHAPTER 8   PAGE 242

1. perfect square   2. not a perfect square
3. perfect square   4. perfect square

5.	perfect square			6.	not a perfect square		
7.	196	8.	25	9.	11.56	10.	.0064	11.	37.21

12.	.0169	13.	.5	14.	9	15.	.09	16.	$\dfrac{2}{5}$

17.	$\dfrac{17}{20}$	18.	$\dfrac{14}{11}$	19.	24.49	20.	77.46	21.	.75
22.	3.30	23.	206.79	24.	.31	25.	10.6	26.	13
27.	1.7	28.	22.6	29.	yes	30.	yes	31.	yes
32.	yes	33.	no	34.	no				

CHAPTER 9

EXERCISES 9.1 PAGE 246

1. A vector is a directed line segment.
2. Two vectors are equal if they have the same magnitude (or length) and the same direction.
3. The opposite of a vector is a vector of the same magnitude with the opposite direction.

5.

6.

7.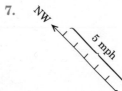

9. (Zero vector)   10. (Zero vector)

11. 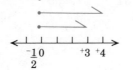   13. 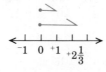   14.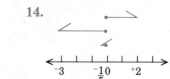

15.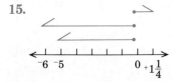

17–23. Any vectors equal to the ones given here.

17. 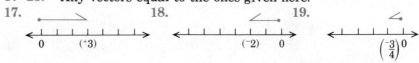   18.   19.

21. 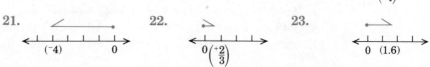   22.   23.

EXERCISES 9.2   PAGE 249

1–19.   Vector diagrams should give the following results.

1. 3	2. 7	3. ⁻5	5. 4	6. ⁻3	7. ⁻2
9. ⁻1	10. ⁻1	11. 0	13. $\frac{-1}{2}$	14. $2\frac{1}{4}$	15. ⁻3
17. ⁻2.6	18. ⁻3.9	19. 1.1	21. 7	22. 32	23. 6
25. 16.7	26. 6.41	27. $22\frac{1}{2}$	29. $17\frac{3}{4}$	30. $4\frac{1}{8}$	31. 13

33. ⁻8   34. ⁻8   35. $\frac{-3}{4}$ 37. neither   38. neither   39. $-\sqrt{3}$

EXERCISES 9.3   PAGE 254

1. 2	2. 1	3. 10	5. 19	6. ⁻10	7. ⁻9
9. 0	10. 25	11. ⁻4	13. 2	14. ⁻3	15. ⁻16
17. ⁻12	18. ⁻9	19. 0	21. 3	22. ⁻6	23. 10
25. ⁻2.36	26. ⁻6	27. 12	29. ⁻12	30. ⁻41	31. ⁻10

33. ⁻48   34. 0   35. 0   37. ⁻52   38. 13   39. $8\frac{17}{20}$

41. ⁻222   42. ⁺141

EXERCISES 9.4   PAGE 256

1. 3	2. 13	3. 11	5. ⁻9	6. 4	7. ⁻10
9. ⁻18	10. 17	11. ⁻31	13. ⁻11	14. ⁻5	15. ⁻2
17. ⁻2	18. ⁻17	19. ⁻131	21. 30	22. 8	23. 4
25. ⁻9	26. 80	27. 24	29. ⁻29	30. 43	

EXERCISES 9.5   PAGE 257

1. 8	2. 6	3. 10	5. −8	6. 7	7. 6
9. −9	10. −4	11. −4	13. −3	14. −10	15. −6
17. 5	18. −102	19. −129	21. 4	22. −25	23. −27
25. 44	26. −15	27. 58	29. −1	30. 0	

EXERCISES 9.6   PAGE 261

1. (a) {1, 5}   (b) {0, 1, 5}   (c) {−3, −1, 0, 1, 5}

    (d) $\{-3, -1, -\frac{1}{2}, 0, 1, \frac{5}{8}, 5\}$ (e) $\{-\sqrt{2}\}$ (f) A

2. 5(−4) = (−4) + (−4) + (−4) + (−4) + (−4)

5. −48	6. 60	7. 16	9. −12	10. −2
11. −3	13. 189	14. −42	15. 0	17. 16

18. −8   19. −.065   21. $\frac{9}{25}$ 22. 4.48   23. 0

25. 1   26. −1   27. 1152   29. −2075   30. 595

31. 2093   33. −11,872   34. 30   35. 105   37. 125

38. 126   39. $\frac{9}{40}$

EXERCISES 9.7   PAGE 265

1. 4   2. −6   3. −2   5. −11   6. 23
7. 21   9. 0   10. 4   11. −1.6   13. 6
14. 1   15. 6   17. −15   18. 18   19. −17
21. −23   22. 33   23. −37   25. −8   26. −48
27. −43   29. −12   30. 11
31. positive   33. positive   34. negative   35. positive

EXERCISES 9.8   PAGE 268

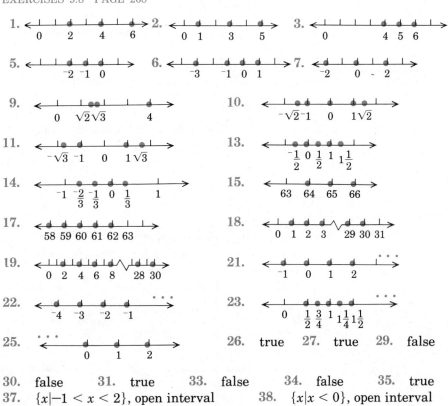

26. true   27. true   29. false
30. false   31. true   33. false   34. false   35. true
37. $\{x|-1 < x < 2\}$, open interval   38. $\{x|x < 0\}$, open interval
39. $\{x|0 \le x \le 3\}$, closed interval
41. open interval

42. open interval

43. closed interval

45. half-open interval

46. half-open interval

47. closed interval

49. open interval

50. open interval

51. open interval

53. half-open interval

54. open interval

55. half-open interval

REVIEW QUESTIONS · CHAPTER 9   PAGE 272

1. directed   2. magnitude; direction
3. magnitude; opposite direction   4. absolute value
5. A variable is a symbol that represents any element in a set that contains more than one element.

6. false	7. true	8. true	9. 7	10. 19	11. −56
12. 6	13. 1.1	14. 9	15. 19	16. −1	17. −3
18. 0	19. 0	20. −82	21. 11	22. 11	23. 37
24. −23	25. 0	26. 6	27. 10	28. −56	29. 5

30.   31.

32.   33.

34.

35. open interval

36. half-open interval

37. $\{x|0 \le x < 3\}$; yes
38. 6   39. 5   40. 2   41. true   42. true   43. false
44. false   45. false   46. false   47. false   48. true   49. true
50. false   51. 48   52. −120   53. 6   54. −11   55. 94

56. $-\dfrac{7}{4}$ 57. 10   58. 0

CHAPTER 10

EXERCISES 10.1   PAGE 278

1. $8x$ 2. x 3. $6x$ 5. $-8z$ 6. $-7y$ 7. $10x$ 9. $-6y$
10. $11p$ 11. $-10x + 5$ 13. $-7, -8, 5, 6$ 14. $0, 1$
15. $-11, -13, 13, 15$ 17. $-11, -14, 25, 28$
18. $-9, -10, 3, 4, -11, -12, 1, 2$ 19. $15, 16, 3, 2$
21. $-3, -4, 9, 10, -2, 11$ 22. $-7, -8, 5, 6, -9, 4$

23. $-26, -31, 34, 39$ 25. $-14\dfrac{1}{2}, -17\dfrac{1}{2}, 21\dfrac{1}{2}, 24\dfrac{1}{2}$

26. $-\dfrac{9}{4}, -\dfrac{11}{4}, \dfrac{15}{4}, \dfrac{17}{4}$ 27. $-\dfrac{2}{3}, -\dfrac{4}{3}$ 29. $8, 11$

30. $-21, -25, 27, 31$ 31. 10   33. 3   34. −4   35. −3

EXERCISES 10.2   PAGE 281

1. A statement is a sentence that can be judged to be true or false.
2. An open sentence is a sentence that contains variables and becomes a statement when the variables are replaced by elements from their replacement sets.
3. The solution set of an open sentence with one variable is that set of numbers from the replacement set of the variable that makes the sentence a true statement when substituted for the variable.
5. true statement   6. false statement   7. open sentence
9. false statement   10. open sentence   11. open sentence
13. ss: $\{1\}$ 14. ss: $\{-2\}$ 15. ss: $\{2\}$

17. ss: $\left\{\dfrac{1}{2}\right\}$ 18. ss: $\{0\}$ 19. ss: $\{-2, -1, 0\}$

21. ss: $\{2, 3\}$ 22. ss: $\left\{-\dfrac{2}{3}\right\}$ 23. ss: $\left\{-\dfrac{2}{3}\right\}$

25. ss: $\{1\}$ 26. ss: $\left\{-\dfrac{1}{3}, -\dfrac{1}{2}, -\dfrac{2}{3}\right\}$ 27. ss: $\{2\}$

29. ss: $\{-2\}$ 30. ss: $\left\{\dfrac{1}{3}\right\}$

EXERCISES 10.3   PAGE 284

1. An equation is either a statement or an open sentence using the equal sign indicating that two expressions represent the same number.
2. Two equations are equivalent if they have the same solution set.
3. (a) If the same number is added to both the left and right members of an equation, the new equation is equivalent to the original equation.
   (b) If both members of an equation are multiplied by the same number (except 0), the new equation is equivalent to the original equation.

5. equivalent	6. not equivalent	7. not equivalent
9. equivalent	10. equivalent	11. equivalent
13. equivalent	14. equivalent	15. not equivalent

17. $x = 7$ 18. $y = 22$ 19. $y = 16$ 21. $y = -25$ 22. $z = 5$

23. $x = 6$ 25. $y = -3$ 26. $a = -7$ 27. $b = -14$ 29. $p = -15$

30. $s = \dfrac{5}{4}$ 31. $s = \dfrac{20}{3}$ 33. $t = -\dfrac{17}{5}$ 34. $x = \dfrac{1}{20}$ 35. $x = -\dfrac{17}{30}$

37. $y = 90$ 38. $y = .5$ 39. $x = -15$ 41. $x = 5$ 42. $y = 5$

43. $x = 20$ 45. $x = \dfrac{8}{7}$

EXERCISES 10.4   PAGE 287

1. $x = 1$ 2. $x = 4$ 3. $y = 2$ 5. $x = -3$ 6. $y = 1$

7. $y = 2$ 9. $x = -\dfrac{11}{2}$ 10. $x = -1$ 11. $z = -3$ 13. $x = 9$

14. $x = 7$ 15. $z = -3$ 17. $x = -1$ 18. $x = \dfrac{5}{6}$ 19. $x = -\dfrac{1}{2}$

21. $x = -5$ 22. $y = 4$ 23. $z = 5$ 25. $x = 0$ 26. $x = 3$

27. $y = -4$ 29. $x = 8$ 30. $y = 5$ 31. $x = -12$ 33. $x = -\dfrac{5}{2}$

34. $y = \dfrac{70}{123}$ 35. $x = \dfrac{260}{3}$ 37. $y = 5$ 38. $x = 0$ 39. $y = 0$

EXERCISES 10.5 PAGE 290

1. $n - 12$ 2. $\dfrac{n}{8}$ 3. $16 + 2n$ 5. $3n - 1$

6. $2(13 + n)$ 7. $10 - \dfrac{1}{2}n$ 9. $8n - 4$ 10. $3n + 12$

11. $7n$ 13. $17 + 2n$ 14. $n - 5$ 15. $20 - n$

17. $n + (n + 2)$ 18. $\dfrac{n}{8}$ 19. $6 - n$ 21. $\dfrac{n}{11}$

22. $50 - n$ 23. $\dfrac{3}{4}n$ 25. $n + (n + 2) + (n + 4)$

26. $n + (n + 2) + (n + 4)$ 27. $n + (n + 1) + (n + 2)$ 29. $4n - 3n$
30. $15n - 6$ 31. a number increased by six
33. five times a number 34. one half of a number
35. two thirds of a number 37. five divided by a number
38. fourteen divided by a number 39. a number divided by eight
41. a number less than twenty
42. two times a number increased by five
43. six less than five times a number
45. three times the sum of a number and eleven

EXERCISES 10.6 PAGE 292

1. 19 2. 12 3. 54 5. 15 6. 25

7. $\dfrac{64}{3}$ 9. 4 10. -24 11. 25 13. 5

14. 18, 19 15. $-13, -14, -15$ 17. 25, 27, 29 18. 8, 10, 12

19. 14, 16, 18, 20 21. $\dfrac{1}{2}$ 22. 5 23. $\dfrac{1}{4}$

25. $-2, -1$ 26. 8, 10 27. $\dfrac{9}{10}$ 29. 6 30. 48

EXERCISES 10.7 PAGE 297

1. $x < 3$ 2. $x > 4$

3. $y \geq 2$

5. $y \leq \dfrac{3}{2}$

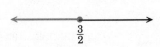

6. $y > -\dfrac{6}{5}$

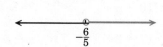

7. $y > \dfrac{2}{3}$

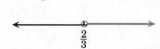

9. $0 < x$

10. $-5 > x$

11. $x \geq -7$

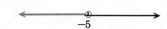

13. $x < 12$

14. $x > -6$

15. $5 < x$

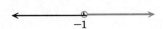

17. $y \geq 1$

18. $x > -1$

19. $0 \geq x$

21. $-3 \leq x \leq -2$

22. $1 \leq x \leq 4$

23. $-1 \leq y \leq -\dfrac{1}{2}$

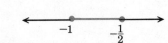

25. $-1 < x \leq 1$

26. $-3 \leq y \leq 1$

27. $-5 \geq x > -6$ 29. $2 \geq y \geq -6$

30. $-\dfrac{5}{6} \leq x \leq -\dfrac{1}{2}$

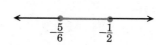

REVIEW QUESTIONS · CHAPTER 10   PAGE 300

1. true or false   2. statement; replacement sets
3. solution set; statement
4. An equation is a mathematical sentence stating that two expressions represent the same real number.
5. solution set   6. 3, 4, 5   7. 5, 6, 7   8. $-1, 0, 1, -2$

9. $-5, -4, -3$ 10. 1, 2, 3, 4, 5, 6   11. $\dfrac{2}{3}, 1, \dfrac{4}{3}$

12. 0   13. 14, 13   14. 11, 12   15. 4, 3
16. statement   17. open sentence
18. It is an open sentence and cannot be judged true or false.
19. $10x$ 20. x 21. $-22y - 2$
22. 6   23. $17p$ 24. equivalent
25. equivalent   26. not equivalent   27. equivalent
28. $x = 2$ 29. $x = -2$ 30. $y = 5$

31. $x = -8$ 32. $y = 1$ 33. $z = -\dfrac{56}{5}$ 34. $a = 1$

35. $a = \dfrac{1}{5}$ 36. $x = 5$ 37. $x = 126$ 38. $y = -11$

39. $x = -\dfrac{1}{3}$ 40. $y = 2$ 41. $3n$ 42. $3n$

43. $n = 13$ 44. $42 - 2n$ 45. $5n + 7 + 2n$ 46. $n + (n + 2)$

47. $\dfrac{28}{n}$ 48. $15 - 3n$ 49. $\dfrac{2n}{8} + \dfrac{3}{8}$

50. $2(n - 5) + 4(n + 7)$ 51. 8   52. $\dfrac{8}{5}$

53. 0   54. $\dfrac{25}{8}$ 55. $-23, -22, -21$

56. $-43, -41, -39$ 57. 2   58. $\dfrac{1}{2}$

59. 0   60. -175

61. $x \leq -9$ 62. $x \geq -3$

-9

-3

63. $x < 3$ 64. $x \leq 2$

3   2

65. $-3 \leq x \leq 0$

-3 0

APPENDIX I

EXERCISES I.1   PAGE 307

1. 254   2. 163,041   3. 21,327   5. 140   6. 256   7. 53
9. 2362   10. 97   11. 744   13. 978
14. (64)   15. (532)   17. (846)

(a) 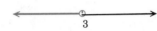   (a)    (a)

(b) ∴∴∴.   (b) $\overset{\cdots}{\underset{\cdots}{=}}$ (b) $\overset{\cdot\cdot}{\div}$

(c) ϜΔIIII   (c) ϜΔΔΔII   (c) ϜHHHΔΔΔΓI
(d) LXIV   (d) DXXXII   (d) DCCCXLVI

EXERCISES I.2   PAGE 310

1. 4983   2. 32   3. 1802   5. 47,900   6. 222   7. 46   9. 999

10. (a) VVVV <<< VV   11. (a) VVVVV <<< VVV
        VVV   <<   VVVV   <<   VVV

(b) $\overline{\text{YOB}}$ (b) $\overline{\phi\text{CF}}$

(c) 四
五
七
十
二

(c) 五
五
九
十
六

13. (a) V V VVV
VV

(b) $\overline{\Gamma \Chi \Xi \Epsilon}$

(c) 三
千
六
五
六
十
五

14. (a) VV V <<<VVV

(b) $\overline{\Zeta \Sigma \mathrm{C} \Gamma}$

(c) 七
千
二
五
九
十
三

15. (a) VVV <<<<VV

(b) A
/MΩNB

(c) 十
千
八
五
五
十
二

APPENDIX II

1. $35 = 3(10^1) + 5(10^0)$ 2. $761 = 7(10^2) + 6(10^1) + 1(10^0)$

3. $8469 = 8(10^3) + 4(10^2) + 6(10^1) + 9(10^0)$

5. $62{,}322 = 6(10^4) + 2(10^3) + 3(10^2) + 2(10^1) + 2(10^0)$

6. $11_{(2)} = 1(2^1) + 1(2^0) = 1(2) + 1(1) = 2 + 1 = 3$

7. $101_{(2)} = 1(2^2) + 0(2^1) + 1(2^0) = 1(4) + 0(2) + 1(1) = 4 + 0 + 1 = 5$

9. $1011_{(2)} = 1(2^3) + 0(2^2) + 1(2^1) + 1(2^0) = 1(8) + 0(4) + 1(2) + 1(1)$
    $= 8 + 0 + 2 + 1 = 11$

10. $1101_{(2)} = 1(2^3) + 1(2^2) + 0(2^1) + 1(2^0) = 1(8) + 1(4) + 0(2) + 1(1)$
    $= 8 + 4 + 0 + 1 = 13$

11. $110{,}111_{(2)} = 1(2^5) + 1(2^4) + 0(2^3) + 1(2^2) + 1(2^1) + 1(2^0)$
    $= 1(32) + 1(16) + 0(8) + 1(4) + 1(2) + 1(1)$
    $= 32 + 16 + 0 + 4 + 2 + 1 = 55$

13. $101{,}011_{(2)} = 1(2^5) + 0(2^4) + 1(2^3) + 0(2^2) + 1(2^1) + 1(2^0)$
    $= 1(32) + 0(16) + 1(8) + 0(4) + 1(2) + 1(1)$
    $= 32 + 0 + 8 + 0 + 2 + 1 = 43$

14. $11{,}010_{(2)} = 1(2^4) + 1(2^3) + 0(2^2) + 1(2^1) + 0(2^0)$
    $= 1(16) + 1(8) + 0(4) + 1(2) + 0(1) = 16 + 8 + 0 + 2 + 0 = 26$

15. $1000_{(2)} = 1(2^3) + 0(2^2) + 0(2^1) + 0(2^0) = 1(8) + 0(4) + 0(2) + 0(1)$
    $= 8 + 0 + 0 + 0 = 8$

17. $11{,}101_{(2)} = 1(2^4) + 1(2^3) + 1(2^2) + 0(2^1) + 1(2^0)$
    $= 1(16) + 1(8) + 1(4) + 0(2) + 1(1) = 16 + 8 + 4 + 0 + 1 = 29$

18. $10{,}110_{(2)} = 1(2^4) + 0(2^3) + 1(2^2) + 1(2^1) + 0(2^0)$
    $= 1(16) + 0(8) + 1(4) + 1(2) + 0(1) = 16 + 0 + 4 + 2 + 0 = 22$

19. $111{,}111_{(2)} = 1(2^5) + 1(2^4) + 1(2^3) + 1(2^2) + 1(2^1) + 1(2^0)$
    $= 1(32) + 1(16) + 1(8) + 1(4) + 1(2) + 1(1)$
    $= 32 + 16 + 8 + 4 + 2 + 1 = 63$

21. 111

1. $24_{(5)} = 2(5^1) + 4(5^0) = 2(5) + 4(1) = 10 + 4 = 14$

2. $13_{(5)} = 1(5^1) + 3(5^0) = 1(5) + 3(1) = 5 + 3 = 8$

3. $10_{(5)} = 1(5^1) + 0(5^0) = 1(5) + 0(1) = 5 + 0 = 5$

5. $104_{(5)} = 1(5^2) + 0(5^1) + 4(5^0) = 1(25) + 0(5) + 4(1) = 25 + 0 + 4 = 29$

6. $312_{(5)} = 3(5^2) + 1(5^1) + 2(5^0) = 3(25) + 1(5) + 2(1) = 75 + 5 + 2 = 82$

7. $32_{(5)} = 3(5^1) + 2(5^0) = 3(5) + 2(1) = 15 + 2 = 17$

9. $423_{(5)} = 4(5^2) + 2(5^1) + 3(5^0) = 4(25) + 2(5) + 3(1) = 100 + 10 + 3$
    $= 113$

10. $444_{(5)} = 4(5^2) + 4(5^1) + 4(5^0) = 4(25) + 4(5) + 4(1) = 100 + 20 + 4$
    $= 124$

11. $1034_{(5)} = 1(5^3) + 0(5^2) + 3(5^1) + 4(5^0) = 1(125) + 0(25) + 3(5) + 4(1)$
    $= 125 + 0 + 15 + 4 = 144$

13. $244_{(5)} = 2(5^2) + 4(5^1) + 4(5^0) = 2(25) + 4(5) + 4(1) = 50 + 20 + 4$
    $= 74$

14. $3204_{(5)} = 3(5^3) + 2(5^2) + 0(5^1) + 4(5^0) = 3(125) + 2(25) + 0(5) + 4(1)$
    $= 375 + 50 + 0 + 4 = 429$

15. $13{,}042_{(5)} = 1(5^4) + 3(5^3) + 0(5^2) + 4(5^1) + 2(5^0)$
    $= 1(625) + 3(125) + 0(25) + 4(5) + 2(1)$
    $= 625 + 375 + 0 + 20 + 2 = 1022$

17. yes; 13   18. yes; 28   19. $\{0, 1, 2, 3, 4, 5, 6, 7\}$

EXERCISES II.3   PAGE 318

1. $1000_{(2)} = 8$ 2. $311_{(5)} = 81$ 3. $11{,}000_{(2)} = 24$

5. $432_{(5)} = 117$ 6. $1000_{(2)} = 8$ 7. $1011_{(2)} = 11$

9. $433_{(5)} = 118$ 10. $1302_{(5)} = 202$ 11. $10{,}100_{(2)} = 20$

13. $10{,}100_{(2)} = 20$ 14. $1222_{(5)} = 187$ 15. $2241_{(5)} = 321$

17. $110{,}111_{(2)} = 55$ 18. $23{,}240_{(5)} = 1695$ 19. $3002_{(5)} = 377$

21. $30{,}141_{(5)} = 1921$ 22. $1{,}101{,}001_{(2)} = 105$ 23. $110{,}001_{(2)} = 49$

25. $10{,}001{,}010_{(2)} = 138$

INDEX

POWERS, ROOTS, AND PRIME FACTORIZATIONS

NO.	SQUARE	SQUARE ROOT	CUBE	CUBE ROOT	PRIME FACTORIZATION
1	1	1.0000	1	1.0000	—
2	4	1.4142	8	1.2599	prime
3	9	1.7321	27	1.4423	prime
4	16	2.0000	64	1.5874	2 · 2
5	25	2.2361	125	1.7100	prime
6	36	2.4495	216	1.8171	2 · 3
7	49	2.6458	343	1.9129	prime
8	64	2.8284	512	2.0000	2 · 2 · 2
9	81	3.0000	729	2.0801	3 · 3
10	100	3.1623	1000	2.1544	2 · 5
11	121	3.3166	1331	2.2240	prime
12	144	3.4641	1728	2.2894	2 · 2 · 3
13	169	3.6056	2197	2.3513	prime
14	196	3.7417	2744	2.4101	2 · 7
15	225	3.8730	3375	2.4662	3 · 5
16	256	4.0000	4096	2.5198	2 · 2 · 2 · 2
17	289	4.1231	4913	2.5713	prime
18	324	4.2426	5832	2.6207	2 · 3 · 3
19	361	4.3589	6859	2.6684	prime
20	400	4.4721	8000	2.7144	2 · 2 · 5
21	441	4.5826	9261	2.7589	3 · 7
22	484	4.6904	10,648	2.8020	2 · 11
23	529	4.7958	12,167	2.8439	prime
24	576	4.8990	13,824	2.8845	2 · 2 · 2 · 3
25	625	5.0000	15,625	2.9240	5 · 5
26	676	5.0990	17,576	2.9625	2 · 13
27	729	5.1962	19,683	3.0000	3 · 3 · 3
28	784	5.2915	21,952	3.0366	2 · 2 · 7
29	841	5.3852	24,389	3.0723	prime
30	900	5.4772	27,000	3.1072	2 · 3 · 5
31	961	5.5678	29,791	3.1414	prime
32	1024	5.6569	32,768	3.1748	2 · 2 · 2 · 2 · 2
33	1089	5.7446	35,937	3.2075	3 · 11
34	1156	5.8310	39,304	3.2396	2 · 17
35	1225	5.9161	42,875	3.2711	5 · 7
36	1296	6.0000	46,656	3.3019	2 · 2 · 3 · 3
37	1369	6.0828	50,653	3.3322	prime
38	1444	6.1644	54,872	3.3620	2 · 19
39	1521	6.2450	59,319	3.3912	3 · 13
40	1600	6.3246	64,000	3.4200	2 · 2 · 2 · 5
41	1681	6.4031	68,921	3.4482	prime
42	1764	6.4807	74,088	3.4760	2 · 3 · 7
43	1849	6.5574	79,507	3.5034	prime
44	1936	6.6333	85,184	3.5303	2 · 2 · 11
45	2025	6.7082	91,125	3.5569	3 · 3 · 5
46	2116	6.7823	97,336	3.5830	2 · 23
47	2209	6.8557	103,823	3.6088	prime
48	2304	6.9282	110,592	3.6342	2 · 2 · 2 · 2 · 3
49	2401	7.0000	117,649	3.6593	7 · 7
50	2500	7.0711	125,000	3.6840	2 · 5 · 5